Extra Class

FCC License Preparation for Element 4 Extra Class Theory

by

Gordon West
WB6NOA

---

with Technical Editor
Eric P. Nichols
KL7AJ

Seventh Edition

Master Publishing, Inc.

This book was developed and published by:
Master Publishing, Inc.
Niles, Illinois

Editing by:
Pete Trotter, KB9SMG and Karin Thompson, K0RTX

Cartoons by:
Jim Massara, N2EST, *based on the original Elmer by* Carson Haring, AC0BU

Thanks to the following for their assistance with this book: Suzy West, N6GLF; Chip K7JA & Janet KL7MF Margelli; Dennis Gonya, K6LIG; Jim Ford, N6JF; Joe Cetinske, KK6GSI; K0UOS; Robert Hawkinson PE, N1RKH; Richard, W4KBX; Mike Collis, WA6SVT; Jodi Lenocker, WA6JL; Mike Wolcott, W4WYI; Dave Dodge, K4CTV; Gary, KC0PBC; Don Wilbanks, AE5DW; George WA6RIK, Mimi and Rosy; Dr. Tamitha Skov; and Ira Wiesenfeld, P.E., WA5GXP;

Photograph Credit:
All photographs that do not have a source identification are either courtesy of the authors, Chip Margelli, K7JA, or Master Publishing, Inc. originals.

Printing by:
Arby Graphic Service
Niles, Illinois

6019 W. Howard St.
Niles, Illinois 60714
847/763-0916 (voice)
847/763-0918 (fax)
E-Mail: MasterPubl@aol.com

Visit Master Publishing
on the Internet at:
www.masterpublishing.com

Seventh Edition
5 4 3 2 1

About the Author

Gordon West, WB6NOA

Gordon West has been a ham radio operator for more than 50 years, holding the top Extra Class license, call sign WB6NOA. He also holds the highest FCC commercial operator license, the First Class General Radiotelephone Certificate with Radar Endorsement.

Gordon teaches evening ham radio classes and offers weekend ham radio licensing seminars on a monthly schedule. These seminars cover entry-level and upgrade licenses in ham radio. He has served on the faculty of Coastline College and Orange Coast College.

Gordon is a regular contributor to amateur radio, marine, and general two-way radio magazines. He is a fellow of the Radio Club of America, and a life member of the American Radio Relay League. The ARRL presented Gordon with its "Instructor of the Year" award. The Dayton Amateur Radio Association named Gordon their "2006 Amateur of the Year" for his efforts in recruiting and training new amateurs, in addition to his lifelong involvement in ham radio. Through his Gordon West Radio School, he has trained eight out of ten newly-licensed hams with his classes, books and audio courses over the past 35 years.

I wish to dedicate this 7th Edition of Extra Class to the memory of young Kevin Guice, KG6MIH, an extraordinary Extra Class ham and Eagle Scout, and to the entire Guice family who make Amateur Radio grow and happen by serving as fellow VEs and Instructors. Thank you!

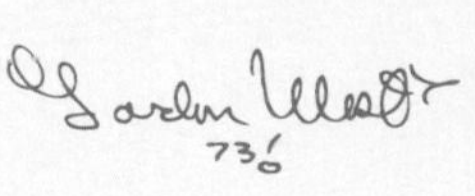

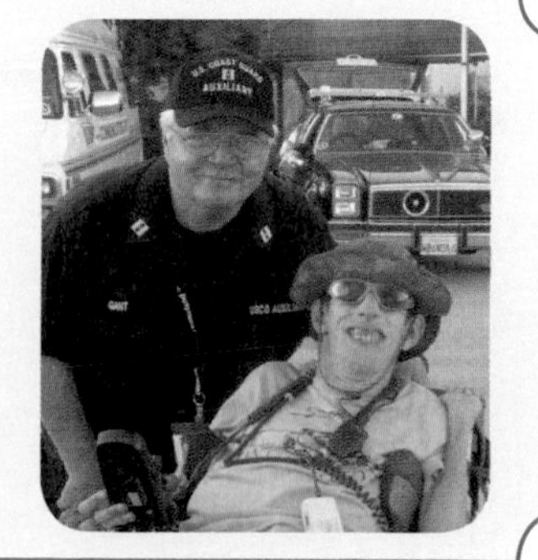

Pictured with Kevin is Pastor Jason Gant, W6AUX.

Eric P. Nichols, KL7AJ
Technical Editor

Joining author Gordon West for this new Seventh Edition of his Extra Class study manual as Technical Editor is Eric P. Nichols, KL7AJ. Eric has been a licensed amateur operator since 1972, holding the top Extra Class license, as well as the FCC GROL commercial license. He has spent his adult career in various aspects of radio research, broadcast communications, and teaching.

A prolific author, he has written many articles for ***QST, QEX,*** and a number of other ham radio magazines, as well as professional journals. In 2010 and again in 2014 he was the recipient of the William Orr, W6SAI, Technical Writing Award conferred by the American Radio Relay League for articles he authored for ***QST*** magazine. He is the author of ***Radio Science for the Radio Amateur***, and ***The Opus of Amateur Radio Knowledge and Lore***. Along with Gordon West, WB6NOA he is Technical Editor of the ***GROL+RADAR*** study manual for the FCC commercial radio licenses.

Eric lives in North Pole, AK. He has operated just about every mode available to the radio amateur, but always returns to CW on the lower HF bands as his "default" mode. He enjoys restoring vintage "boat anchor" radio equipment as well as designing cutting edge radio instrumentation with Arduino, his latest obsession. Being in interior Alaska, he gets a lot of requests from new hams (and some older ones) for a first-time Alaskan radio contact, and is always willing to try a "sked" on any HF band or mode!

Table of Contents

QUESTION POOL NOMENCLATURE
The latest nomenclature changes recommended by the Volunteer Examiner Coordinator's Question Pool Committee (QPC) for the Element 4 question pool effective July 1, 2016 and valid to June 30, 2020 have been incorporated in this book.

FCC RULES, REGULATIONS AND POLICIES
The NCVEC QPC releases revised question pools on a regular cycle, and deletions as necessary. The FCC releases changes to FCC rules, regulations and policies as they are implemented. This book includes the most recent information released by the FCC when this copy was printed.

Preface

Amateur Extra Class is the highest-level Federal Communications Commission license you can hold in the amateur radio service. Achieving the Extra Class license opens up exclusive sub-bands for voice operation on worldwide frequencies, along with other exclusive Extra Class sub-bands for Morse code and data. Your Extra Class license gives you full operating privileges across the complete spectrum of every ham radio band!

On February 23, 2007, all Morse code testing for amateur radio licenses was completely eliminated by the FCC. From that time until now with the elimination of the code "stress test" a strange phenomenon has occurred. More hams than ever are learning the code and practicing CW over the worldwide airwaves. We encourage you to learn the code. It's fun!

Achieving an Extra Class license allows you to administer all amateur radio exams. As an Extra Class licensee, you may be accredited as a Volunteer Examiner to prepare and administer ham exams as a member of a 3-person VE team. You also could qualify to become a contact Volunteer Examiner to help coordinate ham exam teams in your area. This is a great way to help our hobby and service grow!

The new Element 4 Extra Class question pool included in this book is valid for written examinations from July 1, 2016 through June 30, 2020. The total number of questions in the new pool is about the same as in years past – with 50 of these multiple choice questions appearing on the Extra Class exam. No additional math complexity has been added to this fresh new pool, but there are many updated questions on newer operating modes and techniques. Our book explains everything about the Extra Class exam:

- *Chapter 1* reviews the operating privileges you'll earn when you pass the Extra Class exam.
- *Chapter 2* is a brief review of the history of ham radio licensing and explains the exam requirements for all of the current FCC Amateur Radio licenses.
- *Chapter 3* contains all 712 Element 4 questions and answers, along with our description of the correct answer, followed by the correct answer key.
- *Chapter 4* tells you what to expect on examination day and reviews the latest FCC e-filing requirements.

Eric and I welcome Advanced and General Class operators to Extra Class exam preparation! Get set for new privileges and increased VE opportunities. We hope to hear you on our top bands very soon!

Gordon West 73's

Gordon West, WB6NOA

Eric P. Nichols

Eric P. Nichols, KL7AJ

1

Extra Class Privileges

The Extra Class amateur license permits you to use all segments of the worldwide bands designated for amateur communications for long-distance voice, data, and video communications and expands your General or Advanced Class operating privileges. Of course, you keep all of the privileges you enjoyed as a General or Advanced Class operator.

When you pass your Extra Class examination, you will gain a 25-kHz "window" of exclusive CW privileges at the bottom end of each of the 80, 40, 20, and 15 meter bands. You also gain a total of 150 kHz of additional "windows" of exclusive Extra Class phone privileges on 75, 20, and 15 meters. You will gain valuable DX operating privileges on the Extra Class bands.

Extra Class has always been recognized as the "top" amateur radio license because of the unlimited skywave privileges it provides on each of the worldwide ham bands. Earning Extra Class privileges requires passing written exams for Element 2, Technician Class, and Element 3, General Class, prior to taking your Element 4 exam.

Never in the history of amateur radio has it been easier to achieve an Extra Class license. There is no longer a Morse code test required for the Extra Class license. Certainly, the old 20 wpm code requirement was a huge obstacle for many accomplished ham radio operators who wanted to achieve Extra Class but just couldn't get past that tough, tough CW test. In February, 2007, the Federal Communications Commission ended code test requirements for all amateur radio licenses, concluding that "...this change eliminates an unnecessary burden that may have discouraged current amateur radio operators from advancing their skills and participating more fully in the benefits of amateur radio." Most other nations had already dropped their Morse code test requirements when the FCC chose to adopt this change.

Just because the code test has been eliminated, it doesn't mean earning your Extra Class license will be a snap. For most of us, it requires time dedicated to developing our operating skills and means spending some serious time diligently studying the electronic theory and operating techniques covered in the questions. Becoming an Extra Class means you've reached the top of our hobby and service.

The Extra Class license allows you to give something back to the amateur service as a volunteer who conducts examinations as part of a 3-person accredited Volunteer Examiner team. As an accredited volunteer, you are qualified to prepare and administer all levels of amateur radio exams. You can help recruit and organize other hams to serve on your VE team. This is a great opportunity to help our hobby grow and prosper!

As an Extra Class operator, you will be eligible for the 2x1 short-block call signs that are reserved exclusively for those amateurs who have made it to the very

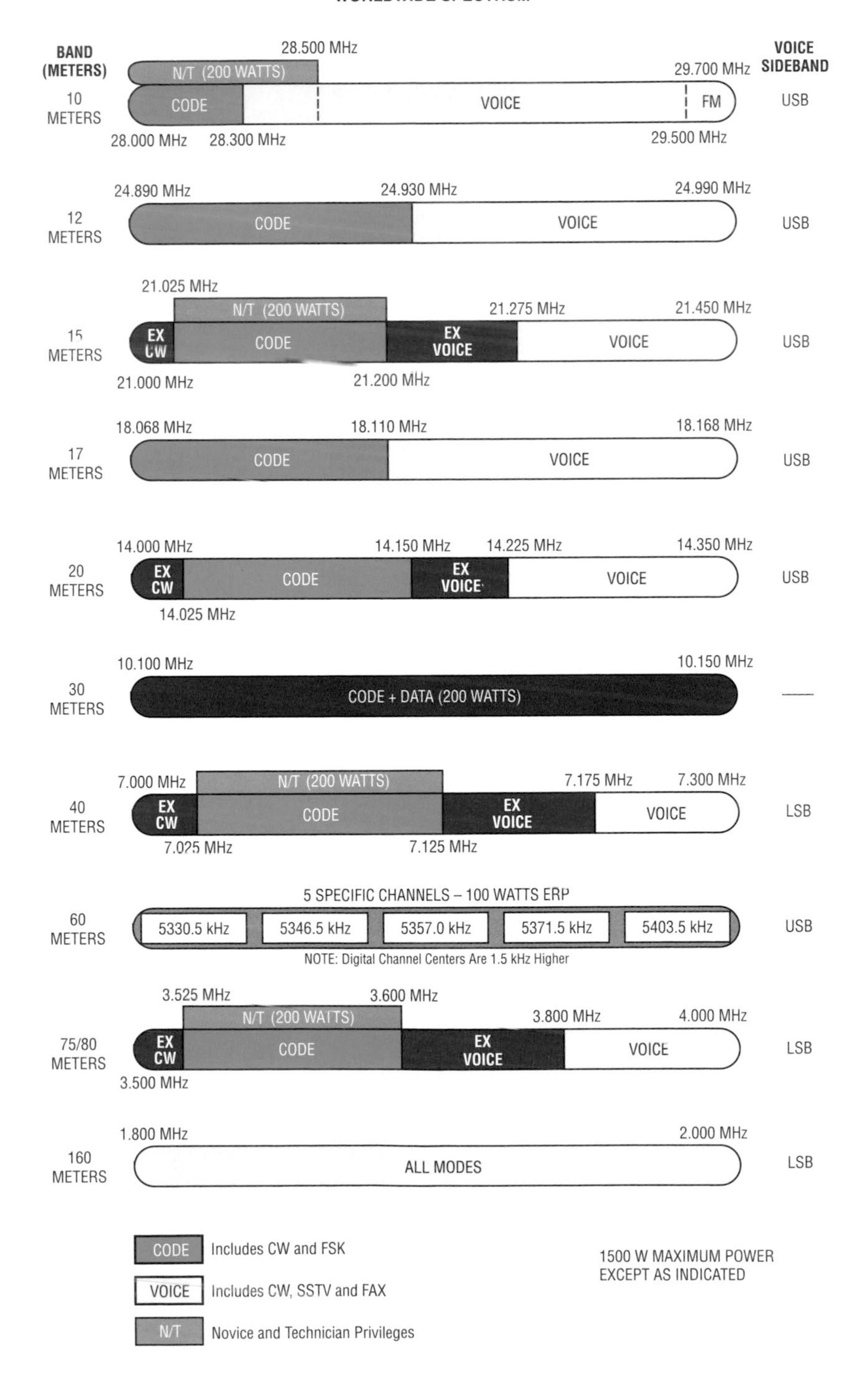

Figure 1-1. Extra Class Privileges

highest level of license in the amateur service. You may wish to take advantage of the Vanity Call Sign program to select specific call letters of your choice (see Chapter 4 for details on this).

Best of all, when you achieve your Extra Class license, you do not have to prepare for anymore license upgrade tests. You're the top!

Now, let's take a look at privileges of the Extra Class license.

THE EXTRA CLASS LICENSE

Figure 1-1 graphically illustrates your Extra Class code, data and voice privileges on the medium-frequency (MF) (1.8 to 2.0 MHz) and the high frequency (HF) (3.0 to 30 MHz) bands.

Code and data privileges are in the designated areas on the left side of each band edge. Voice privileges are on the right side of each band other than 30 meters, which is reserved exclusively for CW and data. In the 80, 20, and 15 meter bands, there are areas exclusively for Extra Class licensees. If you are upgrading from Advanced Class, passing the Extra Class exam gives you privileges in a greater part of these bands.

Someone upgrading from General Class gains privileges on the Extra Class sub-bands and the grandfathered Advanced Class sub-bands. In other words, as an Extra Class operator, *you get everything!*

Figure 1-1 includes "EX" areas. These are parts of the band where you have *exclusive* voice and CW operating privileges shared with no one other than fellow Extra Class operators.

EXTRA CLASS LICENSE PRIVILEGES

160 METERS: 1.8 MHz - 2.0 MHz.

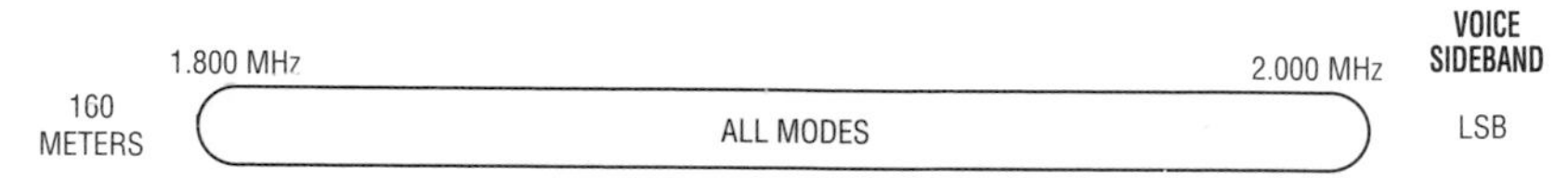

Extra Class privileges let you operate voice and code from one end to the other of the 160 meter band. These privileges are shared with General and Advanced Class operators. The 160 meter band is great for long-distance nighttime communications. At night, you work the world on this band.

75/80 METERS: 3.5 MHz - 4.0 MHz.

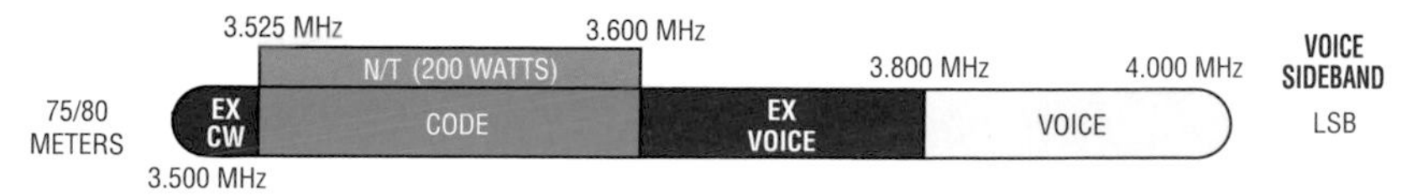

Extra Class privileges on 75/80 meters give you exclusive voice privileges that were recently expanded to include 3.600 to 3.800 MHz. This part of the band is not shared with any other ham license class and gives the Extra Class operator a huge increase for nighttime DX voice contacts; some well over 6000 miles away. Extra Class operators also have exclusive CW privileges at the bottom of the 75/80 meter band.

60 METERS: 5 SPECIFIC CHANNELS

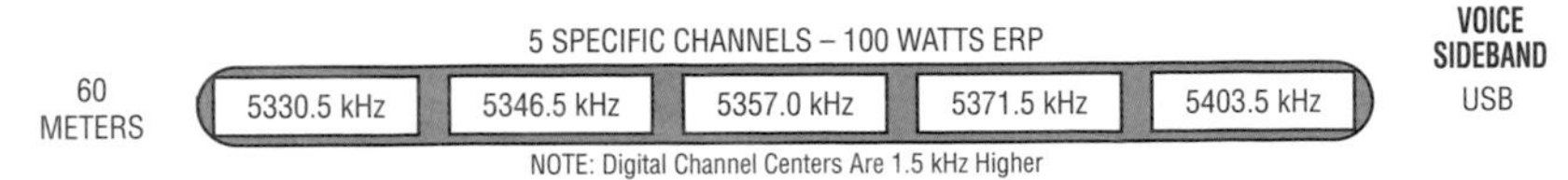

In 2003, a portion of the 60 meter band was allocated by the FCC for amateur radio use. Unlike other ham bands where we have a range of frequencies, 60 meter privileges are assigned to 5 specific channels. Recently, the middle channel on 60 meters was changed from 5366.5 kHz on your dial to 5357.0 kHz, upper sideband. Effective radiated power is limited to 100 watts when using a half-wave dipole antenna. If you use a different type of antenna, you must keep a log and adjust your power to stay within the FCC's rules. On SSB voice, your signal cannot occupy more than 2.8 kHz bandwidth. Narrow bandwidth data, such as PSK-31 and PACTOR III are allowed, as is CW, on any of the channels. Your transmitted signal must end up exactly on the channel CENTER, which is 1.5 kHz ABOVE the carrier frequencies listed on our chart. See page 288 in the Appendix for more operating tips for 60 meters. Also see the explanation to question E1A06 on page 26. It's complicated!

40 METERS: 7 MHz - 7.3 MHz.

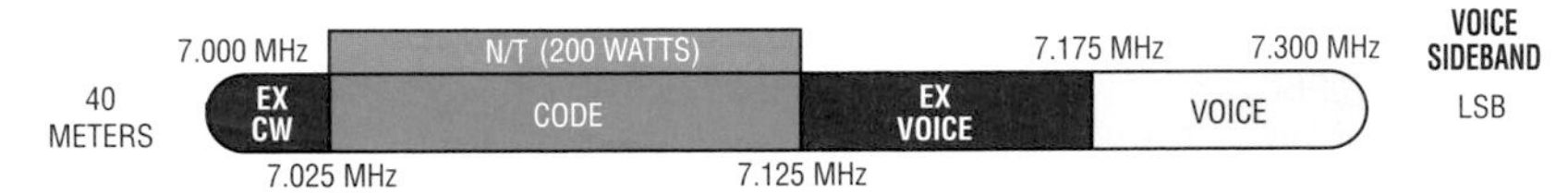

Extra Class operators have an exclusive 25 kHz of CW operating privileges at the bottom of the of the 40 meter band; 7.000 to 7.025 MHz. Voice privileges were recently extended to 7.125 to 7.300 MHz giving more "elbow room" on this band. During daylight hours, 40 meters is a great band for base station and mobile contacts up to 500 miles away. At night, 40 meters skips all over the country, and many times all over the world! Beware, though, since 40 meters is shared with worldwide AM shortwave broadcast stations, at night you should be prepared to dodge megawatt carriers playing everything from rock and roll to political broadcasts.

30 METERS: 10.1 MHz - 10.15 MHz.

Only code, data, and RTTY are permitted on this band. Privileges on 30 meters are shared by General, Advanced and Extra Class amateur operators. Thirty meters is located just above the 10 MHz WWV time broadcasts on shortwave radio so voice transmissions are not allowed on this band by any class of amateur operator. Power is limited to 200 watts PEP for any of the operating modes.

20 METERS: 14 MHz - 14.350 MHz.

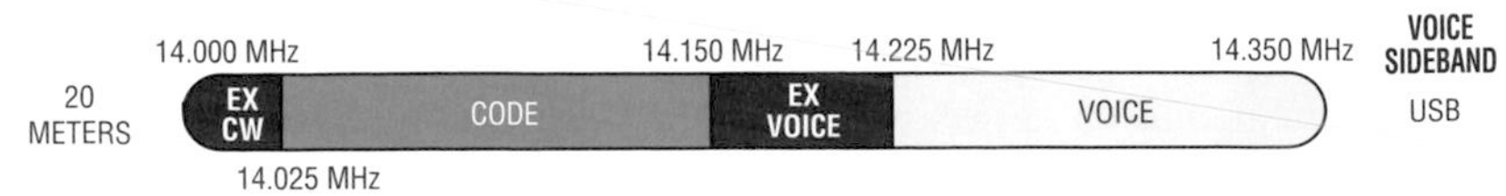

20 meters is where the real DX activity takes place! In addition to the parts of the band shared with the General and Advanced class licensees, Extra Class operators have two band segments of exclusive frequency privileges on this band. For CW, you gain the bottom 25 kHz of the band (14.000 – 14.025 MHz). For voice, you have exclusive use of 14.150 to 14.175 MHz. Happy DXing!

Almost 24 hours a day, you should be able to work stations in excess of 5,000 miles with a modest antenna set up. If you are a mariner, most maritime mobile operations take place on the 20 meter band. If you are into recreational vehicles (RVs), there are nets all over the country especially for you. The "where it's at" band is 20 meters when you want to work the world from your car, boat, RV, or home shack!

17 METERS: 18.068 MHz - 18.168 MHz.

18.068 MHz 18.110 MHz 18.168 MHz VOICE SIDEBAND
17 METERS CODE VOICE USB

All emission types are authorized on this newer Amateur Radio band. There is plenty of elbow room here with lots of foreign DX coming in day and night. The 17 meter band is shared with General and Advanced class operators giving you plenty of opportunity for fun operations and lots of contacts locally as well as from foreign lands.

15 METERS: 21.0 MHz - 21.450 MHz.

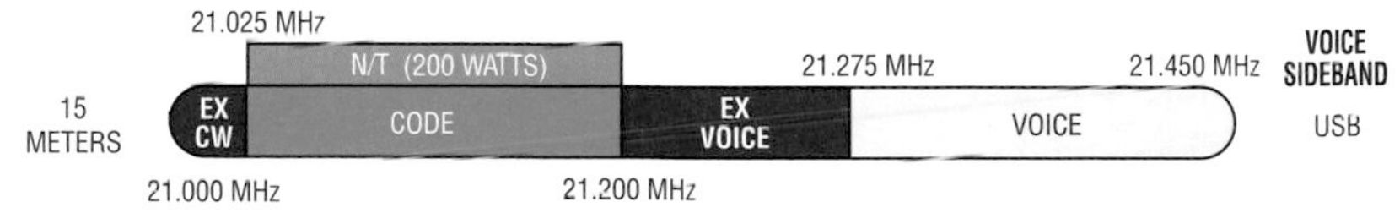

Along with privileges shared with General and Advanced class amateur operators on 15 meters, Extra Class operators have two exclusive operating segments on this band: the bottom 25 kHz from 21.00 to 21.025 MHz and another 25 kHz segment from 21.200 to 21.225 MHz.

The 15-meter band is loved by hams throughout the world because of the extremely low noise and strong signal levels at these frequencies. Band conditions on 15 meters usually favor daytime and evening contacts in the direction of the Sun. Skywave coverage on 15 meters fades as the Sun sets. This is a popular band for mobile operators because antennas are smaller.

12 METERS: 24.890 MHz - 24.990 MHz.

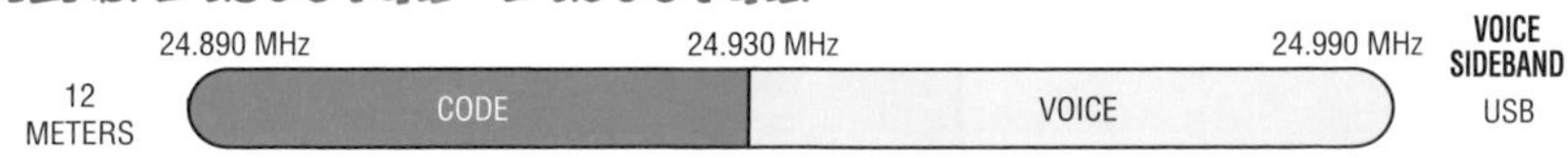

Code, data, and RTTY privileges on the 12 meter band extend from 24.890 to 24.930 MHz. Voice privileges are from 24.930 to 24.990 MHz. Extra Class operators share this band with General and Advanced class amateur operators. Although this is a very narrow band, expect excellent daytime range throughout the world. At night, range is limited to groundwave coverage because the ionosphere is not receiving sunlight to produce skywave coverage on this band.

10 METERS: 28.0 MHz - 29.7 MHz.

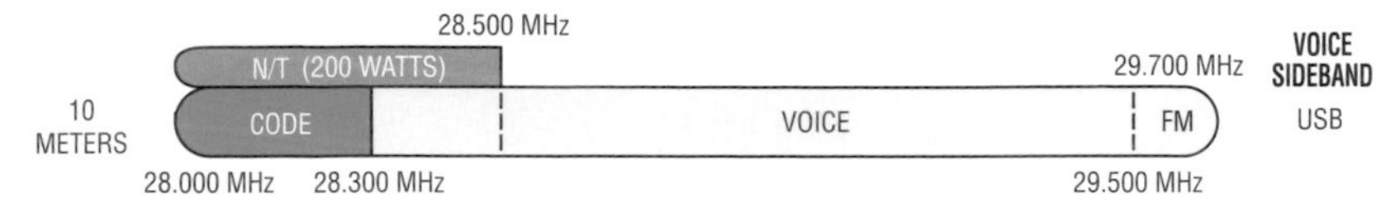

There is plenty of excitement shared by all classes of amateur operators on the 10 meter band. Worldwide and continental band openings may pop up anytime there are extra amounts of solar activity. During the summertime, Sporadic-E ionospheric conditions make 1,500-mile band openings exciting for all operators on "10." The Extra Class operator has the entire 10 meter band from top to bottom (28.000 to 29.700 MHz). There is plenty of CW activity at 28.1 MHz, propagation beacons at 28.2 MHz, phone excitement between 28.3 to 29.0 MHz with all sorts of elbow room. And wait until you try the full fidelity of frequency modulation (FM) on 29.600 MHz! There are even FM repeaters from 29.5 to 29.7 MHz at the top of 10 meters.

VOICE BAND EXPANSION

In late 2006, the FCC issued rules that provide expanded voice privileges and removed the 200 watt power limitation for CW on the worldwide bands for General, Advanced and Extra Class operators. Novice and Technician licensees were given additional CW privileges on the 80, 40, 15, and 10 meter bands. This Report and Order granted ham radio operators more "elbow room" on frequencies that FCC viewed as "grossly underused" and "wasted" spectrum within the CW portion of these bands. Here's what we got:

On 75 meters: General class operators may now operate voice from 3800 to 4000 kHz, an increase of 50 kHz. Advanced Class licensees are able to operate voice from 3700 to 4000 kHz, a 75 kHz increase. Extra Class operators gained 150 kHz of bandwidth for voice operations from 3600 to 4000 kHz.

On 40 meters: Advanced and Extra Class operators are able to use voice from 7125 to 7300 kHz, an increase of 25 kHz. General Class may now use voice from 7175 to 7300 kHz, a 50 kHz increase.

On 15 meters: Generals gained an additional 25 kHz of voice from 21.275 to 21.450 MHz. Novice and Technician operators gained CW privileges in an expanded portion in the lower half of the 10, 15, 40, and 80 meter bands. And Technician Class gained voice communications on 10 meters from 28.300 to 28.500 MHz, giving these brand new ham operators a taste of skywave communications on this worldwide band. As a new Extra, spend some time in this part of "10" to encourage these new hams!

On 60 meters: While the Report did not change the number of frequencies available to all amateur operators on 60 meters, one of the 60 meter frequencies changed. Dial frequency 5366.5 kHz was replaced by dial frequency 5357.0 kHz. Other changes in the Report for this band included an increase of power to 100 watts ERP and digital modes plus CW operation on all 5 channels.

6 Meters and Up: Your Extra Class license allows unlimited band privileges using all emission types on the frequencies detailed in the following chart. These privileges are shared with all amateur operators.

Meters	Frequency
6 meters	50-54 MHz
2 meters	144-148 MHz
1.25 meters	222-225 MHz (219-220 MHz Digital)
70 centimeters	420-450 MHz
35 centimeters	902-928 MHz
23 centimeters	1240-1300 MHz

Microwave Bands: The Extra Class license allows for unlimited privileges using all emission types on the following microwave bands:

Frequency	Frequency
2300-2310 MHz	47.0-47.2 GHz
2390-2450 MHz	75.5-81.0 GHz
3.3-3.5 GHz	119.98-120.02 GHz
5.65-5.925 GHz	142-149 GHz
10.0-10.5 GHz (popular X band)	241-250 GHz
24.0-24.25 GHz	All above 300 GHz

Look for plenty of Extra Class excitement on the 10 GHz band. This may soon become the new frontier for those amateur operators establishing records on microwave frequencies.

BECOME A VOLUNTEER EXAMINER

Earning the Extra Class license provides you with the opportunity to give back to amateur radio! As an Extra Class licensee you may be accredited as a Volunteer Examiner to prepare and administer ham exams as a member of a three-person VE team. You also could qualify to become a contact Volunteer Examiner to help coordinate ham exam teams in your area. This is a great way to help our hobby and service grow and prosper! To learn more, visit ☞ **www.W5YI.org** and click on the W5YI VEC link on the left side of the page, and then select the "Become a VE" link.

BE AN EXTRA CLASS ELMER – TEACH A CLASS

When you pass your Extra Class exam, you will have arrived at the TOP! We encourage you to share your enthusiasm for our amateur radio service by becoming an instructor. The W5YI Ham Instructor program is designed to help you spread the word – and training – to radio enthusiasts and encourage them to earn their ham radio licenses.

Ham Instructors may find kids and scouting a great place to introduce ham radio and develop amateur radio training classes. Boaters, flyers, and RVers are *all* interested in what amateur radio can do to make their traveling safer and more enjoyable. You might become an Elmer to help them locate club or local weekend classes in your area where they can earn their entry-level Technician Class license.

The radio industry supports W5YI Ham Instructors with free band plan charts, frequency guides, and other training materials. Extra Class Ham Instructors help the amateur radio service continually grow. To learn more about the W5YI Ham Instructor program, call 800-669-9594 or visit ☞ **www.HamInstructor.com**. Join the outreach program to bring more kids and radio enthusiasts into our hobby and public service.

Become a Ham Ambassador and help our hobby grow!

SUMMARY

Whether you are upgrading to Extra Class from a grandfathered Advance Class license or a General Class license, you will gain exciting new privileges on the bands for worldwide communications. This is where the rare DX happens!

But your biggest privilege as an Extra Class operator is the ability for you to work closely with a Volunteer Examination Coordinator and serve as a VE team leader. Now that General and Advanced Class operators may assist Extra Class operators in administering license exams, your role as a Volunteer Examiner will be an important one for the testing teams – and the growth of our hobby – throughout the country.

Join the Gordon West / W5YI Ham Instructor Program Today!

Help our hobby and service grow! Sign up today as a Gordon West / W5YI Ham Instructor and join the ranks of teachers, scout leaders, Elmers, and hams using Gordo's classroom techniques to successfully train new hams. W5YI provides FREE support to registered instructors, including:

- ✔ Instructor patch
- ✔ Instructor Guides for all 3 license classes
- ✔ Pre-study homework assignments for your students
- ✔ Downloadable Power Point presentations for Tech, General, and Extra
- ✔ Postings of your ham radio class dates
- ✔ Discounted Gordon West / W5YI training materials

Learn more and register as a Gordon West / W5YI Ham Instructor by visiting ☞ **www.haminstructor.com**

2

A Little Ham History!

Ham radio has changed a lot in the 100-plus years since radio's inception. In the past 25 years, we have seen some monumental changes! So, before we get started preparing for the exam, we're going to give you a little refresher lesson about our hobby, its history, and an overview of how you'll progress through the amateur ranks to the top amateur ticket – the Extra Class license.

In this chapter you'll learn all of the licensing requirements under the FCC rules that became effective April 15, 2000. And you'll learn about the six classes of license that were in effect *prior* to those rules changes. That way, when you run into a Novice or Advanced Class operator on the air you'll have some understanding of their skill level, experience, and frequency privileges.

WHAT IS THE AMATEUR SERVICE?

There are more than 735,000 Americans who are licensed amateur radio operators in the U.S. today. The Federal Communications Commission, the Federal agency responsible for licensing amateur operators, defines our radio service this way:

"The amateur service is for qualified persons of all ages who are interested in radio technique solely with a personal aim and without pecuniary interest."

Ham radio is first and foremost a fun hobby! In addition, it is a service. And note the word "qualified" in the FCC's definition – that's the reason you're studying for an exam; by passing the exam you prove that you're qualified to get on the air.

Millions of operators around the world exchange ham radio greetings and messages by voice, radioteletype, telegraphy, facsimile, and television worldwide. Japan, alone, has more than a million hams! It is very commonplace for U.S. amateurs to communicate with Russian amateurs. China is just getting started with its amateur service. Being a ham operator is a great way to promote international good will.

The benefits of ham radio are countless! Ham operators are probably known best for their contributions during times of disaster. In recent years, many recreational sailors in the Caribbean who were attacked by modern-day pirates had their lives saved by hams directing rescue efforts. Following the 9/11 terrorist attacks on the World Trade Center and Pentagon, literally thousands of local hams assisted with emergency communications. After Katrina blew through New Orleans, ham radio operators were the first to report the levee breach and to warn that flood waters were rushing into the city.

The ham community knows no geographic, political or social barrier. If you study and make the effort, you are going to earn your upgrade to Extra Class with its expanded privileges on the HF worldwide bands. Follow the suggestions in our book and your chances of passing the written exam are excellent. Learning what you need to know to be an Extra Class operator will be easy and fun!

A BRIEF HISTORY OF AMATEUR RADIO LICENSING

Before government licensing of radio stations and amateur operators was instituted in 1912, hams could operate on any wavelength they chose and could even select their own call letters. The Radio Act of 1912 mandated the first Federal licensing of all radio stations and assigned amateurs to the short wavelengths of less than 200 meters. These "new" requirements didn't deter them, and within a few years there were thousands of licensed ham operators in the United States.

Since electromagnetic signals do not respect national boundaries, radio is international in scope. National governments enact and enforce radio laws within a framework of international agreements that are overseen by the International Telecommunications Union (ITU). The ITU is a worldwide United Nations agency headquartered in Geneva, Switzerland. It divides the radio spectrum into a number of frequency bands, with each band reserved for a particular use. Amateur radio is fortunate to have many bands allocated to it all across the radio spectrum.

In the U.S., the Federal Communications Commission is the government agency responsible for the regulation of wire and radio communications. The FCC allocates frequency bands in accordance with the ITU plan to the various services – including the amateur service – and regulates stations and operators.

By international agreement in 1927, the alphabet was apportioned among various nations for basic call sign use. The prefix letters K, N and W were assigned solely to the United States. The letter A is shared by the United States and other countries.

In the early years of amateur radio licensing in the U.S., the classes of licenses were designated by the letters "A," "B," and "C." The highest license class with the most privileges was "A." In 1951, the FCC dropped the letter designations and gave the license classes names – names that are familiar to us today. Class A became Advanced; Class B became General; and Class C became Conditional. At the same time, the FCC also created the Extra Class and Technician Class licenses. They also added a new Novice Class – a one-year, non-renewable license for beginners that required a 5-wpm Morse code speed proficiency test and a 20-question written examination on elementary theory and regulations, with both tests taken in the presence of one licensed ham.

The General exam required 13-wpm code speed, and Extra required 20-wpm. Each of the five written exams was progressively more comprehensive and formed what came to be known as the *Incentive Licensing System.*

In the '70s, the Technician Class license became very popular because of the number of repeater stations appearing on the air that extended the range of VHF and UHF mobile and handheld radios. It also was very fashionable to be able to patch your mobile radio into the telephone system, which allowed hams to make telephone calls from their automobiles long before the advent of cell phones.

In 1979, the international amateur service regulations were changed to permit all countries to waive the manual Morse code proficiency requirement for "...stations making use exclusively of frequencies above 30 MHz." This set the stage for creation of the Technician "no-code" license in 1991, when the 5-wpm Morse code requirement was eliminated. New "no-code" Technician licensees were permitted to operate on all amateur bands above 30 MHz. Applicants for the "no-code" Technician license had to pass the 35-question Novice and 30-question Technician

Class written examinations, but no Morse code test.

"No-code" Technician Class amateurs who passed a 5-wpm code test were awarded a Technician-Plus license. Besides their 30 MHz and higher no-code frequency privileges, Tech-Plus licensees gained the Novice CW privileges and a sliver of the 10 meter voice spectrum.

With these changes in 1991, there were six amateur service license classes – Novice, Technician, Technician-Plus, General, Advanced, and Extra – along with five written exams and three Morse code tests used to qualify hams for their various licenses.

Then in 1984 the FCC made another important change to amateur radio rules, when the length of term for amateur radio licenses was increased from five to ten years.

The Amateur Service is Restructured

As you can see, through the years ham radio licensing and privileges underwent a myriad of changes. In 1998, the FCC began working with the ham radio community to restructure our amateur radio service to make it more effective. The objective of this restructuring was to streamline the license process, eliminate unnecessary and duplicated rules, and reduce the emphasis on the Morse code test. The result of this review was a complete restructuring of the U.S. amateur service, which became effective April 15, 2000. The restructuring included a change in the Morse code requirement for General and Extra classes to 5-wpm and the reduction from six to three amateur radio license classes. Since 2000 there have been only 3 amateur radio license classes:

- Technician Class, the HF/VHF/UHF entry level license.
- General Class, the HF entry level license, which required a 5-wpm code test.
- Amateur Extra Class, a technically oriented senior license, based on-5wpm code credit.

The next major change to FCC Amateur Radio rules occurred on February 23, 2007, when the FCC completely eliminated the Morse code examination for all amateur radio licenses! The FCC action was based on 6,200 written comments, with most supporting the elimination of all code tests. At the same time, the FCC also "upgraded" No-code Technician Class operators, awarding them use of the Novice and Technician-Plus high frequency privileges on 80, 40, 15, and 10 meters.

In its public notice announcing elimination of the Morse code test requirement, the FCC Commissioners wrote: "We believe that because the international requirement for telegraphy proficiency has been eliminated, we should treat Morse code telegraphy no differently than other amateur service communication techniques. This change eliminates an unnecessary regulatory burden that may discourage current amateur radio operators from advancing their skills and participating more fully in the benefits of amateur radio."

Eliminating the Morse code test in no way diminishes the enthusiasm many ham operators have for the technique of sending dits and dahs over the air. In fact, we believe we have more hams than ever before learning the code now that they can do it on high frequency where practicing code with other hams is fun!

Self-Testing in the Amateur Service

Prior to 1984, all amateur radio exams were administered by FCC personnel at FCC Field Offices around the country. In 1984, the FCC adopted a two-tier system beneath it called the Volunteer Examiner Coordinator (VEC) system to handle amateur radio license exams. The VEC System was formed after Congress passed laws that allowed the FCC to accept the services of Volunteer Examiners (VEs) to prepare and administer amateur service license examinations. Testing activity of the VEs is managed by the VECs. The VEC acts as the administrative liaison between the VEs, who administer the various ham examinations, and the FCC, which grants the license.

A team of three VEs, who must be accredited by a VEC, is required to conduct amateur radio examinations. General Class amateurs may serve as examiners for the Technician class. Advanced Class and Extra Class amateurs may administer exams for Technician and General class. The Extra Class exam may only be administered by VEs who hold an Extra Class license.

In 1986, the FCC turned over responsibility for maintenance of the examination question pools for amateur radio licenses to the National Conference of Volunteer Examiner Coordinators (NCVEC). The Conference appoints a Question Pool Committee (QPC) to develop and revise the various question pools according to a schedule. As a rule, each of the three question pools is changed once every four years. The FCC requires at least ten times as many questions in each of the pools as may appear on an examination. Both the Technician Class Element 2 and General Class Element 3 written examinations contain 35 multiple-choice questions. The Extra Class Element 4 written examination has 50 questions.

OPERATOR LICENSE CLASSES AND EXAM REQUIREMENTS

Anyone is eligible to become a U.S. licensed amateur operator (including foreign nationals, if they are not a representative of a foreign government). There is no age limitation – if you can pass the examinations, you can become a ham!

Today, there are three amateur operator licenses issued by the FCC – Technician, General, and Extra. Each license requires progressively higher levels of learning and proficiency, and each gives you additional operating privileges. This is known as incentive licensing – a method of strengthening the amateur service by offering more radio spectrum privileges in exchange for more operating and electronic knowledge.

There is no waiting time required to upgrade from one amateur license class to another. Similarly, there is no required waiting time to retake a failed exam. You can even take all three examinations at one sitting if you're really brave! *Table 2-1* details the amateur service license structure and required examinations.

Table 2-1: Current Amateur License Classes and Exam Requirements

License Class	Exam Element	Type of Examination
Technician Class	2	35-question, multiple-choice written examination. Minimum passing score is 26 questions answered correctly (74%).
General Class	3	35-question, multiple-choice written examination. Minimum passing score is 26 questions answered correctly (74%).
Extra Class	4	50-question, multiple-choice written examination. Minimum passing score is 37 questions answered correctly (74%).

ABOUT THE WRITTEN EXAMS

What is the focus of each of the written examinations, and how does it relate to gaining expanded amateur radio privileges as you move up the ladder toward your Extra Class license? *Table 2-2* summarizes the subjects covered in each written examination element.

Table 2-2. Question Element Subjects

Exam Element	License Class	Subjects
Element 2	Technician	Elementary operating procedures, radio regulations, and a smattering of beginning electronics. Emphasis is on VHF and UHF operating.
Element 3	General	HF (high-frequency) operating privileges, amateur practices, radio regulations, and a little more electronics. Emphasis is on HF bands.
Element 4	Extra	A technical examination covering specialized operating procedures, more radio regulations, formulas and heavy math. Also covers the specifics on amateur testing procedures.

No Jumping Allowed

You may not by-pass a required examination as you upgrade from Technician to General to Extra. For example, to obtain a General Class license, you must first take and pass the Element 2 written examination for the Technician Class license then take and pass the Element 3 written examination. To obtain an Extra Class license, you must first pass the Element 2 (Technician) and Element 3 (General) written examinations and then successfully pass the Element 4 (Extra) written examination.

TAKING THE ELEMENT 4 EXAM

Here's a summary of what you can expect when you go to the session to take the Element 4 written exam for your Extra Class license. Detailed information about how to find an exam session, what to expect at the session, what to bring to the session, and more is included in Chapter 4.

Examination Administration

All amateur radio examinations are administered by a team of at least three Volunteer Examiners (VEs) who have been accredited by a Volunteer Examiner Coordinator (VEC). The VEs are licensed hams who volunteer their time to help our hobby grow.

How to Find an Exam Session

Examination sessions are organized under the auspices of an approved VEC. A list of VECs is located in the Appendix on page 283. The W5YI-VEC and the ARRL-VEC are the two largest VECs in the country. These organizations test in all 50 states. Their 3-member, accredited examination teams are just about everywhere. So when you call the VEC, you can be assured they probably have an examination team only a few miles from where you are reading this book right now!

Taking the Exam

The Element 4 exam is a multiple-choice format. The VEs will give you a test paper that contains the 50 questions and multiple choice answers and an answer sheet for you to complete. Take your time! Make sure you read each question carefully and select the correct answer. Once you're finished, double check your work before handing in your exam papers.

The VEs will score your exam immediately. You'll know before you leave the exam site whether you've passed. Chances are very good that, if you've studied hard, you'll get that passing grade!

GETTING/KEEPING YOUR CALL SIGN

Once the VE team scores your exam and you've passed, the process of upgrading your official FCC Amateur Radio License begins – usually that same day.

At the exam site, you will complete NCVEC Form 605, which is your application to the FCC for your license. If you pass the exam, the VE team will send on the required paperwork to their VEC. The VEC reviews the paperwork then files your application with the FCC. This filing is done electronically and your license upgrade will be granted and posted on the FCC's website within a few days.

Should you keep your old call sign or ask for a new one? If you check the box on the Form 605 asking for a new call sign, the FCC's computer will assign one to you. But you might not like what you get. So it is best to stick with your old call sign, and if you want a new, specific, call file an application for a Vanity Call Sign. See Chapter 4 for more details.

As soon as you pass your exam and have the CSCE, you are permitted to go on the air with your Extra Class HF privileges – even before your license upgrade appears on the FCC's Universal License System database. See Chapter 4 for more details on this process.

Vanity Call Signs

Once you've upgraded to Extra Class, you may want to get a personalized call sign that has special meaning to you. If so, you can apply for a Vanity Call Sign with a different letter combination. This process is explained in more detail In Chapter 4.

HOW MANY CLASSES OF LICENSES?

Once you've passed your Element 4 exam and go on the air as a new Extra Class operator, you'll be talking to more hams throughout the U.S. and around the world. Here's a summary of the new and "grandfathered" licenses that your fellow amateurs may hold and a recap of the level of expertise they have demonstrated in order to gain their licenses.

New License Classes

Following the FCC's restructuring of Amateur Radio licensing that took effect April 15, 2000, there are just three license classes – Technician, General, and Extra. Persons who hold licenses issued prior to April 15, 2000 may continue to hold onto their license class and renew it every 10 years for as long as they wish.

"Grandfathered" Licensees

Once you get on the air with your new Extra Class privileges, every now and then you might meet a Novice operator while yakking on 10 meters, or sending CW on 15, 40, or 80 meters. And then there are the Advanced Class operators who may continue to hold onto their license class designation until they finally decide to move up to Extra Class. When you meet an operator holding an Advanced Class license, you know without a doubt that he had to successfully complete a 13 word per minute Morse code test along with the written exam to earn that license.

When you look at the Frequency Charts in this book, you'll see that we have updated them to reflect the expanded privileges on many high frequency bands for Extra Class operators. We also list the sub-band privileges for Extra Class, Advanced Class, General Class, Technician Class and even the Novice.

"Paper Upgrade" to General

The FCC Rules (Part 97.505) provides credit for General Class Element 3, without examination, to those applicants who can prove they held an unexpired Technician Class license granted before March 21, 1987. No other class of license receives this element credit.

To see if a candidate qualifies for any of these opportunities to regain their former license privileges, or for a "paper upgrade" to General Class, contact the W5YI VEC at 800-669-9594.

Element Credit for Prior Licenses

Recent rule changes by the FCC make it possible for hams whose licenses have expired to regain their privileges simply by taking and passing the Element 2 Technician Class written exam. If you were licensed as a General, passing the Tech exam will earn you a new General Class license. If you were Advanced, passing the Tech exam will also get you a new General Class license. And if you were an Extra, passing the Tech exam will earn you a new Extra Class license!

This rule applies to those with expired General, Advanced, and Extra Class licenses that are beyond the two-year grace period. In order to take advantage of this, you must show proof of holding your expired license. If you do not have a copy of your old license, one resource to help verify your former status is to e-mail call sign historian Pete Varounis, NL7XM, at twelveVDC@aol.com.

Paperless Licenses

Effective February 17, 2015 the FCC no longer routinely issues/mails a paper license document to amateur radio licensees and other WTB licensed services. The Commission maintains that the official Amateur Radio license authorization is the electronic record that exists in its Universal Licensing System database. Licensees may access their current, official authorization, and print out an unofficial "reference copy" of their license, using the FCC ULS License Search feature at: ☞ **http://wireless.fcc.gov/uls**.

The FCC will provide a paper license to those who notify the Commission that they prefer to receive one. To notify the FCC that you want a license mailed to you, contact the FCC at 877-480-3201 between 8:00 a.m. to 6:00 p.m. Eastern Time Monday through Friday (except Federal holidays). You can also request a paper license by logging into the ULS license record using your FRN and CORES password and selecting to receive a paper copy by mail. If you do not know your CORES password, you can reset it on the FCC website:
☞ **https://apps.fcc.gov/coresWeb**

NOTE: The renewal period on your new Extra Class license is NOT 10 years. It remains the SAME renewal date that is now on your General or Advanced Class license. Make sure your address is up-to-date, too! An upgrade to your license class does NOT change the original grant date of your previous license class.

IT'S EASY!

The primary pre-requisite for passing any amateur radio operator license exam is the "will" to do it. If you follow our suggestions in this book, your chances of passing the Extra Class exam are excellent.

The 2007 FCC rule changes that eliminated the Morse code test requirement for Extra Class make it even easier for you to earn your Extra Class license and enjoy voice privileges on the high frequency bands that were expanded in those same changes.

When you become an Extra Class operator, GET ON THE AIR! Operating is an important step to becoming a good ham. So GET ON THE AIR and enjoy your new Extra Class worldwide HF privileges. Remember, you are now at the top of the hobby!

Want to find a test site fast?
Visit the W5YI-VEC website at ☞ **www.W5YI.org**, or call 800-669-9594.

3

Getting Ready for the Exam

The complete new Element 4 Question Pool of 712 question and answers that will be used to make-up your 50-question Extra Class written theory examination is included in this chapter. The questions and answers are presented *exactly as they will appear on your exam.* Each question has a correct answer and three distracters (the incorrect answers - those that distract you from the truth!). A score or 74%, or no more than 13 wrong answers, will earn you a passing grade.

THIS EXTRA CLASS QUESTION POOL IS VALID FROM JULY 1, 2016, THROUGH JUNE 30, 2020.

The Volunteer Examiners who make up the exam are not permitted to reword the questions, nor are they permitted to change any of the answers (but they can change the A B C D *order* of the answer choices). *Every question in this book, letter for letter, word for word, number for number, and every right and wrong answer, will be exactly the same on your upcoming 50-question Extra Class Element 4 exam.*

This new seventh edition study guide is designed to educate you about Extra Class operating techniques, rules and regulations, propagation, electrical theory, practical circuits, antennas, and more. We do this by including an explanation of the correct answer for each question. We also offer some memory tricks to help you remember the correct answer. Key words that help you associate the question and correct answer are highlighted in ***blue*** in the explanations. For those pesky distracters, we provide hints on how to identify wrong answers that may look very much like the correct one.

The new Element 4 Extra Class question pool was developed by the Question Pool Committee of the National Conference of Volunteer Exam Coordinators. They began by carefully reviewing the most recent Extra Class question pool, editing out more than 128 obsolete or duplicate questions, and then adding in about 138 brand new questions designed to keep the pool up-to-date with newer technology and current FCC rules and regulations. This new 2016-2020 question pool was carefully developed by the QPC under the leadership of chairman Roland Anders, K3RA, of the Laurel Amateur Radio Club VEC, and committee members Perry Green, WY1O, of the ARRL VEC, Larry Pollock, NB5X, of the W5YI VEC, and Jim Wiley, KL7CC, of the Anchorage Amateur Radio Club VEC.

Three dozen additional volunteers (including Gordo and Eric) recommended potential new questions to the Question Pool Committee. We all want to thank the members of the Question Pool Committee for their dedication and the hundreds of hours spent updating Extra Class pool. Thanks, Team, for a fresh, new Extra Class question pool!

WHAT THE EXAM CONTAINS

Your exam will be comprised of 50 questions taken from the current Element 4 question pool. The pool is divided into 10 subelements. *Table 3-1* shows you the subelement topics, the total number of questions in each, and the number of questions from each that will appear on your exam. Carefully study *Table 3-1* to gain an understanding of how your exam will be constructed.

You will find that each subelement is divided into topic groups. There will be one question on your exam taken from each topic group. If there is a specific group of questions that are giving you a hard time, study them later. Only one question will be taken from any one group. If you just can't get a good understanding of that one group, remember that will cause you to miss only one question.

Table 3-1. Question Distribution for the Extra Class Element 4 Exam

Subelement Number	Subelement Topic	No. of Questions in Pool	No. of Questions on Exam
E1	***Commission's Rules*** (FCC rules for the Amateur Radio services)	75	6
E2	***Operating Procedures*** (Amateur station operating procedures)	73	5
E3	***Radio Wave Propagation*** (Radio wave propagation characteristics)	46	3
E4	***Amateur Practices***	79	5
E5	***Electrical Principles*** (Electrical principles as applied to amateur station equipment)	65	4
E6	***Circuit Components*** (Amateur station equipment circuit components)	87	6
E7	***Practical Circuits*** (Practical circuits employed in amateur station equipment)	117	8
E8	***Signals and Emissions*** (Signals and emissions transmitted by amateur stations)	47	4
E9	***Antennas and Feed Lines*** (Amateur station antennas and feed lines)	112	8
E0	***Safety***	11	1
Total		712	50

NOTE: The Element 4 Question Pool originally contained 713 questions. One question was deleted by the QPC prior to publication of this book, resulting in a pool with 712 active questions. The deleted question does not appear in this book

QUESTION POOL VALID THROUGH JUNE 30, 2020

The Element 4 question pool in this book is valid from July 1, 2016, until June 30, 2020. During this period, the questions and answers will not be changed. *Every question in this book, letter for letter, word for word, number for number, and every right and wrong answer, will be exactly the same on your 50-question Extra Class Element 4 exam.* If FCC rules change or advancements in technology cause any question to become obsolete, that question will be deleted from the question pool and no longer appear on an exam. There will be no surprises in the exam room between what you study here and the questions and answers you see on your test.

The new 2016-2020 Extra Class Element 4 pool covers many new subject areas designed to keep the Extra Class questions fresh:

- *Ham Rules and Regulations on band edge SSB operation*
- *Satellite operation and modes*
- *ATV and SSTV technicalities*
- *Contest operating*
- *Contest log files*
- *PSK and QPSK*
- *Antenna takeoff patterns*
- *Antenna analyzer versus SWR Bridge*
- *NPN transistors*
- *Impedance calculations*
- *Q for a series tuned circuit*
- *dBm noise floor calculations*
- *Roofing filter topics*
- *Receiver blocking dynamic range*
- *Intermodulation interference*
- *Third order intercept levels*
- *Noise blankers*
- *RF attenuators*
- *Schottky barrier diodes*
- *Emitter amplifier bias*
- *Third order intermodulation distortion*
- *Parametric amplifiers*
- *High voltage power supply filters*
- *Frequency counters*
- *Antenna modeling*
- *Safety practices, RF radiation hazards, hazardous materials*

QUESTION CODING

Each question in the Element 4 Extra Class pool is assigned a coded number using a system of letters and numbers. *Figure 3-1* explains the coding for question E5G09. Understanding how the questions are numbered in the pool will help you know when you encounter a question on your exam that is from a section that was giving you trouble. Not to panic… there will be only one question from that subelement group on the exam.

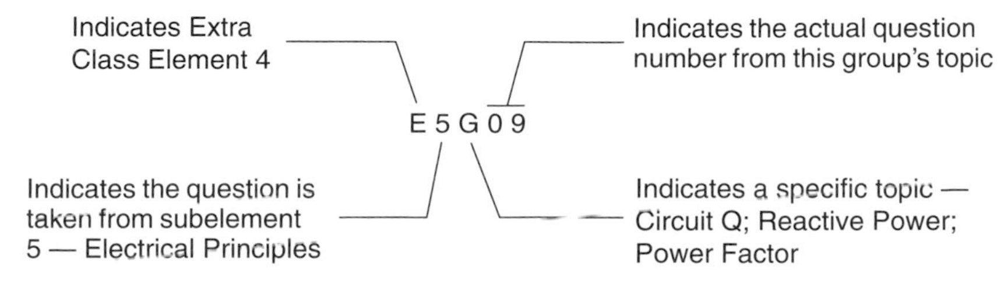

Figure 3-1. Examination Question Coding

STUDY HINTS

You're probably saying: "712 questions and answers! This is an overwhelming number to learn for the Extra Class exam!" But good news – almost every question is asked at least twice with just slightly different wording. Many of the math questions are based on a single formula. Once you know it, you can easily answer the 4 or 5 questions based on that formula

The mathematical formulas that you need for the Extra Class exam are noted with an E▶ in the questions that follow and are summarized in the Appendix on pages 290 to 294. You should learn the formulas, understand how to use them, and make sure you know the correct units to use. Many times you may have to convert kilohertz (kHz) to megahertz (MHz), or vice versa. Understand the formulas and learn how to apply them to arrive at a correct answer. In many cases, we will give you handy shortcuts to help you through the longer formulas. These shortcuts and a little humor now and then should keep you on track.

Work your book! Put a check mark beside questions you have down cold. Circle those that may need a little bit more study. Highlight those that appear to be brain-busters and figure out ways to remember the correct answers. Remember, do not memorize the answer key (A, B, C, or D) since the order of the answers can be changed for your exam.

If you study the questions and answers for a half hour in the morning and a half hour in the evening, it should take you approximately 30 days to prepare for the Extra Class exam. If you are an Advanced Class operator, it may take you only 15 days to prepare for your upgrade. One week before the exam, highlight the correct answer to each question and spend that week looking at the question and only the highlighted correct answer. The day before the exam, study the ***BLUE KEY WORDS*** in each explanation.

QUESTIONS REARRANGED FOR SMARTER LEARNING

The first thing you'll notice when you look at how the Element 4 question pool is presented in the *Extra Class* book is that the entire question pool is arranged into logical topic areas for easier learning. This arrangement is the same as Gordo uses when he teaches his 4-day Extra Class seminars. Rather than jumping back and forth between subelement topic areas to study similar-topic questions, he and Eric take you LOGICALLY through the entire question pool in 16 topic areas. Here is a list of the topic areas and the page numbers on which each begins in this book:

A numerical cross-reference of all of the Element 4 questions is found in the Appendix, on page 284. If you need to look up a specific question by its number, we tell you precisely the page that it is on. This logical arrangement of questions is an exclusive training method found in all Gordon West ham radio test preparation books.

Every single question in the Element 4 pool has been placed within the proper topic area for logical learning as a group. Gordo's classroom students love this approach to learning, and you will, too!

ADDITIONAL STUDY MATERIALS

Passing your Extra Class exam will require plenty of study. There are additional resources that will help you learn – and understand – the material presented in this book.

First, Gordo and Eric have recorded a six CD audio course for Extra Class, which is an excellent addition to this book. If you spend a lot of time in your car or pickup, or if you just learn better by listening, the audio CD course is for you. It follows the same outline as this book, and includes many sounds of Extra Class operation to help bring the material to life.

Next, if you like learning at your PC, study software is available from the W5YI Group to accompany this book. This W5YI program includes the answer explanations from the book along with the questions and answers to study. When you take a practice exam, the software shows you the correct answer, scores your work, and tells you where you need to focus your study efforts.

When you're ready to learn more about the electronic principles introduced in the questions of the Element 4 pool, we offer several books on "basic" electronics – *Basic Electronics, Basic Communications Electronics,* and *Basic Digital Electronics* – available where you purchased this book, or from The W5YI Group. If you are extra technical, we also have 4 Engineer's Mini Notebooks written by Forrest M. Mims III, which include great electronic projects with step by step instructions. A listing of these additional study materials can be found in the back of this book.

Of course, the real learning of the material covered in these questions happens after you pass your exam and get on the air as an Amateur Extra Class operator!

EXAMINATION QUESTIONS

Let's get started reviewing the questions and answers. Read over the question and see if you already know the answer. Next, spot the likely answer and then read the explanation of the correct answer. Finally, at the end of the explanation, we give you the correct A, B, C, or D answer. Use a small card to cover up the explanations and then slowly reveal the 4 possible answers, the explanation, and then the correct answer. Put a check mark beside the question number if you answered it correctly. Put an X if you need more study with a specific question.

And one more time, be assured that every question in this book, letter for letter, word for word, number for number, and every right and wrong answer, will be exactly the same on your upcoming 50-question Extra Class Element 4 exam.

Work the book! In Gordo's classes, he can usually tell who will pass the exam the first time by looking at their marked up textbooks.

Ready to get started? Here we go with your first look into the world of the amateur Extra Class license.

Rules & Regs

E1A01 When using a transceiver that displays the carrier frequency of phone signals, which of the following displayed frequencies represents the highest frequency at which a properly adjusted USB emission will be totally within the band?

A. The exact upper band edge.
B. 300 Hz below the upper band edge.
C. 1 kHz below the upper band edge.
D. 3 kHz below the upper band edge.

On 20 meters, stay 3 kHz below 14.350 MHz.

Worldwide high frequency transceivers may automatically select the correct upper or lower sideband mode when you move around the bands. But they will NOT automatically stop you from tuning to the edge of the band and beyond. It is up to you to know your limits in each band for each operating mode. Say you are on 20 meters dialed in to the maritime mobile net frequency of 14.300 MHz. This is the CARRIER frequency from which the voice emission will modulate UP approximately 2.8 kHz. Your transmitted voice will occupy just under 3 kHz of bandwidth, 14.300 to 14.303 MHz. To meet this requirement, the carrier frequency displayed on your dial would be at least ***3 kHz below the upper band edge***. 14.347 MHz is the closest you would want to dial your carrier frequency to stay within the upper limit of the 20 meter band. [97.301, 97.305] **ANSWER D.**

E1A03 With your transceiver displaying the carrier frequency of phone signals, you hear a station calling CQ on 14.349 MHz USB. Is it legal to return the call using upper sideband on the same frequency?

A. Yes, because you were not the station calling CQ.
B. Yes, because the displayed frequency is within the 20 meter band.
C. No, the sideband will extend beyond the band edge.
D. No, U.S. stations are not permitted to use phone emissions above 14.340 MHz.

No! 14.347 MHz is the closest we would want to dial our carrier frequency to stay within the upper limit of the 20 meter band. Remember, band limit frequencies vary for other countries. You must remain within the limits for U.S. operators. Don't be tempted to operate out of band to contact this DX station. ***Band edge operation may result in sidebands extending beyond the U.S. band edge limits***, which is a rules violation for us. [97.301, 97.305] **ANSWER C.**

E1A02 When using a transceiver that displays the carrier frequency of phone signals, which of the following displayed frequencies represents the lowest frequency at which a properly adjusted LSB emission will be totally within the band?

A. The exact lower band edge.
B. 300 Hz above the lower band edge.
C. 1 kHz above the lower band edge.
D. 3 kHz above the lower band edge.

This question now has us on lower sideband, maybe 40 meters, so we need to ***stay 3 kHz ABOVE the lower band edge*** to stay within the band. It would be considered poor operating practice to intentionally hug a band edge and switch to the non-conforming sideband. [97.301, 97.305] **ANSWER D.**

E1A04 With your transceiver displaying the carrier frequency of phone signals, you hear a DX station calling CQ on 3.601 MHz LSB. Is it legal to return the call using lower sideband on the same frequency?

A. Yes, because the DX station initiated the contact.
B. Yes, because the displayed frequency is within the 75 meter phone band segment.
C. No, the sideband will extend beyond the edge of the phone band segment.
D. No, U.S. stations are not permitted to use phone emissions below 3.610 MHz.

Lower sideband is used on 75 meters. The lowest frequency near the voice band edge that we would select as our dial carrier frequency is 3.603 MHz. ***U.S. operators on 3.601 MHz*** would have ***sideband emissions extending illegally beyond the band edge*** of the phone segment. [97.301, 97.305] **ANSWER C.**

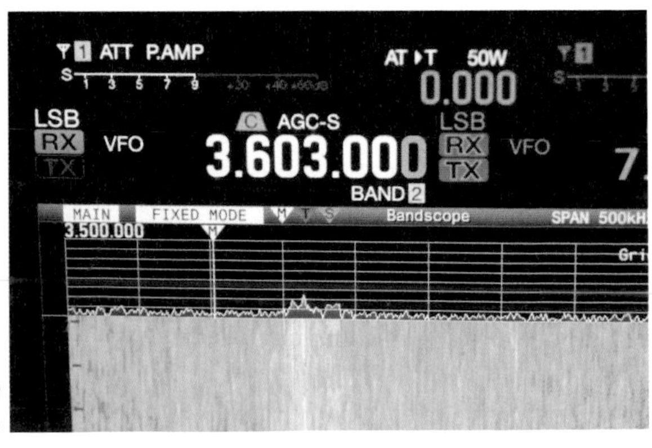

Stay 3 kHz above the lower voice limit on 75m.

E1A12 With your transceiver displaying the carrier frequency of CW signals, you hear a DX station's CQ on 3.500 MHz. Is it legal to return the call using CW on the same frequency?

A. Yes, the DX station initiated the contact.
B. Yes, the displayed frequency is within the 80 meter CW band segment.
C. No, one of the sidebands of the CW signal will be out of the band.
D. No, U.S. stations are not permitted to use CW emissions below 3.525 MHz.

Be cautious NOT to operate precisely at the band edge of your privileges. Emissions of your ***CW signal might extend beyond band edge*** limitations. While quite narrow, a CW signal still occupies a certain bandwidth which is dependent on both keying speed and keying wave shape. [97.301, 97.305] **ANSWER C.**

☞ **http://www.comportco.com/~w5alt/cw/cwindex.php?pg=5**

E1A07 Which amateur band requires transmission on specific channels rather than on a range of frequencies?

A. 12 meter band.
B. 17 meter band.
C. 30 meter band.
D. 60 meter band.

It is only on the ***60 meter band*** where ham radio transmissions are limited to five specific channels. [97.303] **ANSWER D.**

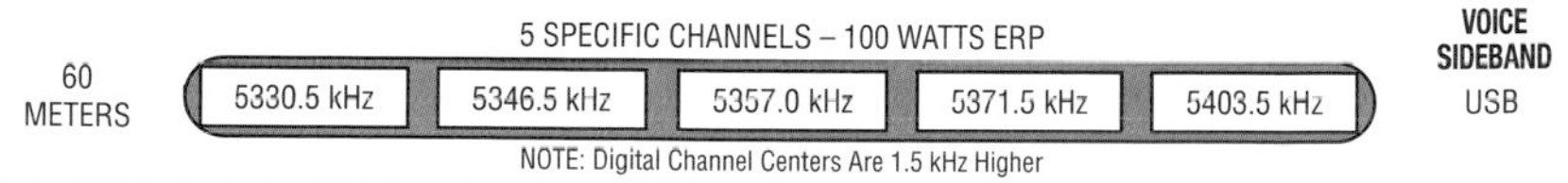

E1A05 What is the maximum power output permitted on the 60 meter band?

A. 50 watts PEP effective radiated power relative to an isotropic radiator.
B. 50 watts PEP effective radiated power relative to a dipole.
C. 100 watts PEP effective radiated power relative to the gain of a half-wave dipole.
D. 100 watts PEP effective radiated power relative to an isotropic radiator.

The FCC recently granted higher power output on the 60 meter band. We can now transmit ***100 watts*** effectively radiated power ***(ERP) relative to the gain of a half-wave dipole*** on that band. Since most HF rigs output just ***100 watts***, you are good to go with a 60 meter dipole antenna. However, running an amplifier on 60 meters to exceed 100 watts output is not allowed. [97.313] **ANSWER C.**

CW DXer Chip Margelli, K7JA, can now run code on the 60 meter band!

E1A14 What is the maximum bandwidth for a data emission on 60 meters?

A. 60 Hz.
B. 170 Hz.
C. 1.5 kHz.
D. 2.8 kHz.

FCC Rule 97.303 states that amateur radio operators must ensure that ***emissions from their 60 meter stations do not occupy more than 2.8 kHz bandwidth***, centered on the channel center frequency. PSK31 audio is on channel center frequencies, generally 1500 Hz above the suppressed carrier frequency. However, CW 150HA1A must be centered on channel, where channel sharing and cooperation are of paramount importance. [97.303] **ANSWER D.**

E1A06 Where must the carrier frequency of a CW signal be set to comply with FCC rules for 60 meter operation?

A. At the lowest frequency of the channel.
B. At the center frequency of the channel.
C. At the highest frequency of the channel.
D. On any frequency where the signal's sidebands are within the channel.

The CW designator 150HA1A precludes tone modulated CW where different tones could help sort out different CW stations transmitting on the same channel. The new rules read: "For ***CW emissions, the carrier frequency is set to the CENTER frequency***." This means that if all ham sets sending CW have zero error on the center channel display, all the CW signals will come out with the same exact tone! Band planning may help solve the problem of multiple CW signals sharing a specific channel sounding the same! [97.15] **ANSWER B.**

CARRIER TUNING FREQUENCY	CENTER FREQUENCY
5330.5 kHz	5332.0 kHz
5346.5kHz	5348.0 kHz
5357.0 kHz (New replacement)	5358.5 kHz
5371.5 kHz	5373.0 kHz
5403.5 kHz	5405.0 kHz

Recently, the previously authorized channel 5368 kHz was replaced with a new center frequency, 5358.5 kHz. Amateur radio operators also now have multiple modes authorized on the five 60-meter channels – voice, CW, and data modes using PACTOR III and RTTY using the PSK31 technique. Here's the hard part – we are required to share all of this on these five 60 meter channels. The ARRL recommends digital modes PSK31 and PACTOR III only. Time-sharing the channels and cooperating with other stations that also have access to these channels may allow these multiple modes to co-exist on the 60 meter band, which is perfectly situated in the spectrum for effective emergency communications. Learn more about the 60 meter regs on the FCC website: www.fcc.gov/document/amateur-radio-service-5-mhz.

E1A13 Who must be in physical control of the station apparatus of an amateur station aboard any vessel or craft that is documented or registered in the United States?

A. Only a person with an FCC Marine Radio.
B. Any person holding an FCC issued amateur license or who is authorized for alien reciprocal operation.
C. Only a person named in an amateur station license grant.
D. Any person named in an amateur station license grant or a person holding an unrestricted Radiotelephone Operator Permit.

If you go offshore aboard your U.S. documented or registered boat, ***anyone*** on board ***holding an FCC issued amateur radio license*** of the correct grade can work the ham radio equipment. This also holds true for your non-U.S.-citizen friend who holds an ***FCC issued reciprocal license***. Just because you may be in international waters, non-licensed personnel (the ship's Captain, for instance) would not be permitted to operate on the ham bands, even though they may hold a Marine Radiotelephone Operator's Permit. [97.5] **ANSWER B.**

E1A11 Which of the following describes authorization or licensing required when operating an amateur station aboard a U.S.-registered vessel in international waters?

A. Any amateur license with an FCC Marine or Aircraft endorsement.
B. Any FCC-issued amateur license.
C. Only General class or higher amateur licenses.
D. An unrestricted Radiotelephone Operator Permit.

This is a great question dealing with maritime mobile operation when out on the high seas. If the vessel is flying the American flag, registered in the United States, ***any ham radio transmissions MUST be from an FCC-licensed amateur operator or person holding a reciprocal permit for an alien amateur license***. Years ago there was the false belief that once someone was in international waters, anyone could transmit over ham radio without a license. Not so! Your ham license is required and is separate from the ship's marine radio license or the captain's Marine Radio Operator Permit. [97.5] **ANSWER B.**

☞ **www.arrl.org/regulations**

Maritime Mobile (M/M) is found on 14.300 MHz.

E1A10 If an amateur station is installed aboard a ship or aircraft, what condition must be met before the station is operated?

A. Its operation must be approved by the master of the ship or the pilot in command of the aircraft.
B. The amateur station operator must agree not to transmit when the main radio of the ship or aircraft is in use.
C. The amateur station must have a power supply that is completely independent of the main ship or aircraft power supply.
D. The amateur operator must have an FCC Marine or Aircraft endorsement on his or her amateur license.

If you are planning a long cruise, or taking a ride in a friend's airplane, make sure you have received ***approval by the captain (master of the ship) or aircraft pilot before you begin to transmit***. Look at the three incorrect answers – they sound logical, but only the answer we just gave you is correct. [97.11] **ANSWER A.**

E1B03 Within what distance must an amateur station protect an FCC monitoring facility from harmful interference?

A. 1 mile.
B. 3 miles.
C. 10 miles.
D. 30 miles.

The Federal Communications Commission recently automated its monitoring stations. Now there are remote-controlled, unmanned monitoring stations, and all amateur ***operators within one mile of those stations must contact the local FCC engineer in charge*** to ensure they do not cause harmful interference to the remote-controlled monitoring units. [97.13] **ANSWER A.**

E1B05 What is the National Radio Quiet Zone?

A. An area in Puerto Rico surrounding the Arecibo Radio Telescope.
B. An area in New Mexico surrounding the White Sands Test Area.
C. An area surrounding the National Radio Astronomy Observatory.
D. An area in Florida surrounding Cape Canaveral.

A national ***radio quiet zone*** covering parts of Maryland, West Virginia, and Virginia has been established near the ***National Radio Astronomy Observatory***. The quiet zone restriction applies to "constant-on" stations like beacons and repeaters. Operations from your home or vehicle would not be affected unless you actually enter the Observatory property in Greenbank, WV. [97.3] **ANSWER C.**

E1B02 Which of the following factors might cause the physical location of an amateur station apparatus or antenna structure to be restricted?

A. The location is near an area of political conflict.
B. The location is of geographical or horticultural importance.
C. The location is in an ITU Zone designated for coordination with one or more foreign governments.
D. The location is of environmental importance or significant in American history, architecture, or culture.

The ***environmental impact*** of a proposed installation is now a big consideration on anything that may have significant ***impact on American history, architecture, or culture***. For instance, you would not want to place your station on an Indian burial ground. [97.13] **ANSWER D.** ☞ **www.fcc.gov/mb/facts**

E1B04 What must be done before placing an amateur station within an officially designated wilderness area or wildlife preserve, or an area listed in the National Register of Historical Places?

A. A proposal must be submitted to the National Park Service.
B. A letter of intent must be filed with the National Audubon Society.
C. An Environmental Assessment must be submitted to the FCC.
D. A form FSD-15 must be submitted to the Department of the Interior.

If you plan to install your station on a wildlife preserve, or an area listed as an historical place, you will need to ***submit an environmental assessment to the FCC*** for its review and approval. Be prepared for an extremely long wait before your permit may be granted! [97.13, 1.1305-1.1319] **ANSWER C.**

E1C12 What types of communications may be transmitted to amateur stations in foreign countries?

A. Business-related messages for non-profit organizations.
B. Messages intended for connection to users of the maritime satellite service.
C. Communications incidental to the purpose of the amateur service and remarks of a personal nature.
D. All of these choices are correct.

With your new Extra Class privileges you will likely make even more contacts in foreign countries. Remember, no business communications allowed, so keep your communicatios to purely ***personal remarks or communications of a technical nature***. [97.117] **ANSWER C.**

E1F02 What privileges are authorized in the U.S. to persons holding an amateur service license granted by the Government of Canada?

A. None, they must obtain a U.S. license.
B. All privileges of the Extra Class license.
C. The operating terms and conditions of the Canadian amateur service license, not to exceed U.S. Extra Class privileges.
D. Full privileges, up to and including those of the Extra Class License, on the 80, 40, 20, 15, and 10 meter bands.

When operating in a foreign country, you need to be sure you comply with both your own license terms *and* those of the country in which you operate. This works both ways, as in the example of Canada. While amateur radio privileges are approximately matched in most countries, in many cases the "band borders" are not exactly the same... even those of Canada and the U.S. ***So be sure you comply with both countries' rules at all times***. [97.107] **ANSWER C.**

Going "down under?" The Australian Communications and Media Authority (ACMA) has issued a new class of license to allow visiting amateurs to operate in Australia for up to 90 days. All you have to do is use your home call sign followed by the suffix "VK" followed by "portable" and then announce the location of your station, without doing anything more.

E1C04 What is meant by IARP?

A. An international amateur radio permit that allows U.S. amateurs to operate in certain countries of the Americas.
B. The internal amateur radio practices policy of the FCC.
C. An indication of increased antenna reflected power.
D. A forecast of intermittent aurora radio propagation.

Amateur Radio is an international hobby, and we should take advantage of every opportunity to enhance our international good will. This often involves traveling to foreign countries. ***The International Amateur Radio Permit (IARP) allows you to operate in many countries in the Americas.*** **ANSWER A.**

☞ **www.oas.org/juridico/english/sigs/a-62.html** for a list of participating IARP countries.

E1C11 Which of the following operating arrangements allows an FCC-licensed U.S. citizen to operate in many European countries, and alien amateurs from many European countries to operate in the US?

A. CEPT agreement.
B. IARP agreement.
C. ITU reciprocal license.
D. All of these choices are correct.

The ***CEPT agreement*** (European Conference of Postal & Telecommunications Administrations – CEPT) is great news for U.S. Advanced and Extra Class amateur operators ***when visiting many countries in Europe*** – no more reciprocal paperwork! Before leaving for Europe, pack a copy of FCC Public Notice DA 11-211 along with the original and copies of your U.S. ham license. This paperwork is necessary in case you are asked about the equipment you are bringing into that country at customs. Have fun in Europe with ham radio! [97.5] **ANSWER A.**

E1C13 Which of the following is required in order to operate in accordance with CEPT rules in foreign countries where permitted?

A. You must identify in the official language of the country in which you are operating.
B. The U.S. embassy must approve of your operation.
C. You must bring a copy of FCC Public Notice DA 11-221.
D. You must append "/CEPT" to your call sign.

The Central European Postal and Telecommunications (CEPT) authority is the "FCC" for much of Europe. If you travel there, you may operate in participating countries by presenting proof of U.S. Citizenship, your FCC license, and a copy of ***FCC Public Notice DA 11-221***. Note that CEPT for much of Europe is only *automatically* available to Advanced and Extra Class amateurs, which is yet another incentive to go for the Extra Class license. **ANSWER C.**

☞ **http://www.arrl.org/files/file/Regulatory/DA-11-221A1.pdf** for a copy of the Public Notice

Participating CEPT countries, as of February, 2015, are:

Albania
Austria
Belgium
Bosnia & Herzegovina
Bulgaria
Croatia
Cyprus
Czech Republic
Denmark
Estonia
Finland
France & its possessions
Germany
Greece
Hungary
Iceland
Ireland
Italy
Latvia
Liechtenstein
Lithuania
Luxembourg
Macedonia
Moldova
Monaco
Montenegro
Netherlands
Netherland Antilles
Norway
Poland
Portugal
Romania
Russian Federation
Serbia
Slovak Republic
Slovenia
Spain
Sweden
Switzerland
Turkey
Ukraine
United Kingdom & its possessions

ELMER E

Good To Know:

European Reciprocal Licenses Now Limited to Advanced & Extra Class Licensees

The European Conference of Postal and Telecommunications Administrations (CEPT) has revised its table of equivalence between FCC amateur licenses and the CEPT license. Effective February 4, 2008, Recommendation T/R 61-01 (as amended) *now grants* ***full CEPT privileges only to those U.S. citizens who hold an FCC-issued Advanced or Amateur Extra Class license****. This means that those U.S. licensees who hold an FCC-issued General or Technician license are no longer eligible for full operating privileges in countries where CEPT-reciprocal operation had previously been permitted. U.S. Novice class licensees have had no reciprocal operating privileges under the CEPT provisions.*

E1F04 Which of the following geographic descriptions approximately describes "Line A"?

A. A line roughly parallel to and south of the US-Canadian border.
B. A line roughly parallel to and west of the US Atlantic coastline.
C. A line roughly parallel to and north of the US-Mexican border and Gulf coastline.
D. A line roughly parallel to and east of the US Pacific coastline.

The ***United States and Canada*** have adopted the ***"Line A" area as a buffer zone*** for certain types of radio stations. This keeps normally "line of sight" radio energy from interfering with the other country's communications. Amateur transmissions are restricted on the 70-cm band between 420-430 MHz if the station is located north of "Line A." This is because Canada has a different type of radio allocation in the bottom part of the 70-cm band. [97.3] **ANSWER A.**

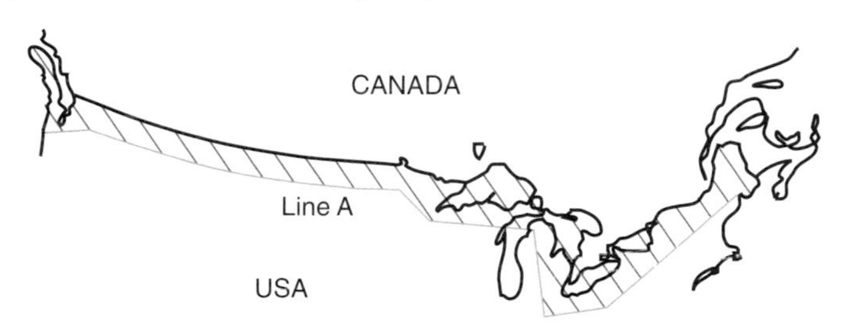

Amateur Radio Restrictions

E1F05 Amateur stations may not transmit in which of the following frequency segments if they are located in the contiguous 48 states and north of Line A?

A. 440 MHz - 450 MHz.
B. 53 MHz - 54 MHz.
C. 222 MHz - 223 MHz.
D. 420 MHz - 430 MHz.

Frequencies between 420-430 MHz are normally used for digital modes, fast-scan television, and auxiliary links. ***Within the "Line A" area*** along the Canadian border, ***U.S. hams are prohibited from operating below 430 MHz*** to keep their transmissions from interfering with the different radio allocations Canada has in this frequency range. [97.303] **ANSWER D.**

E1B06 Which of the following additional rules apply if you are installing an amateur station antenna at a site at or near a public use airport?

A. You may have to notify the Federal Aviation Administration and register it with the FCC as required by Part 17 of FCC rules.
B. No special rules apply if your antenna structure will be less than 300 feet in height.
C. You must file an Environmental Impact Statement with the EPA before construction begins.
D. You must obtain a construction permit from the airport zoning authority.

Live right next to an airport? If you do, you will need to ***consult Part 17 of the FCC Rules*** regarding the installation of that antenna tower. Likely, you will be ***working with the Federal Aviation Administration*** to get that tower sanctioned, as well as getting it documented, under Part 17 Rules. [97.15] **ANSWER A.**

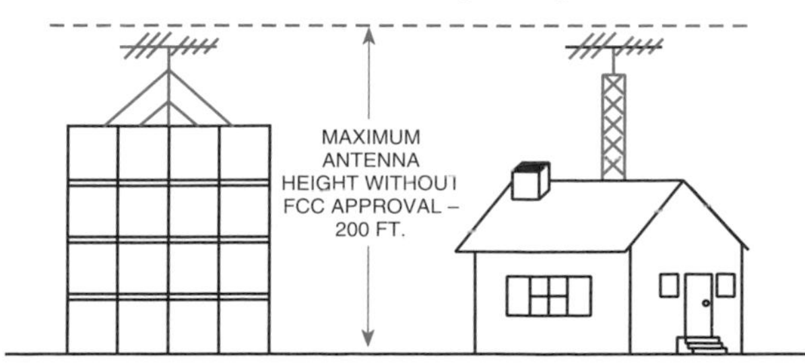

Antenna Height near a public use airport is governed by both the FCC and the FAA.

E1F07 When may an amateur station send a message to a business?

A. When the total money involved does not exceed $25.
B. When the control operator is employed by the FCC or another government agency.
C. When transmitting international third-party communications.
D. When neither the amateur nor his or her employer has a pecuniary interest in the communications.

In 1993, the prohibited business communications rules were redefined by the FCC to allow an amateur to ***send a message to a business as long as neither the amateur nor his or her employer had a pecuniary interest in the communications***. If you run a tree-trimming business, it would not be legal to use an autopatch to call your answering machine to find out where your next customer is waiting. On the other hand, if you are a school teacher and discover that a big pine tree has fallen down across the school's driveway, it would be perfectly acceptable to place an autopatch call to a tree-trimmer and have them come out to do their job. But could the school teacher call into the school district and tell them that she is going to be late for work? Probably not legal. [97.113] **ANSWER D.**

E1F08 Which of the following types of amateur station communications are prohibited?

A. Communications transmitted for hire or material compensation, except as otherwise provided in the rules.
B. Communications that have a political content, except as allowed by the Fairness Doctrine.
C. Communications that have a religious content.
D. Communications in a language other than English.

This question deals with permissible communications from one ham to another. Political, religious, and foreign language communications are permitted, but ***communications transmitted for hire or material compensation are normally prohibited***. [97.113] **ANSWER A.**

E1A08 If a station in a message forwarding system inadvertently forwards a message that is in violation of FCC rules, who is primarily accountable for the rules violation?

A. The control operator of the packet bulletin board station.
B. The control operator of the originating station.
C. The control operators of all the stations in the system.
D. The control operators of all the stations in the system not authenticating the source from which they accept communications.

A few years ago, dozens of amateur operators were sent "greetings" from the FCC for taking part in forwarding a packet radio message that contained materials specifically prohibited by the rules and regulations. The hams forwarding the messages claimed everything was in the automatic mode and it was not up to them to censor what was coming down the electronic data line! Hence, new rules hold the ***control operator of the station originating the message as primarily accountable*** and the control operator of the first forwarding station accountable for any violation of the rules. More specifically, everyone else "down the line" is no longer responsible for the message content, but must shut down their packet bulletin boards if informed that there is a message being automatically forwarded that violates FCC rules. [97.219] **ANSWER B.**

E1A09 What is the first action you should take if your digital message forwarding station inadvertently forwards a communication that violates FCC rules?

A. Discontinue forwarding the communication as soon as you become aware of it.
B. Notify the originating station that the communication does not comply with FCC rules.
C. Notify the nearest FCC Field Engineer's office.
D. Discontinue forwarding all messages.

If you are advised that your packet station is automatically forwarding communications in violation of the rules, get into the computer and ***discontinue forwarding this specific communication*** as soon as you become aware of it. You may continue to forward all other messages. [97.219] **ANSWER A.**

E1C07 What is meant by local control?

A. Controlling a station through a local auxiliary link.
B. Automatically manipulating local station controls.
C. Direct manipulation of the transmitter by a control operator.
D. Controlling a repeater using a portable handheld transceiver.

Local control is the actual place where you turn on your equipment, press the microphone PTT, and transmit. This is called ***"direct manipulation of the transmitter"*** by a control operator. [97.3] **ANSWER C.**

This is local control – the operator in front of his transmitter and nearby antenna.

E1C01 What is a remotely controlled station?

A. A station operated away from its regular home location.
B. A station controlled by someone other than the licensee.
C. A station operating under automatic control.
D. A station controlled indirectly through a control link.

Popular amateur radio IRLP stations located high atop mountains or buildings do not have the control operator sitting right in the repeater room. The station is controlled through a control link station. Most control link frequencies are found on UHF 420-430 MHz, or up on 1.2 GHz. ***"Control link" is the key for remote control*** of an amateur station or repeater. [97.3] **ANSWER D.**

E1C08 What is the maximum permissible duration of a remotely controlled station's transmissions if its control link malfunctions?

A. 30 seconds
B. 3 minutes.
C 5 minutes.
D. 10 minutes.

If a distant, remotely-controlled station is undergoing remote control tone commands from an auxiliary station, and the control link on microwave fails, that remotely-controlled station must ***shut down within three minutes***. [97.213] **ANSWER B.**

E2C13 What indicator is required to be used by U.S.-licensed operators when operating a station via remote control where the transmitter is located in the U.S.?

A. / followed by the USPS two letter abbreviation for the state in which the remote station is located.
B. /R# where # is the district of the remote station.
C. The ARRL section of the remote station.
D. No additional indicator is required.

In the past, non-home-based amateur communications were much more restrictive and required specific identifiers. ***Mobile designators are no longer required*** during your ID, but you are certainly free to indicate your mobile or remote status. **ANSWER D.**

E1C08 Which of the following statements concerning remotely controlled amateur stations is true?

A. Only Extra Class operators may be the control operator of a remote station.
B. A control operator need not be present at the control point.
C. A control operator must be present at the control point.
D. Repeater and auxiliary stations may not be remotely controlled.

To ***remotely control*** another ham station, the ***control operator must be present at the control point***. Only under AUTOMATIC control, like a repeater, may the control operator get some sleep! [97.109] **ANSWER C.**

E2E12 Which type of control is used by stations using the Automatic Link Enable (ALE) protocol?

A. Local.
B. Remote.
C. Automatic.
D. ALE can use any type of control.

Automatic Link Enable (ALE) methods are more commonly used in the military or in MARS (Military Auxiliary Radio System) operations, but they are gradually working their way into the amateur ranks. ALE protocol allows your station to ***choose automatically the best band or frequency*** for optimum communications under changing conditions. It's actually quite a fun thing to watch in operation, and it works with nearly any mode – voice or digital. Expect to see more ALE operation in the near future as the solar cycle declines. **ANSWER C.**

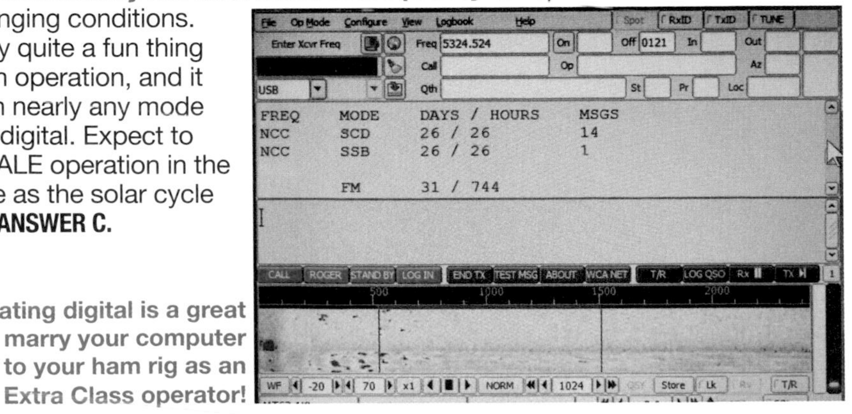

Operating digital is a great way to marry your computer to your ham rig as an Extra Class operator!

E1C02 What is meant by automatic control of a station?

A. The use of devices and procedures for control so that the control operator does not have to be present at a control point.
B. A station operating with its output power controlled automatically.
C. Remotely controlling a station's antenna pattern through a directional control link.
D. The use of a control link between a control point and a locally controlled station.

Ham operators have come up with ***devices and procedures to automatically control a station*** so they don't have to stand guard at the repeater block house on a mountain top. [97.3, 97.109] **ANSWER A.**

E1C03 How do the control operator responsibilities of a station under automatic control differ from one under local control?

A. Under local control there is no control operator.
B. Under automatic control the control operator is not required to be present at the control point.
C. Under automatic control there is no control operator.
D. Under local control a control operator is not required to be present at a control point.

Repeaters under ***automatic control don't have anyone sitting*** at that distant control point. [97.3, 97.109] **ANSWER B.** ☞ www.winlink.org

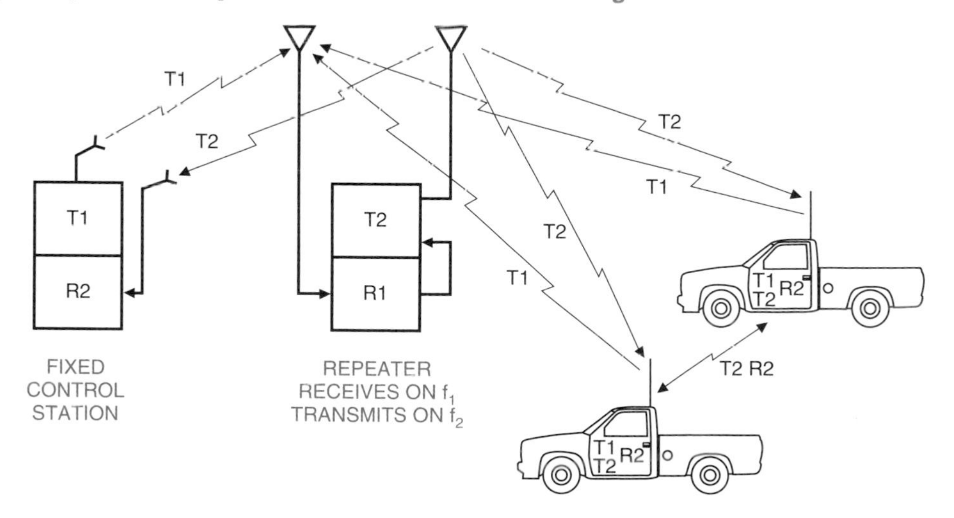

Repeater
Copyright ©1992 Master Publishing, Inc., Niles, IL

E1C09 Which of these ranges of frequencies is available for an automatically controlled repeater operating below 30 MHz?

A. 18.110 MHz - 18.168 MHz.
B. 24.940 MHz - 24.990 MHz.
C. 10.100 MHz - 10.150 MHz.
D. 29.500 MHz - 29.700 MHz.

Repeater operation on HF is limited to the top of the 10 meter band, between ***29.5 to 29.7 MHz***. No other HF (3-30MHz) band permits repeater operation. [97.205] **ANSWER D.**

E1C05 When may an automatically controlled station originate third party communications?

A. Never.
B. Only when transmitting RTTY or data emissions.
C. When agreed upon by the sending or receiving station.
D. When approved by the National Telecommunication and Information Administration.

This question asks, "may an automatically controlled station ORIGINATE third party communications." The answer is ***NEVER***, because a third party without a ham license may not be designated by the first control operator to assume the control operator function required to ORIGINATE that third party communication. It takes a licensed control operator to initiate an unattended "auto-start" station that may be handling third party traffic. [97.221(c)(1) and 97.115(c)] **ANSWER A.**

E1C10 What types of amateur stations may automatically retransmit the radio signals of other amateur stations?

A. Only beacon, repeater or space stations.
B. Only auxiliary, repeater or space stations.
C. Only earth stations, repeater stations or model craft.
D. Only auxiliary, beacon or space stations.

Automatically retransmitting the radio signals of other ham stations takes place through a common ***repeater***, through ***satellite and space stations*** that are like repeaters in orbit around the Earth, and through an ***auxiliary station*** such as one handling IRLP. [97.113] **ANSWER B.** ☞ **www.winsystem.org**

E1F12 Who may be the control operator of an auxiliary station?

A. Any licensed amateur operator.
B. Only Technician, General, Advanced or Amateur Extra Class operators.
C. Only General, Advanced or Amateur Extra Class operators.
D. Only Amateur Extra Class operators.

An amateur radio auxiliary station is a transceiver under your local control which remotely controls another station, like a mountaintop repeater or a distant high-ridge cross-band system. ***Any amateur radio operator, EXCEPT NOVICE CLASS, may set up and be the control operator of an auxiliary station***. While Novice and Advanced Class licenses are no long being issued, there are hams who still hold valid Novice and Advanced Class licenses. [97.201] **ANSWER B.**

E1F06 Under what circumstances might the FCC issue a Special Temporary Authority (STA) to an amateur station?

A. To provide for experimental amateur communications.
B. To allow regular operation on Land Mobile channels.
C. To provide additional spectrum for personal use.
D. To provide temporary operation while awaiting normal licensing.

Ham radio operators are always on the prowl for new types of emissions and vacant frequencies where the ***FCC may allow us special temporary authority***. While obtaining the FCC's STA requires diligence and numerous filings, ham operators have been successful in obtaining permission ***to experiment with new emissions*** that are not even mentioned in the rules and to operate on unused very low frequencies (VLF) for specific propagation studies. These experimental communications may lead to future rule makings on our behalf! [1.931] **ANSWER A.**

E1B09 Which amateur stations may be operated under RACES rules?

A. Only those club stations licensed to Amateur Extra class operators.
B. Any FCC-licensed amateur station except a Technician class.
C. Any FCC-licensed amateur station certified by the responsible civil defense organization for the area served.
D. Any FCC-licensed amateur station participating in the Military Auxiliary Radio System (MARS).

All amateur radio operators are eligible for RACES membership once they are properly ***certified by the civil defense organization responsible for the area served.*** You must plan ahead if you want to be part of RACES. It takes official certification and training before you can help. In an emergency the best an uncertified and untrained operator can do is listen and stay out of the way. [97.407] **ANSWER C.**

☞ **www.fema.gov/pdf/plan/slg101.pdf**

The RACES logo.

E1B10 What frequencies are authorized to an amateur station operating under RACES rules?

A. All amateur service frequencies authorized to the control operator.
B. Specific segments in the amateur service MF, HF, VHF and UHF bands.
C. Specific local government channels.
D. Military Auxiliary Radio System (MARS) channels.

Anyone with a valid amateur radio license, including Novice Class, is eligible to be certified by a civil defense organization as a RACES operator. However, ***you do not gain any out-of-band privileges as a RACES operator***. [97.407] **ANSWER A.**

☞ **www.training.fema.gov/emiweb/is/crslist.asp**

E1E03 What is a Volunteer Examiner Coordinator?

A. A person who has volunteered to administer amateur operator license examinations.
B. A person who has volunteered to prepare amateur operator license examinations.
C. An organization that has entered into an agreement with the FCC to coordinate amateur operator license examinations.
D. The person who has entered into an agreement with the FCC to be the VE session manager.

A Volunteer Examiner Coordinator is ***an organization that has an agreement with the FCC to coordinate amateur radio license exams***. Two of the biggest VECs are the American Radio Relay League and the W5YI-VEC. [97.521] **ANSWER C.**

E1E02 Where are the questions for all written US amateur license examinations listed?

A. In FCC Part 97.
B. In a question pool maintained by the FCC.
C. In a question pool maintained by all the VECs.
D. In the appropriate FCC Report and Order.

The National Conference of Volunteer Examiner Coordinators (NCVEC) has an internal Question Pool Committee that ensures all exam questions are valid. Each of the three amateur radio question pools is updated every four years. ***The question pools are maintained by all of the VECs***. Hams are encouraged to work with their VECs to help keep the pools stocked with fresh questions. The questions do not come from nor are they maintained by the FCC! [97.523] **ANSWER C.** ☞ www.ncvec.org

---

E1E04 Which of the following best describes the Volunteer Examiner accreditation process?

A. Each General, Advanced and Amateur Extra Class operator is automatically accredited as a VE when the license is granted.
B. The amateur operator applying must pass a VE examination administered by the FCC Enforcement Bureau.
C. The prospective VE obtains accreditation from the FCC.
D. The procedure by which a VEC confirms that the VE applicant meets FCC requirements to serve as an examiner.

VE accreditation is received from a Volunteer Examiner Coordinator ***(VEC)*** that ***makes sure each VE applicant meets specific FCC requirements*** and has a clean record. [97.509, 97.525] **ANSWER D.**

National Volunteer Examiner Coordinator

Authorized Examiner

VE ~ 12345

Pete Trotter

KB9SMG

This accreditation is granted under and

The accredited Extra Class Volunteer Examiner may test all levels of ham radio licensing.

E1E01 What is the minimum number of qualified VEs required to administer an Element 4 amateur operator license examination?

A. 5.
B. 2.
C. 4.
D. 3.

It takes a ***team of 3*** Extra Class accredited examiners to administer an Element 4 Amateur Extra Class exam. [97.509] **ANSWER D.**

E1E08 To which of the following examinees may a VE not administer an examination?

A. Employees of the VE.
B. Friends of the VE.
C. Relatives of the VE as listed in the FCC rules.
D. All of these choices are correct.

As a volunteer examiner, ***you may not test your own close relatives, such as kids, grandkids, brothers and sisters, and parents***, as indicated in the FCC Rules. [97.509] **ANSWER C.**

Working with kids as an Elmer or VE is both fun and a great way to help ham radio grow

E1E06 Who is responsible for the proper conduct and necessary supervision during an amateur operator license examination session?

A. The VEC coordinating the session.
B. The FCC.
C. Each administering VE.
D. The VE session manager.

Each one of the administering Volunteer Examiners must continuously supervise everyone in the room taking the exams. The VEs are the final authority for any issue that might arise during the testing session. [97.509] **ANSWER C.**

E1E13 Which of these choices is an acceptable method for monitoring the applicants if a VEC opts to conduct an exam session remotely?

A. Record the exam session on video tape for later review by the VE team.
B. Use a real-time video link and the Internet to connect the exam session to the observing VEs.
C. The exam proctor observes the applicants and reports any violations.
D. Have each applicant sign an affidavit stating that all session rules were followed.

In June, 2014 the FCC modified Part 97 rules to allow VEs to administer examinations by "remote" means, provided that their sponsoring VEC permits testing in this manner. In its Report & Order, the FCC concluded that "remotely" could mean "some method other than being present in person." Some VECs provide their administering VEs with the option to ***use an internet connection and live streaming video to observe applicants as they take their examination***. Most of these "remote exams" are in remote areas, either in distance or where there may be only one licensed amateur accredited as a VE in a local area who can be present to administer an exam. The new rules allow for some or all three administering VEs to observe exam sessions by remote methods without being physically present at the exam site. [97.509] **ANSWER B.**

E1E10 What must the administering VEs do after the administration of a successful examination for an amateur operator license?

A. They must collect and send the documents to the NCVEC for grading.
B. They must collect and submit the documents to the coordinating VEC for grading.
C. They must submit the application document to the coordinating VEC according to the coordinating VEC instructions.
D. They must collect and send the documents to the FCC according to instructions.

If you are the head VE for an exam session, called the Administering VE, ***you must submit the application documents to the coordinating VEC***. Follow the instructions of the VEC for how to submit the paperwork. [97.509] **ANSWER C.**

E1E05 What is the minimum passing score on amateur operator license examinations?

A. Minimum passing score of 70%. C. Minimum passing score of 80%.
B. Minimum passing score of 74%. D. Minimum passing score of 77%.

As an accredited volunteer examiner, you need to know the number of questions an applicant can miss while still achieving a ***minimum passing grade of 74%***. Technician Class and General Class cannot miss more than 9 questions, and pass with 26 correct out of 35. Extra Class cannot miss more than 13 questions, and pass with 37 correct out of 50. [97.503] **ANSWER B.**

E1E11 What must the VE team do if an examinee scores a passing grade on all examination elements needed for an upgrade or new license?

A. Photocopy all examination documents and forward them to the FCC for processing.
B. Three VEs must certify that the examinee is qualified for the license grant and that they have complied with the administering VE requirements.
C. Issue the examinee the new or upgrade license.
D. All these choices are correct.

When an examinee upgrades to a new, higher license, the 3-member VE team must ***certify the exam paperwork and sign the CSCE*** (Certificate of Successful Completion of Examination). The applicant holds onto the CSCE as proof of their successful upgrade until the license is issued by the FCC. [97.509] **ANSWER B.**

W5YI-VEC

National Volunteer Examiner Coordinator

This certifies that:

NAME STATION CALL SIGN

DATE OF ISSUE: ______

CITY / STATE (Session Site)

NUMBER AND STREET

CITY STATE ZIP

has SUCCESSFULLY PASSED the following elements:

☐ Element 2, Technician Class ☐ Element 3, General Class ☐ Element 4, Amateur Extra Class

and will be given credit for this examination element when the appropriate additional examination element is passed at a subsequent examination session within 365 days from the date of issue of this certificate. *(See Section 97.505(a)6)).*

has SUCCESSFULLY PASSED all elements for the following operator license class:

☐ Technician Class Operator ☐ General Class Operator ☐ Amateur Extra Class Operator

If you already hold an FCC-issued amateur radio license, this certificate validates temporary (interim) operation with the rights and privileges of your new operator class (see Section 97.301 of the Commission's Rules) until you receive the license of your new operator class, or for a period not to exceed 365 days from the date of issue of this certificate. (See Section 97.9(b)). When operating on an interim basis in the telegraphy / data / image /RTTY mode, you must append your call sign with /KT (Technician), /AG (General) or /AE (Amateur Extra Class). Use the word "Temporary" before the identifier (KT, AG or AE) when operating in the voice mode. *(See Section 97.119(f)).*

THIS CERTIFICATE IS NOT A LICENSE PERMIT OR ANY OTHER KIND OF OPERATING AUTHORITY

Sample CSCE.

E1E12 What must the VE team do with the application form if the examinee does not pass the exam?

A. Return the application document to the examinee.
B. Maintain the application form with the VEC's records.
C. Send the application form to the FCC and inform the FCC of the grade.
D. Destroy the application form.

You will ***return the application to the examinee*** and inform the candidate of the failing grade. Try to do this compassionately, urging the candidate to continue to study and try again soon. [97.509] **ANSWER A.**

E1E07 What should a VE do if a candidate fails to comply with the examiner's instructions during an amateur operator license examination?

A. Warn the candidate that continued failure to comply will result in termination of the examination.
B. Immediately terminate the candidate's examination.
C. Allow the candidate to complete the examination, but invalidate the results.
D. Immediately terminate everyone's examination and close the session.

If an examinee fails to comply with your instructions, you ***immediately terminate that candidate's examination***. [97.509] **ANSWER B.**

E1E14 For which types of out-of-pocket expenses do the Part 97 rules state that VEs and VECs may be reimbursed?

A. Preparing, processing, administering and coordinating an examination for an amateur radio license.
B. Teaching an amateur operator license examination preparation course.
C. No expenses are authorized for reimbursement.
D. Providing amateur operator license examination preparation training materials.

Volunteer examiners and VECs are allowed to recoup out-of-pocket ***expenses for preparing, processing, administering, and coordinating an examination*** for an amateur radio license. That is what some of the funds from the set fee that each applicant pays for the exam session are used for. [97.527] **ANSWER A.**

E1E09 What may be the penalty for a VE who fraudulently administers or certifies an examination?

A. Revocation of the VE's amateur station license grant and the suspension of the VE's amateur operator license grant.
B. A fine of up to $1000 per occurrence.
C. A sentence of up to one year in prison.
D. All of these choices are correct.

Any VE who is caught administering a fraudulent exam may have his or her ***station license revoked and the VE's amateur radio license grant suspended***. [97.509] **ANSWER A.**

The Effect of the Ionosphere on Radio Waves

To help you with the questions on radio wave propagation, here is a brief explanation on the effect the ionosphere has on radio waves.

The ionosphere is the electrified atmosphere from 40 miles to 400 miles above the Earth. You can sometimes see it as "northern lights." It is charged-up daily by the Sun, and does some miraculous things to radio waves that strike it. Some radio waves are absorbed during daylight hours by the ionosphere's D layer. Others are bounced back to Earth. Yet others penetrate the ionosphere and never come back again. The wavelength of the radio waves determines whether the waves will be absorbed, refracted, or will penetrate. Here's a quick way to memorize what the different layers do during day and nighttime hours:

The D layer is about 40 miles up. The D layer is a Daylight layer; it almost disappears at night. D for Daylight. The D layer absorbs radio waves between 1 MHz to 7 MHz. These are long wavelengths. All others pass through.

The E layer is also a daylight layer, and it is very Eccentric. E for Eccentric. Patches of E layer ionization may cause some surprising reflections of signals on both high frequency as well as very-high frequency. The E layer height is usually 70 miles.

The F1 layer is one of the layers farthest away. The F layer gives us those Far away signals. F for Far away. The F1 layer is present during daylight hours, and is up around 150 miles. The F2 layer is also present during daylight hours, and it gives us the Furthest range. The F2 layer is 250 miles high, and it's the best for the Farthest range on medium and short waves. The F2 layer is strongest in the summer months. During winter months, both the F1 and F2 layers may become unpredictable, but always strong enough to support exciting skywaves! At nighttime, the F1 and F2 layers combine to become just the F layer at 180 miles. This F layer at nighttime will usually bend radio waves between 1 MHz and 15 MHz back to earth. At night, the D and E layers disappear.

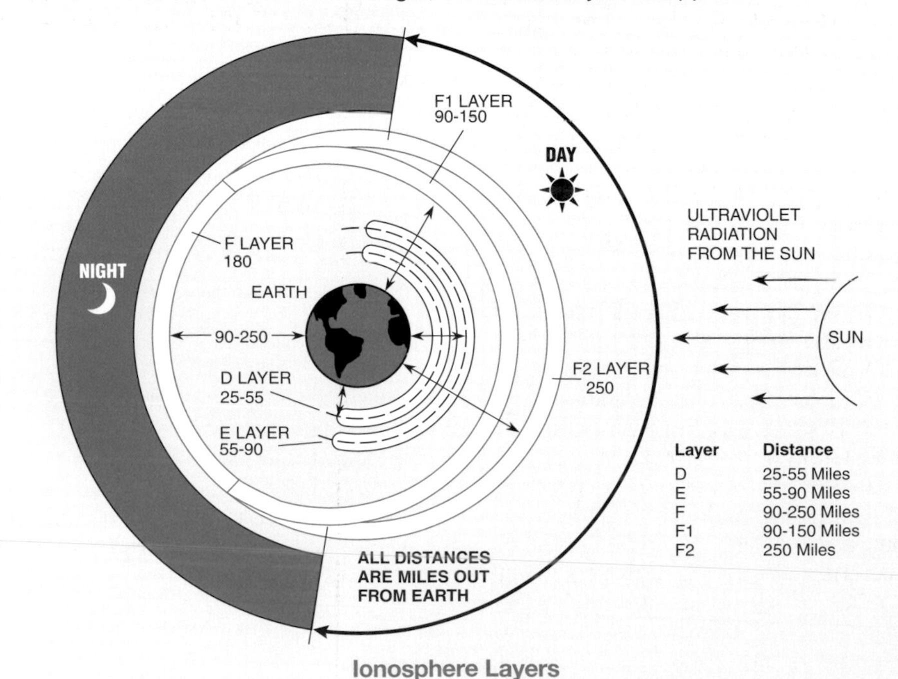

Layer	Distance
D	25-55 Miles
E	55-90 Miles
F	90-250 Miles
F1	90-150 Miles
F2	250 Miles

Ionosphere Layers

Source: *Antennas — Selection and Installation*, © 1986, Master Publishing, Inc., Niles, Illinois

Skywaves & Contesting

E3A15 What is an electromagnetic wave?

A. A wave of alternating current, in the core of an electromagnet.
B. A wave consisting of two electric fields at parallel right angles to each other.
C. A wave consisting of an electric field and a magnetic field oscillating at right angles to each other.
D. A wave consisting of two magnetic fields at right angles to each other.

When you think of an ***electromagnetic radio wave***, think of it as a ***combination of an electric field and a magnetic field*** oscillating at right angles to one another while traveling in waves through the atmosphere at the same time. One field is vertically polarized and the other is horizontally polarized. If the antenna is vertically polarized, the electric field will be vertical and the magnetic field will be horizontal. **ANSWER C.**

E3A16 Which of the following best describes electromagnetic waves traveling in free space?

A. Electric and magnetic fields become aligned as they travel.
B. The energy propagates through a medium with a high refractive index.
C. The waves are reflected by the ionosphere and return to their source.
D. Changing electric and magnetic fields propagate the energy.

Electromagnetic radio waves, traveling in free space, contain both an ***electric field and a magnetic field propagating the energy***. The fields are at a right angle to each other, and it is both the electric field and the magnetic field that make up the radio wave energy traveling at the speed of light in a vacuum. **ANSWER D.**

☞ **www.ips.gov.au/hf_systems/6/9/1**

E3A17 What is meant by circularly polarized electromagnetic waves?

A. Waves with an electric field bent into a circular shape.
B. Waves with a rotating electric field.
C. Waves that circle the Earth.
D. Waves produced by a loop antenna.

Some satellite enthusiasts will use circular polarization to improve their transmission and reception through an orbiting satellite. With ***circular polarization, the electric field rotates***. It could be right-hand circular or left-hand circular, depending on the antenna feed system. **ANSWER B.**

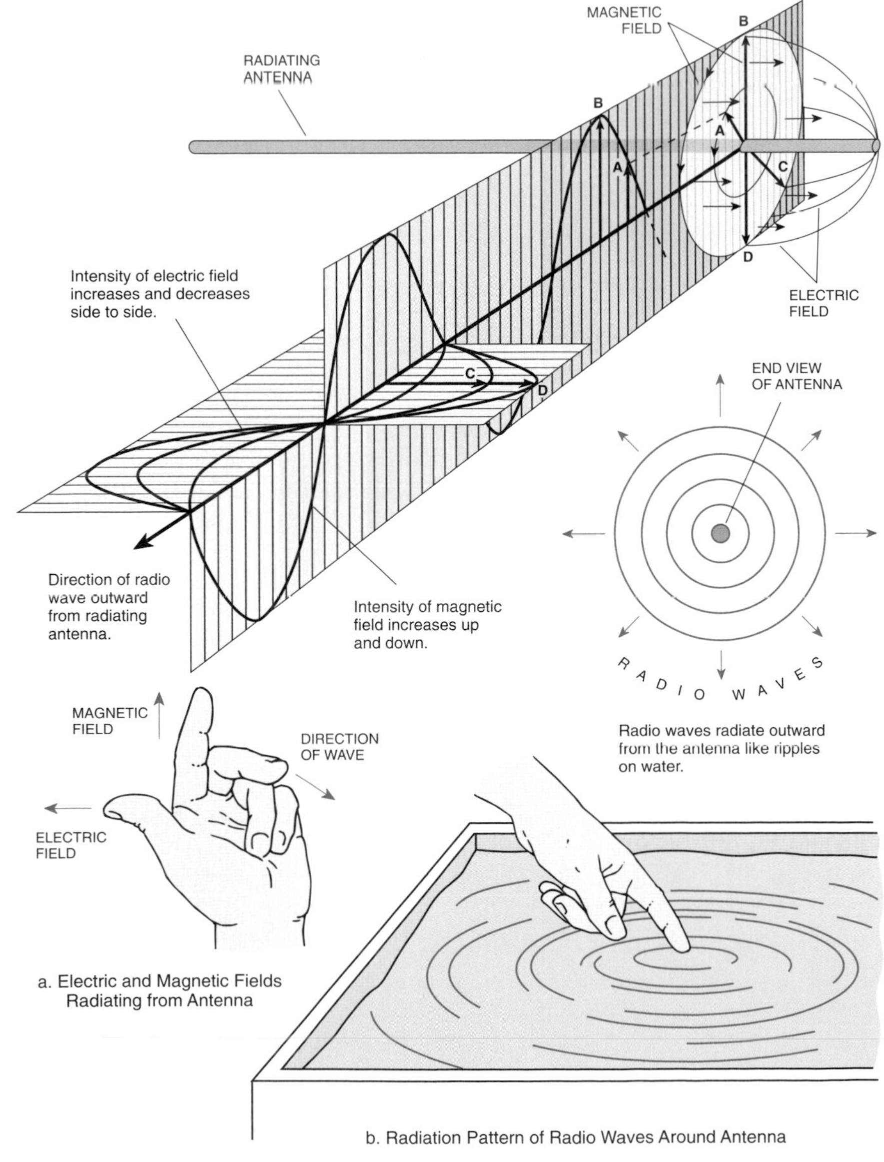

a. Electric and Magnetic Fields Radiating from Antenna

b. Radiation Pattern of Radio Waves Around Antenna

Radio Waves

Source: Basic Electronics © 1994, 2000 Master Publishing, Inc., Niles, Illinois

E3B05 Which amateur bands typically support long-path propagation?

A. 160 meters to 40 meters.
B. 30 meters to 10 meters.
C. 160 meters to 10 meters.
D. 6 meters to 2 meters.

Long-range, ***long-path*** propagation can be found on the worldwide bands from ***10 meters down to 160 meters***. Expect almost daily long-path propagation on 20 meters, which is best in the late evening or early morning hours. **ANSWER C.**

☞ **www.arrl.org/tis/info/propagation.html**

E3B06 Which of the following amateur bands most frequently provides long-path propagation?

A. 80 meters.
B. 20 meters.
C. 10 meters.
D. 6 meters.

The amateur radio ***20 meter band is the best one for long-path*** DX with a relatively modest directional antenna system. When you pass your upcoming Extra Class exam, you will gain an exclusive 20 meter CW window from 14.000 to 14.025 MHz and an exclusive Extra Class window from 14.150 to 14.175 MHz for CW, voice and data. These are great for working rare 20 meter DX on a band shared only with other Extra Class operators in the United States! **ANSWER B.**

E3C01 What does the term ray tracing describe in regard to radio communications?

A. The process in which an electronic display presents a pattern.
B. Modeling a radio wave's path through the ionosphere.
C. Determining the radiation pattern from an array of antennas.
D. Evaluating high voltage sources for X-Rays.

The best ***propagation*** prediction tools involve the use of ***ray tracing for modeling a radio wave's path through the ionosphere***. While ray tracing programs have been around for a long time, only recently have they become readily available to the radio amateur – and they still use considerable processing power. If you use a program such as Proplab Pro 3 to generate a detailed 3D ray trace, it's best to make a new pot of coffee or watch a sitcom while it does its number crunching, even if you have a screaming fast computer. Hopefully the propagation hasn't changed too much by the time your computer issues forth the "answer!" **ANSWER B.**

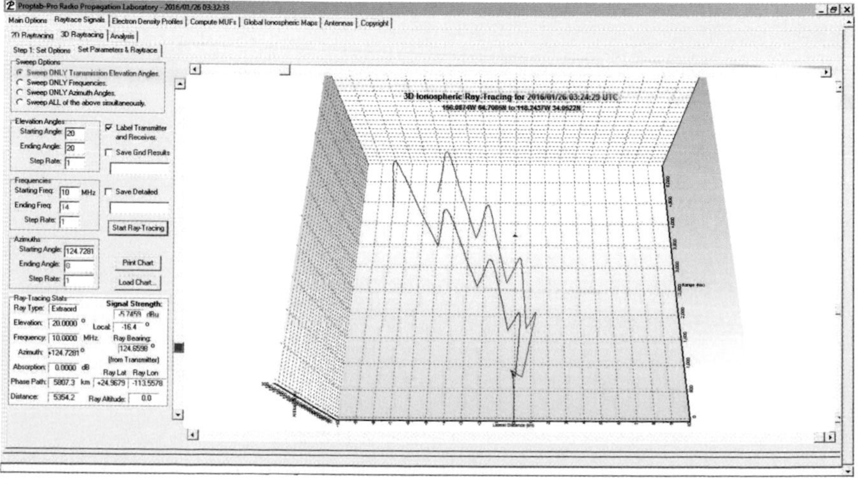

Ray Tracing can give you a 3D view of the path your signal travels around the world.

E3C15 What might a sudden rise in radio background noise indicate?

A. A meteor ping.
B. A solar flare has occurred.
C. Increased transequatorial propagation likely.
D. Long-path propagation is occurring.

While having a steady bath of UV radiation is good for the ionosphere, being blasted with hard X-rays and charged particles from a ***solar flare*** is not. During such an event, radio absorption in the D layer can be almost total. In addition, gigantic blobs of fast-moving charged particles tend to wrinkle the ionosphere to the point that it's no longer a reflective or refractive layer. An ***increase in noise*** also results as ionized particles, primarily in the D layer, recombine and emit vast quantities of random radio waves (and even visible light). **ANSWER B.**

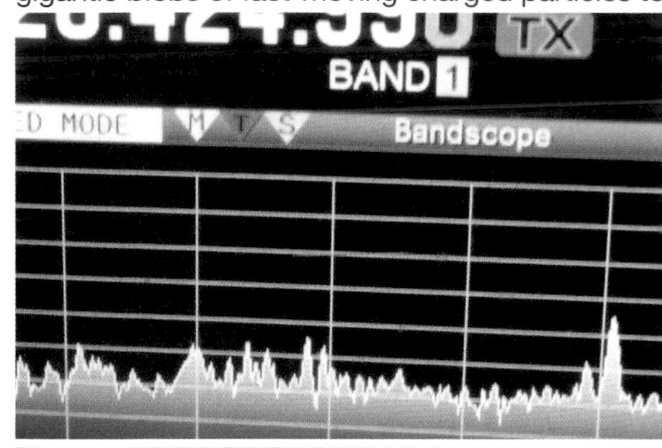

All those band wiggles on the scope are noise pulses from the ionosphere during a solar storm.

E3C02 What is indicated by a rising A or K index?

A. Increasing disruption of the geomagnetic field.
B. Decreasing disruption of the geomagnetic field.
C. Higher levels of solar UV radiation.
D. An increase in the critical frequency.

A and K indices are provided by data from terrestrial *magnetometers.* They are sensitive indicators of solar activity, but need to be interpreted carefully to be truly useful. Learn to read the A and K indices; as they rise they are telling you there is ***increasing disruption of the geomagnetic field***.
ANSWER A.

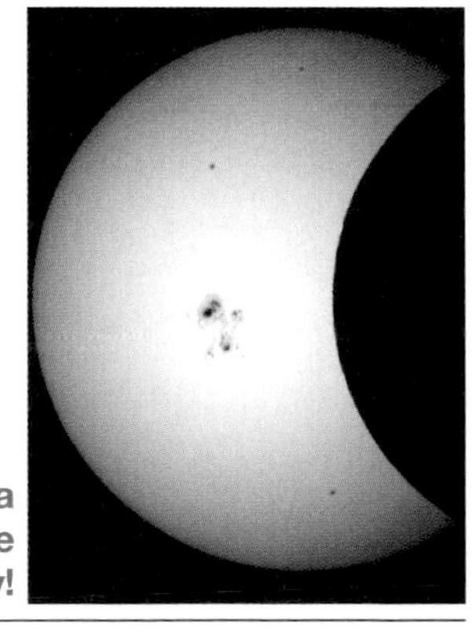

The Sun is seen with several large spots during a rare solar eclipse. An event like this can cause the 75 meter band to come alive during the day!

E3C03 Which of the following signal paths is most likely to experience high levels of absorption when the A index or K index is elevated?

A. Transequatorial propagation.
B. Polar paths.
C. Sporadic-E.
D. NVIS.

Since the Earth's magnetic field lines are concentrated near the magnetic poles (not the geographic poles), magnetic disruptions are also mostly concentrated in these regions. The ionosphere near the polar regions is a magnetized plasma which has some interesting properties. Not only is it absorptive to radio waves, but it is *directionally* absorptive. Don't let anyone tell you there's no such thing as one-way propagation. It is the *norm* in ***polar regions***! **ANSWER B.**

E3C04 What does the value of B_Z (B sub Z) represent?

A. Geomagnetic field stability.
B. Critical frequency for vertical transmissions.
C. Direction and strength of the interplanetary magnetic field.
D. Duration of long-delayed echoes.

The Sun and Earth have an ***interplanetary magnetic field*** which contributes to the character of the Earth's ionosphere. Just like a familiar bar magnet, the magnets embedded in the Sun and the Earth have poles that either attract or repel depending on their orientation. The greatest effect occurs when the Sun's north pole magnetically connects with the Earth's south pole. The magnetic field orientation is defined relative to the interplanetary, which is the plane of the orbit of the Earth around the Sun. The Z axis of B (the magnetic field) is the north/south axis of the interplanetary field, perpendicular to the ecliptic, which is roughly the same direction as the Earth's north/south axis. The X and Y axes have minimal effect on the Sun/Earth interactions and are generally ignored. **ANSWER C.**

E3C05 What orientation of B_Z (B sub Z) increases the likelihood that incoming particles from the Sun will cause disturbed conditions?

A. Southward.
B. Northward.
C. Eastward.
D. Westward.

It is standard practice to designate the type of field with a subscript denoting the AXIS. In this case B designates a magnetic field, the interplanetary magnetic field, and Z designates a Z axis, or UP/DOWN orientation relative to the ecliptic, the plane of the Earth's orbit around the Sun. Just as two bar magnets interact most strongly when the opposite poles are near each other, ***the Earth has the greatest interaction with a magnetic field when that field is oriented north-to-south (or southward).*** Now the real interesting part is that, unlike the Earth's magnetic field, which is fairly fixed, an interplanetary solar field is carried by solar wind, which means it's as variable as the weather! Interplanetary magnetism can range from almost nothing to huge, and it can change very suddenly. Any magnetic field that is not parallel to the Earth's field causes very little interaction. **ANSWER A.**

E3C07 Which of the following descriptors indicates the greatest solar flare intensity?

A. Class A.
B. Class B.
C. Class M.
D. Class X.

Think of X class as being Xtreme. (As we've noted before, physicists can't spell). In any case, ***X class solar flares are extremely disruptive*** as they emit large quantities of charged particles and, as you might suspect, a lot of X-rays, which can certainly create a lot of ionization, but not the type that you want. They penetrate the atmosphere so far that they excite the absorptive D layer far beyond what is normal. The nice thing is that solar flares don't last forever. **ANSWER D.**

E3C09 How does the intensity of an X3 flare compare to that of an X2 flare?

A. 10 percent greater.
B. 50 percent greater.
C. Twice as great
D. Four times as great

Like just about every other radio phenomenon (or even non-radio phenomena like earthquakes) X-class solar flares are best measured on a logarithmic scale. Each numerical suffix represents a doubling of the number before, so an X2 is twice as intense as an X1, ***an X3 is twice as intense as an X2***, and so on. **ANSWER C.**

E3C08 What does the space weather term G5 mean?

A. A severe geomagnetic storm.
B. Very low solar activity.
C. Moderate solar wind.
D. Waning sunspot numbers.

Geomagnetic storms are classified with numbers G1 to G5, with ***G5 being the most severe***. Geomagnetic storms are a *result* of solar storms and can be delayed considerably from the time solar storms are observed; after all the Sun is 93 million miles away, and the ionosphere is just a few hundred miles at the farthest! **ANSWER A.**

☞ **www.swpc.noaa.gov/noaa-scales-explanation**

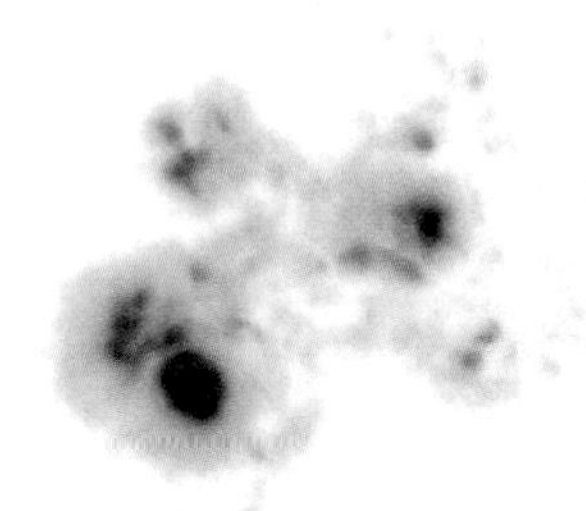

These blotches on the Sun are gigantic solar flares that influence HF propagation.

Good To Know:

WWV AND WWVH SPACE WEATHER BROADCASTS

The National Oceanic and Atmospheric Administration (NOAA) uses WWV and WWVH to broadcast geophysical alert messages that provide information about solar terrestrial conditions. Geophysical alerts are broadcast from WWV at 18 minutes after the hour and from WWVH at 45 minutes after the hour. The messages are less than 45 seconds in length and are updated every 3 hours (typically at 0000, 0300, 0600, 0900, 1200, 1500, 1800, and 2100 UTC). More frequent updates are made when necessary.

The geophysical alerts provide information about the current conditions for long distance HF radio communications. The alerts use a standardized format and terminology that requires some explanation. Before looking at a sample message, let's define some of the terminology:

Solar flux is a measurement of the intensity of solar radio emissions with a wavelength of 10.7 cm (a frequency of about 2800 MHz). The daily solar flux measurement is recorded at 2000 UTC by the Dominion Radio Astrophysical Observatory of the Canadian National Research Council located at Penticton, British Columbia. The value broadcast is in solar flux units that range from a theoretical minimum of about 50 to numbers larger than 300. During the early part of the 11-year sunspot cycle, the flux numbers are low; but they rise and fall as the cycle proceeds. The numbers will remain high for extended periods around sunspot maximum.

The A and K indices are a measurement of the behavior of the magnetic field in and around the Earth. The K index uses a scale from 0 to 9 to measure the change in the horizontal component of the geomagnetic field. A new K index is determined and added to the broadcast every 3 hours based on magnetometer measurements made at the Table Mountain Observatory, north of Boulder, Colorado, or an alternate middle latitude observatory. The A index is a daily value on a scale from 0 to 400 to express the range of disturbance of the geomagnetic field. It is obtained by converting and averaging the eight, 3-hour K index values of a given day. An estimate of the A index is first announced at 2100 UTC, based on 7 measurements and 1 estimated value. At 0000 UTC, the announced A index consists entirely of known measurements, and the word "estimated" is dropped from the announcement.

E3C10 What does the 304A solar parameter measure?

A. The ratio of X-Ray flux to radio flux, correlated to sunspot number.
B. UV emissions at 304 angstroms, correlated to solar flux index.
C. The solar wind velocity at 304 degrees from the solar equator, correlated to solar activity.
D. The solar emission at 304 GHz, correlated to X-Ray flare levels.

Ultraviolet emissions are a good indication of sunspot activity. UV is the primary ingredient necessary for the normal ionosphere to form. ***The 304A solar parameter measures UV emissions at 304 angstroms.*** Ultraviolet actually encompasses a large chunk of the electromagnetic spectrum, as can be seen here: ☞ **http://csep10.phys.utk.edu/astr162/lect/light/spectrum.html** Because of its wide range of wavelengths, ultraviolet is usually subdivided into two or more regions. **ANSWER B.**

Space Weather describes the conditions in space that affect Earth and its technological systems. Space weather is a consequence of the behavior of the Sun, the nature of Earth's magnetic field and atmosphere, and our location in the solar system.

Space Weather storms observed and expected are characterized using the NOAA Space Weather scales. The abbreviated table below shows the levels of activity that are included in the announcements and the associated terminology. The descriptor used to identify observed or expected conditions is the maximum level reached or predicted. The NOAA Space Weather Scales are further described at the Space Environment Center's web site.

Space Weather Storm Levels

The NOAA Space Weather Scales were introduced as a way to communicate to the general public the current and future space weather conditions and their possible effects on people and systems. The scales describe the environmental disturbances for three event types: geomagnetic storms G1 thru G5), solar radiation storms (S1 thru S5), and radio blackouts (R1 thru R5). The scales have numbered levels, analogous to hurricanes, tornadoes, and earthquakes that convey severity. In addition, the scales for radio blackouts also show the peak X-ray levels and flux generated by these solar events.

Read the full explanation of these scales and learn more about space weather at ☞ **swpc.noaa.gov/noaa-scales-explanation**.

To hear current space weather levels dial 303-497-3235.

STORM LEVELS	G Scale	S Scale	R Scale	Peak X-Ray Levels & Flux	K Index
Minor	G1	S1	R1	M1 (10^{-5})	K5 or less
Moderate	G2	S2	R2	M5 (5×10^{-5})	K6
Strong	G3	S3	R3	X1 (10^{-4})	K7
Severe	G4	S4	R4	X10 (10^{-3})	K8
Extreme	G5	S5	R5	X20 (2 x 10-3)	K9

E3C11 What does VOACAP software model?

A. AC voltage and impedance.
B. VHF radio propagation.
C. HF propagation.
D. AC current and impedance.

The Voice of America Coverage Analysis Program, or ***VOACAP***, is one of the oldest and most familiar ***HF radio propagation*** tools. Other programs such as IONCAP were derived from VOACAP. If you're an avid contester or casual DXer, you can benefit from learning how to use VOACAP and its kin. **ANSWER C.**

E3B07 Which of the following could account for hearing an echo on the received signal of a distant station?

A. High D layer absorption.
B. Meteor scatter.
C. Transmit frequency is higher than the MUF.
D. Receipt of a signal by more than one path.

When long-path conditions exist, European ham stations and even European shortwave stations will be received with a noticeable echo on the signal. The ***echo is the slight delay of the signal coming in long-path following what you are also hearing sooner as a short-path signal***. **ANSWER D.**

E3B12 What is the primary characteristic of chordal hop propagation?

A. Propagation away from the great circle bearing between stations.
B. Successive ionospheric reflections without an intermediate reflection from the ground.
C. Propagation across the geomagnetic equator.
D. Signals reflected back toward the transmitting station.

Chordal hop propagation can result in some surprisingly loud signals, especially on 80 meters. Since there are ***no ground reflections*** in between sky reflections, the losses are much less than "normal" multi-hop propagation. One additional indication that chordal hop is in play is that there's a huge dead "skip zone" in the middle of the path, which is along the dark side of the Earth. For further information and some great diagrams showing chordal hop propagation, check out this website:

☞ **http://www.antennex.com/prop/prop0806/prop0806.pdf**

ANSWER B.

Chip's home brew 10 meter beam gets a shot at chordal hop propagation on the 10 meter band for loud DX signals!

E3B13 Why is chordal hop propagation desirable?

A. The signal experiences less loss along the path compared to normal skip propagation.
B. The MUF for chordal hop propagation is much lower than for normal skip propagation.
C. Atmospheric noise is lower in the direction of chordal hop propagation.
D. Signals travel faster along ionospheric chords.

Chordal hop signals can be unusually strong, experiencing ***less loss*** compared to skip propagation. There is even some tantalizing evidence that *amplification* can take place along a chordal hop path. (This process can be duplicated in the plasma physics laboratory, while it is unknown if all the proper conditions indeed exist "in the wild." This is wide open for further amateur experimentation.) Chordal hop signals always occur along the "dark path" opposite the Sun. **ANSWER A.**

E2C12 What might help to restore contact when DX signals become too weak to copy across an entire HF band a few hours after sunset?

A. Switch to a higher frequency HF band.
B. Switch to a lower frequency HF band.
C. Wait 90 minutes or so for the signal degradation to pass.
D. Wait 24 hours before attempting another communication on the band.

Now that we are cresting on the peak of solar cycle 24, many bands like 15 and 20 meters may remain "open" for skip signals well into the evening hours. When the band finally gives up and goes into noise with no more DX heard in the background, ***switch to a lower frequency HF band*** to see if that band is still open for skywave excitement. **ANSWER B.** ☞ **http://propagation.hfradio.org**

When we get news of a flare on the Sun like this, expect some unsettled band conditions about 40 hours later.

E3B08 What type of HF propagation is probably occurring if radio signals travel along the terminator between daylight and darkness?

A. Transequatorial.
B. Sporadic-E.
C. Long path.
D. Gray-line.

Once you earn your Extra Class license, you will stay up for long hours working rare DX. Just as the Sun is rising here, it is setting over Europe or Asia. Working stations in the twilight at both ends of the circuit may lead to some extraordinary DX contacts that can last up to 30 minutes. This is called ***gray-line propagation***. **ANSWER D.**

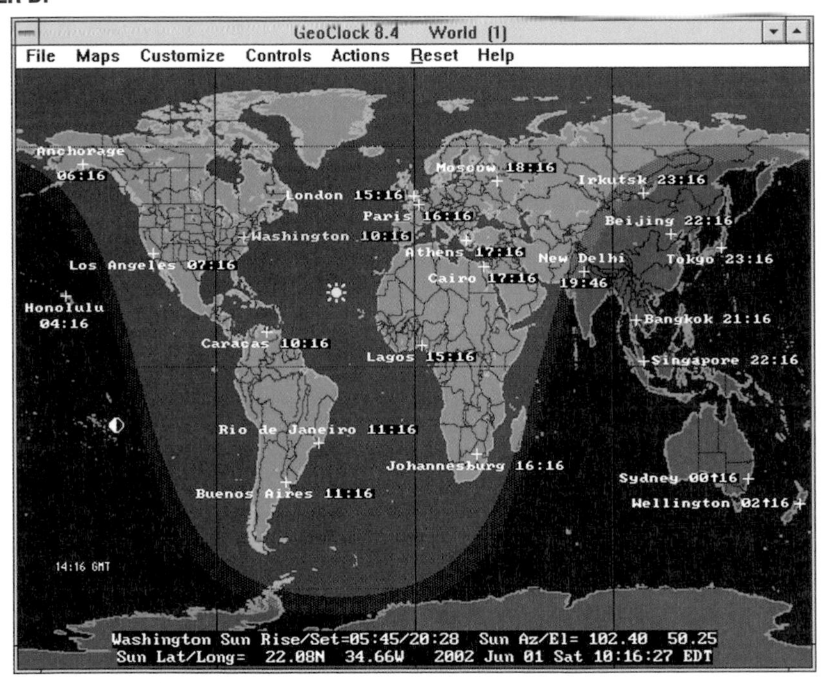

Gray line conditions follow the Sun's terminator. It's a 10 minute opportunity at great DX for stations at each end of the gray line terminator.

E3B10 What is the cause of gray-line propagation?

A. At midday, the Sun super heats the ionosphere causing increased refraction of radio waves.
B. At twilight and sunrise, D-layer absorption is low while E-layer and F-layer propagation remains high.
C. In darkness, solar absorption drops greatly while atmospheric ionization remains steady.
D. At mid-afternoon, the Sun heats the ionosphere decreasing radio wave refraction and the MUF.

During twilight, the D layer quickly disappears resulting in less absorption, ***while the E and F layers continue relatively strong***. This same condition holds true during solar eclipses. **ANSWER B.**

E3B11 At what time of day is Sporadic-E propagation most likely to occur?

A. Around sunset.
B. Around sunrise.
C. Early evening.
D. Any time.

While Sporadic-E appears to be a seasonal phenomenon, it doesn't seem to care much about the time of day. This further adds to the mystery and excitement of the mode. What this means is, if you want to work ***Sporadic E***, it just plain helps to be on the air a lot since it ***can happen any time!*** In all his sporadic E log notes, including the largest sunspot event in 1958, Gordo says he has NEVER encountered any E skip on VHF bands from 10 pm to 6 am in the dead of night! But for the correct answer, go with ***any time*** and let him know if you every make any midnight contacts! **ANSWER D.**

E3B09 At what time of year is Sporadic E propagation most likely to occur?

A. Around the solstices, especially the summer solstice.
B. Around the solstices, especially the winter solstice.
C. Around the equinoxes, especially the spring equinox.
D. Around the equinoxes, especially the fall equinox.

Sporadic E is, as the name suggests, *sporadic*. It can actually occur any time of the year, and is often a surprise when it does. However, it is most likely to occur during the dead of winter or the ***dead of summer***. The reasons for this are not yet fully understood, but the phenomenon is well documented. **ANSWER A.** Here is a link dedicated to Sporadic E "sightings" and some possible predictions:
☞ **http://www.uksmg.org/content/sporade.htm**

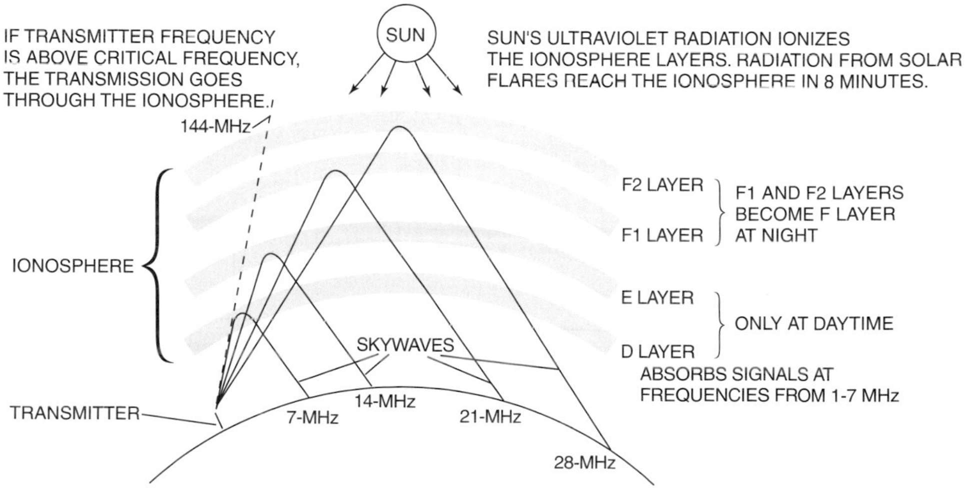

Ionosphere and Its Layers

E3A08 When a meteor strikes the Earth's atmosphere, a cylindrical region of free electrons is formed at what layer of the ionosphere?

A. The E layer.
B. The F1 layer.
C. The F2 layer.
D. The D layer.

It is within the ***E-layer*** of the ionosphere that 6 and 2 meter excitement like Sporadic-E and ***meteor scatter contacts*** take place. Think mEtEor! **ANSWER A.**

E3A09 Which of the following frequency ranges is most suited for meteor scatter communications?

A. 1.8 MHz - 1.9 MHz.
B. 10 MHz - 14 MHz.
C. 28 MHz - 148 MHz.
D. 220 MHz - 450 MHz.

During a recent meteor shower, Gordo was able to maintain 30-second meteor-scatter QSOs on the 10 meter band; 10-second QSOs on the 6 meter band, and 3-second QSOs on the 2 meter band. The higher you go in frequency, the shorter the contact time. ***28-148 MHz are the best*** areas to hear the effects of ***meteor scatter***. **ANSWER C.** ☞ **http://earthsky.org/astronomy-essentials/earthskys-meteor-shower-guide**

Making meteor contacts is a great way to work CW during a 3-second "burn."

E2D01 Which of the following digital modes is especially designed for use for meteor scatter signals?

A. WSPR.
B. FSK441.
C. Hellschreiber.
D. APRS.

If you are into computers, ham radio digital signaling is just the thing for you! There are many software programs that work with your computer's sound card and there are excellent add-on radio signal controllers that let the more sophisticated transceivers both transmit and receive ham radio digital calls. Up on VHF and UHF, the most popular digital mode for ***meteor scatter*** contacts is ***FSK 441 (Frequency Shift Keying)***. **ANSWER B.** ☞ **http://spaceweather.com**

E2D02 Which of the following is a good technique for making meteor scatter contacts?

A. 15 second timed transmission sequences with stations alternating based on location.
B. Use of high speed CW or digital modes.
C. Short transmission with rapidly repeated call signs and signal reports.
D. All of these choices are correct.

There are several different ways of conducting ***meteor scatter contacts***. If you and your friend 500 miles away want to set a meteor scatter schedule, you first make sure your clocks are perfectly in sync with WWV at 10,000 kHz. Agree on who takes the first and third ***15 second calls***, and then each listen for the other's fifteen second transmission. Transmitting west to east takes the first and third 15 second period. Another way for random meteor scatter contacts is through ***high speed CW, digital modes, or rapidly giving your call sign for a few seconds***. Listen to Gordo's audio CDs to hear exciting meteor scatter contacts. **ANSWER D.**
☞ **www.VHFDX.DE/WSJT**

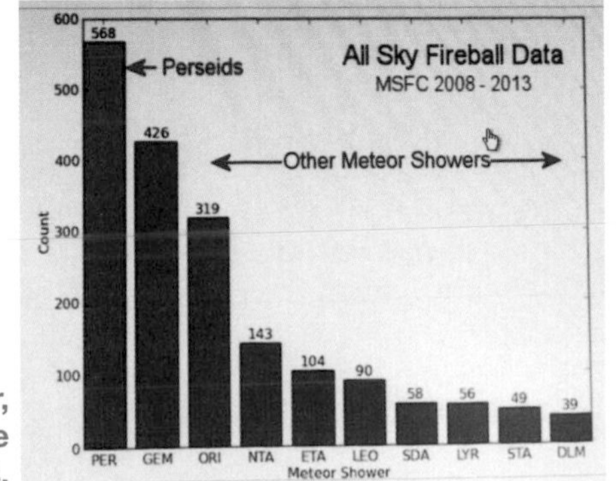

Perseids in August, Geminids in December, and Orionids in October are favorite ham radio meteor shower events.

E3B01 What is transequatorial propagation?

A. Propagation between two mid-latitude points at approximately the same distance north and south of the magnetic equator.
B. Propagation between any two points located on the magnetic equator.
C. Propagation between two continents by way of ducts along the magnetic equator.
D. Propagation between two stations at the same latitude.

A TE (transequatorial) signal is propagated between stations north and south of the magnetic Equator. Some of the best contacts have been between Florida and countries in South America. Any ***stations*** in contact with each other are ***about the same distance away from the Equator*** – ideal conditions for a TE contact. CW as well as SSB are the normal modes for TE on VHF and UHF bands. **ANSWER A.**

E3B02 What is the approximate maximum range for signals using transequatorial propagation?

A. 1000 miles.
B. 2500 miles.
C. 5000 miles.
D. 7500 miles.

The maximum total distance for a TE (transequatorial) contact is about ***5,000 miles***. **ANSWER C.**

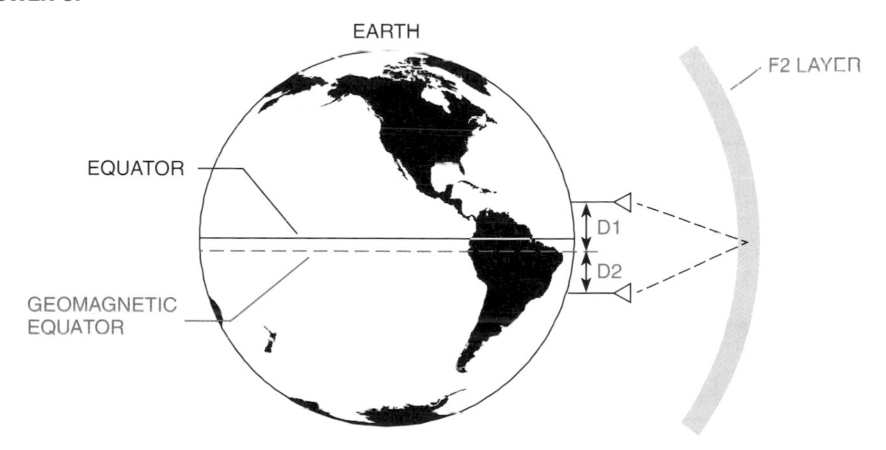

For transequatorial propagation to occur, stations at D1 and D2 must be equidistant from the geomagnetic equator. Popular theory is that reflections occur in and off the F2 layer during peak sunspot activity.

Transequatorial Propagation

E3B03 What is the best time of day for transequatorial propagation?

A. Morning.
B. Noon.
C. Afternoon or early evening.
D. Late at night.

The ***best time of day for TE*** propagation is ***mid-afternoon or early evening***. By this time of day, the Sun's ultraviolet radiation has had time to "warm up" the possible path. **ANSWER C.**

E3A12 What is the cause of auroral activity?

A. The interaction in the F2 layer between the solar wind and the Van Allen belt.
B. A low sunspot level combined with tropospheric ducting.
C. The interaction in the E layer of charged particles from the Sun with the Earth's magnetic field.
D. Meteor showers concentrated in the extreme northern and southern latitudes.

The ionization levels necessary to create a visible aurora are far higher than what is necessary to create "normal" F layer propagation. ***Charged particles*** with enough energy to ***ionize the E layer usually create*** havoc with the upper ionosphere. As beautiful as ***auroras*** are to watch, they are not generally an indication of good radio conditions! The Earth's magnetic field tends to cause auroral activity to concentrate near the Earth's magnetic poles, but this is not always the case. Extreme solar activity can "warp" the magnetic field lines far south of their normal "homes" causing auroras to be visible far south of the Arctic. **ANSWER C.**

Solar flares and sunspots affect radiowave propagation.
Photo courtesy of N.A.S.A.

E3A13 Which emission mode is best for aurora propagation?

A. CW.
B. SSB.
C. FM.
D. RTTY.

Bouncing a signal off of an aurora is difficult. CW always provides the best opportunity for someone hearing raspy sounds coming into their receiver. On many ***auroral contacts on CW***, we don't even hear a tone – only short and long rushes of super-imposed noise. Play Gordo's audio CD Extra Class course and hear for yourself. **ANSWER A.**

This aurora photo is from Chip and Janet Margelli, K7JA and KL7MF, during their week of radio fun in Coldfoot, Alaska.

E3A14 From the contiguous 48 states, in which approximate direction should an antenna be pointed to take maximum advantage of aurora propagation?

A. South.
B. North.
C. East.
D. West.

You would ***aim your directional antenna North*** at the aurora in order to pick up VHF and UHF fluttery auroral CW sounds. Also, with big auroras, you can sometimes even transmit SSB. Forget about FM or RTTY, these modes usually will not work off of an aurora. **ANSWER B.**

E3B04 What is meant by the terms extraordinary and ordinary waves?

A. Extraordinary waves describe rare long skip propagation compared to ordinary waves which travel shorter distances.
B. Independent waves created in the ionosphere that are elliptically polarized.
C. Long path and short path waves.
D. Refracted rays and reflected waves.

Any time you launch an ***HF signal*** into the ionosphere, it ***splits into two elliptically polarized "characteristic waves,"*** one being clockwise (O mode or "ordinary"), and one being counterclockwise (X mode or "extraordinary"). These two waves follow very different paths, especially near the polar regions where the Earth's magnetic field is strong. **ANSWER B.**

A great description of X and O propagation can be found on the Lowell Digisonde website: ☞ **http://digisonde.com/instrument-description.html**

E3B14 What happens to linearly polarized radio waves that split into ordinary and extraordinary waves in the ionosphere?

A. They are bent toward the magnetic poles.
B. Their polarization is randomly modified.
C. They become elliptically polarized.
D. They become phase-locked.

In the northern magnetic hemisphere, the clockwise elliptical wave is known as the O, "ordinary," wave and the counterclockwise ***elliptical wave*** is known as the X, "extraordinary," wave. These two waves or "rays" can take vastly different paths around the Earth. Folks who operate in the Arctic for the first time are usually surprised to discover that Great Circle paths mean almost nothing! **ANSWER C.**

☞ **http://www.swpc.noaa.gov/**

E4E06 What is a major cause of atmospheric static?

A. Solar radio frequency emissions.
B. Thunderstorms.
C. Geomagnetic storms.
D. Meteor showers.

Thunderstorms carry lightning activity, and lightning noise can be received from as far away as 800 miles on the 80 and 40 meter bands. **ANSWER B.**

A sudden lightning bolt like this will create noise on HF, and can toast your gear (and you) with a direct strike!

E3C14 Why does the radio-path horizon distance exceed the geometric horizon?

A. E-region skip.
B. D-region skip.
C. Downward bending due to aurora refraction.
D. Downward bending due to density variations in the atmosphere.

The VHF/UHF radio horizon is approximately 15% farther away than the geometric horizon. This is because the ***lower portion of the VHF/UHF wave front is slightly slowed by our denser atmosphere***, allowing the higher altitude wave front to travel slightly faster, ***causing the entire wave front to "bend"*** slightly beyond the geometric horizon. Listen to Gordo's audio CD course for the sound of long range VHF/UHF over-the-horizon range!
ANSWER D.

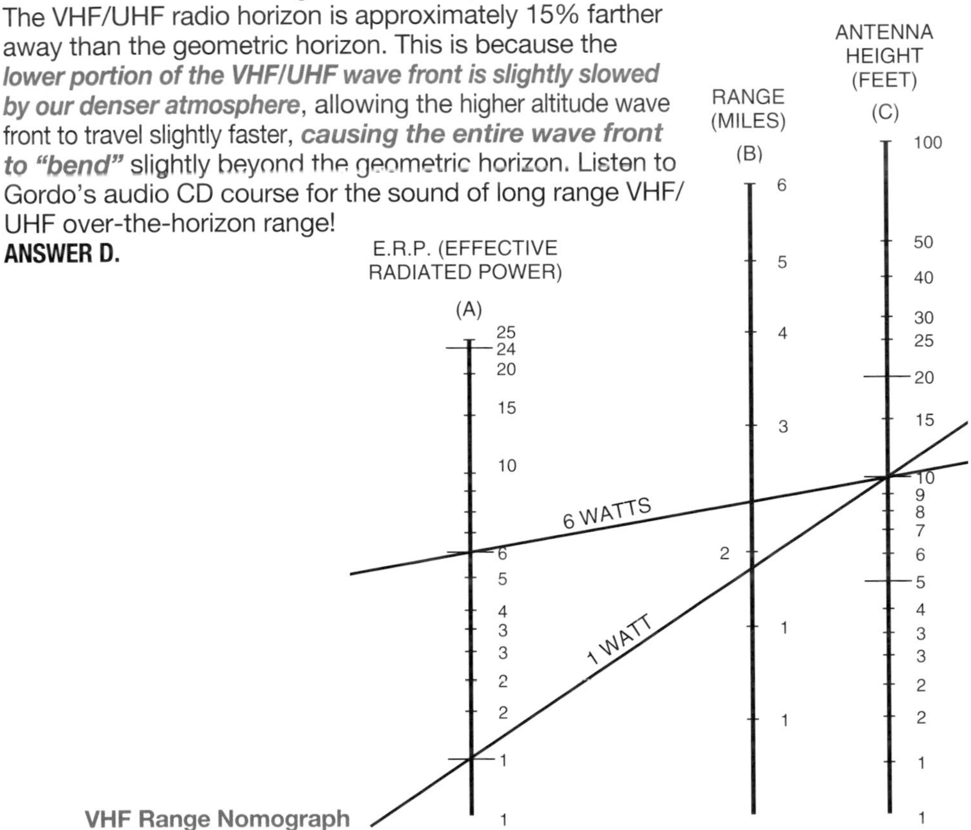

VHF Range Nomograph

E3C06 By how much does the VHF/UHF radio horizon distance exceed the geometric horizon?

A. By approximately 15 percent of the distance.
B. By approximately twice the distance.
C. By approximately 50 percent of the distance.
D. By approximately four times the distance.

Radio waves bend slightly over the horizon because of the difference in the air's refractive index at higher altitudes. You can calculate your approximate VHF range to the horizon with the formula: D (miles) = 1.415 x $\sqrt{2H}$ (in feet). Determine the height (in feet) of the station you are wishing to communicate with, calculate its distance to the horizon, and add the two answers together for your rock-solid communications range. Depending on local weather conditions, a ***15%*** to 30% ***range enhancement over the geometric horizon*** will usually take place at VHF and UHF radio frequencies. **ANSWER A.**

E3A05 Tropospheric propagation of microwave signals often occurs along what weather related structure?

A. Gray-line.
B. Lightning discharges.
C. Warm and cold fronts.
D. Sprites and jets.

Warm and cold weather fronts form large-surface reflectors for VHF and microwave signals (as well as plain old light). **ANSWER C.**

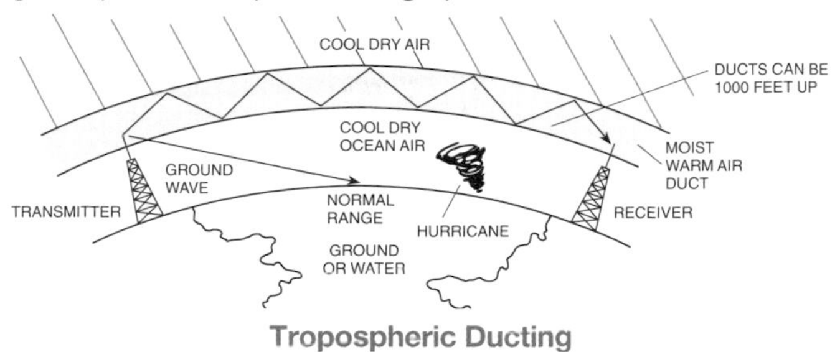

Tropospheric Ducting

E3A07 Atmospheric ducts capable of propagating microwave signals often form over what geographic feature?

A. Mountain ranges.
B. Forests.
C. Bodies of water.
D. Urban areas.

Ducting works best when the two "walls" forming the duct are flat and parallel. This is most likely to happen over a large flat surface such as a ***body of water***, but can also occur over a large expanse of desert. **ANSWER C.**

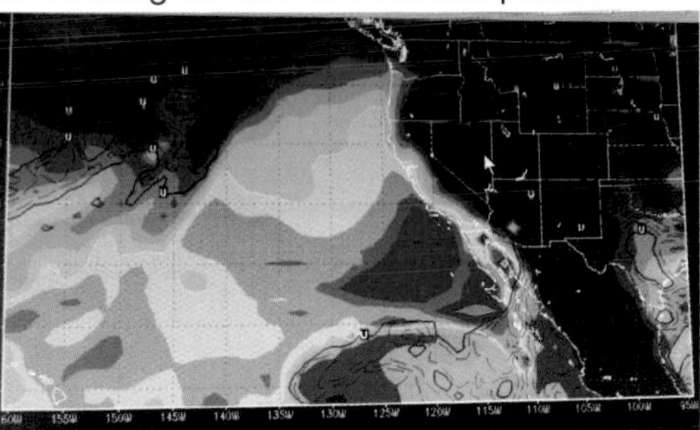

VHF and UHF signals will travel freely within the red and orange areas of troposphere ducting. See Hawaii at bottom left? We work it tropo every summer for days on end on 2 meters through 5.6 GHz!

E3A10 Which type of atmospheric structure can create a path for microwave propagation?

A. The jet stream.
B. Temperature inversion.
C. Wind shear.
D. Dust devil.

Temperature inversions generally form large *stable* structures, unlike wind-shear structures or the jet stream. Just like skipping a stone on water, microwaves are better reflected from a flat smooth surface than a rough, wrinkly surface. **ANSWER B.**

Here is a visual of a tropo duct—a mirage as images are refracted from above to the roadway below.

E3A11 What is a typical range for tropospheric propagation of microwave signals?

A. 10 miles to 50 miles.
B. 100 miles to 300 miles.
C. 1200 miles.
D. 2500 miles.

Because the troposphere is so low compared to ionospheric regions, it stands to reason that "skip distances" will be a lot less for tropospheric modes than, say F2 ionospheric reflections. ***Tropo propagation is typically on the order of 100 to 300 miles***, but can occasionally be much farther because of multiple hops, or "mixed mode" propagation. **ANSWER B.**

E3A04 What do Hepburn maps predict?

A. Sporadic E propagation.
B. Locations of auroral reflecting zones.
C. Likelihood of rain-scatter along cold or warm fronts.
D. Probability of tropospheric propagation.

Since the troposphere is where weather happens, it makes sense that weather prediction can help a lot with tropospheric propagation prediction – and it does. The ***Hepburn map*** is a specialized weather map geared just ***for predicting radio phenomena in the tropo "weather zone."***

ANSWER D.

You can find first-hand Hepburn data here: ☞ **www.dxinfocentre.com**

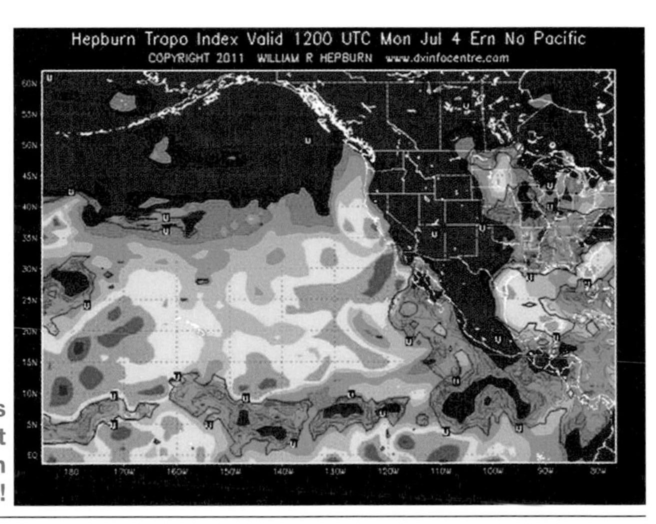

Great tropo conditions seen on the West Coast and in the Gulf between Texas and Florida!

E3A06 Which of the following is required for microwave propagation via rain scatter?

A. Rain droplets must be electrically charged.
B. Rain droplets must be within the E layer.
C. The rain must be within radio range of both stations.
D. All of these choices are correct.

Like any other "scattering" mode, the process is rather inefficient, but can sometimes be surprisingly reliable. Microwave signals can scatter from rain droplets, but in a rather incoherent fashion. This mode is possible whenever there is a ***region of sky containing a large number of water droplets situated between the transmitter and receiver.*** Scatter modes usually benefit from lots of transmitter power. **ANSWER C.**

E3C12 How does the maximum distance of ground-wave propagation change when the signal frequency is increased?

A. It stays the same.

B. It increases.

C. It decreases.

D. It peaks at roughly 14 MHz.

Ground waves travel further on lower frequencies, so as the ***frequency*** of a signal is ***increased, ground wave propagation decreases***. This same propagation phenomena occurs with audio frequencies. Lower audio frequencies travel further than higher audio frequencies. That's why fog horns always blast out a very low audio tone. **ANSWER C.**

E3C13 What type of polarization is best for ground-wave propagation?

A. Vertical.

B. Horizontal.

C. Circular.

D. Elliptical.

We use ***vertical polarization for ground-wave propagation***. Knife-edge bending is a good example of vertically-polarized ground waves that drag the bottom half of the wave over a mountain top, resulting in the ground-wave signal getting into valleys where normally there would be no reception. **ANSWER A.**

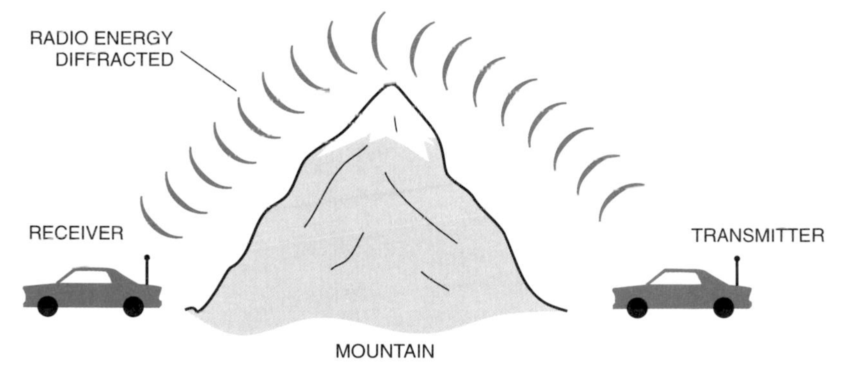

Knife-Edge Diffraction

E2C01 Which of the following is true about contest operating?

A. Operators are permitted to make contacts even if they do not submit a log.

B. Interference to other amateurs is unavoidable and therefore acceptable.

C. It is mandatory to transmit the call sign of the station being worked as part of every transmission to that station.

D. Every contest requires a signal report in the exchange.

Many of you are preparing for Extra Class to gain valuable, exclusive Extra Class frequencies for contesting. Contesting hones our ability to pick weak signals out of the noise and enhances speedy operating techniques essential in some emergency communication situations. ***There is no requirement to submit a contest log*** when participating in a contest. However, sending in a log lets contest officials know the popularity of their specific event. **ANSWER A.**

E2C11 How should you generally identify your station when attempting to contact a DX station during a contest or in a pileup?

A. Send your full call sign once or twice.
B. Send only the last two letters of your call sign until you make contact.
C. Send your full call sign and grid square.
D. Send the call sign of the DX station three times, the words "this is," then your call sign three times.

Always make sure to ***send your full call sign*** when operating DX, not just the last letter or two in your call sign, which is technically not legal. Rules also require that you momentarily listen to your offset transmit frequency to ensure the frequency is not in use. **ANSWER A.**

Veteran DXers operate split to work rare DX foreign stations.

E2C10 Why might a DX station state that they are listening on another frequency?

A. Because the DX station may be transmitting on a frequency that is prohibited to some responding stations.
B. To separate the calling stations from the DX station.
C. To improve operating efficiency by reducing interference.
D. All of these choices are correct.

As a new Extra Class operator, learn from listening during your first contest in the Extra Class portion of the band. If it is an international contest, you will hear rare DX stations calling for a contact "listening up 10." They are not listening on their own transmit frequency, but are listening 10 kHz higher for a response. An example might be a foreign DX station operating just below your Extra Class band limits on 20 meters, at 14.145 MHz, listening "up 10." This means you would transmit back to them on 14.155 MHz. To do this, you tune VFO-A to the receive frequency of 14.145 MHz, and put VFO-B on your transmit frequency of 14.155 MHz. Now engage SPLIT VFO OPERATION and make sure you are transmitting on VFO-B, while receiving on VFO-A. You must be in SPLIT for this to occur. If you transmit back to the foreign station outside of your Extra Class band edge you are illegally out of band and the foreign station can't hear you because they are listening "up 10 kHz." ***Split operation reduces interference, separates calling stations from the DX station, and lets the DX station stay in the clear, out of our U.S. phone band.***
ANSWER D. ☞ www.arrl.org/contests

E2C07 What is the Cabrillo format?

A. A standard for submission of electronic contest logs.
B. A method of exchanging information during a contest QSO.
C. The most common set of contest rules.
D. The rules of order for meetings between contest sponsors.

The various computer logging programs (such as CT by K1EA, TR Log, N1MM, etc.) have built-in utilities that *automatically generate a Cabrillo-formatted log* after the contest is over. It's one of those features where you click on a box and the program automatically ker-chunks out a Cabrillo file for submission to the contest sponsor. **ANSWER A.**

E2C02 Which of the following best describes the term self-spotting in regards to HF contest operating?

A. The generally prohibited practice of posting one's own call sign and frequency on a spotting network.
B. The acceptable practice of manually posting the call signs of stations on a spotting network.
C. A manual technique for rapidly zero beating or tuning to a station's frequency before calling that station.
D. An automatic method for rapidly zero beating or tuning to a station's frequency before calling that station.

Most avid contesters, looking for a high contest score, will stay glued to a computer call sign spotting network. This allows them to see rare stations and where they are operating on the radio dial. It is considered *poor practice to spot yourself on a computer posting* of contest activity, called "self spotting." **ANSWER A.**

Band	USB Calling Frequency
6 meters	50.125 MHz
2 meters	144.200 MHz
1.25	222.1 MHz
70 cm	432.100 MHz
35 cm	902.100 MHz
23 cm	1296.100 MHz

E2C06 During a VHF/UHF contest, in which band segment would you expect to find the highest level of activity?

A. At the top of each band, usually in a segment reserved for contests.
B. In the middle of each band, usually on the national calling frequency.
C. In the weak signal segment of the band, with most of the activity near the calling frequency.
D. In the middle of the band, usually 25 kHz above the national calling frequency.

During a VHF/UHF contest, *most of the weak signal single sideband activity is found on each side of the calling frequency*. **ANSWER C.**

These two frequencies are where to hang out looking for a random DX call. Once you make contact, move at least 5 kHz from these calling channels.

E2C03 From which of the following bands is amateur radio contesting generally excluded?

A. 30 meters.
B. 6 meters.
C. 2 meters.
D. 33 centimeters.

The 30 meter, 10 MHz, band allows CW and data only with a maximum power of 200 watts output. ***There is no contesting allowed on the 30 meter band***. **ANSWER A.**

E2C05 What is the function of a DX QSL Manager?

A. To allocate frequencies for DXpeditions.
B. To handle the receiving and sending of confirmation cards for a DX station.
C. To run a net to allow many stations to contact a rare DX station.
D. To relay calls to and from a DX station.

It's fun going out on a DXpedition. When Gordo was at Christmas Island, operating T32GW, the QSL cards poured in for months! Sometimes, on a big scale DXpedition, minding the paper QSL cards and the electronic QSL requests is best handled by a ***DX QSL manager***. They carefully ***check the logs to insure all of the contact information is valid, and then send out the coveted QSL card confirmation***. It is a huge job, so if you know of a DX QSL manager, give them a good pat on the back! **ANSWER B.**

Some of Gordo's QSL cards go back to the 3 cent stamp! Collecting paper QSL cards is a fun hobby.

E2C08 Which of the following contacts may be confirmed through the U.S. QSL bureau system?

A. Special event contacts between stations in the U.S.
B. Contacts between a U.S. station and a non-U.S. station.
C. Repeater contacts between U.S. club members.
D. Contacts using tactical call signs.

QSL Bureaus are very busy! Don't load down the U.S. QSL bureau with domestic QSLs! These bureaus are set up to handle the ***special mail requirements between U.S. and non-U.S. DX stations***, both incoming and outgoing. **ANSWER B.**

HERE ARE SOME WEB ADDRESSES OF USEFUL SITES RELATED TO SKYWAVES & CONTESTING:

http://www.ncdxf.org/beacon/beaconprograms/html
http://www.dxatlas.com/
http://dx-world.net/
http://www.voacap.com/prediction.html

CD 2 TRACK 1

Outer Space Comms

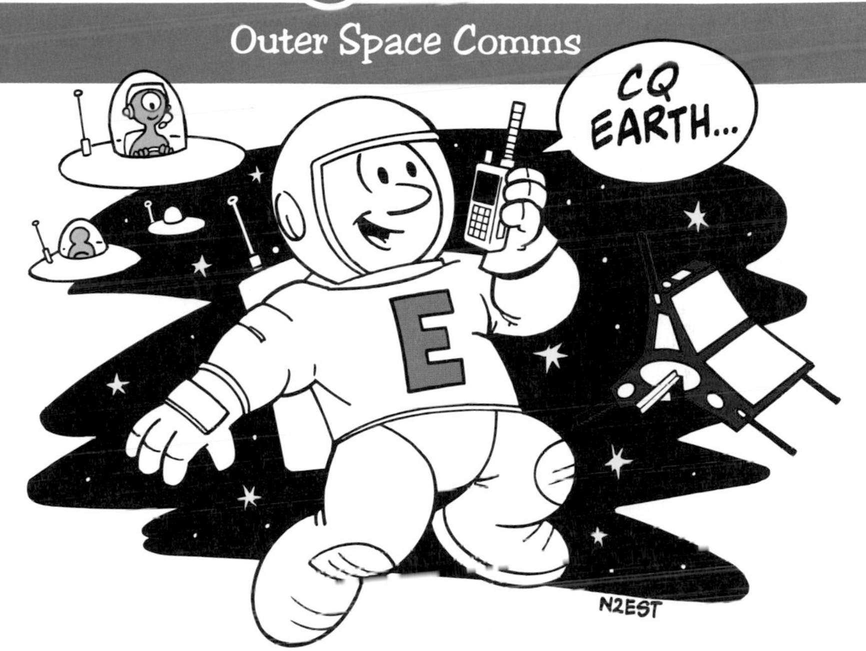

E1D02 What is the amateur satellite service?

A. A radio navigation service using satellites for the purpose of self training, intercommunication and technical studies carried out by amateurs.
B. A spacecraft launching service for amateur-built satellites.
C. A radio communications service using amateur radio stations on satellites.
D. A radio communications service using stations on Earth satellites for public service broadcast.

The amateur satellite service is a ***radio communication service for the purpose of self-training, intercommunication, and technical investigations*** that allows hams to communicate through amateur stations located on orbiting satellites. [97.3] **ANSWER C.**

E2A02 What is the direction of a descending pass for an amateur satellite?

A. From north to south.
B. From west to east.
C. From east to west.
D. From south to north.

There are several factors to consider when attempting to communicate through an amateur satellite. One is the direction in which the satellite travels. A satellite travelling with a ***descending*** path will pass over the Equator going ***from north to south***. This will get you started for spotting a satellite. Tracking programs available on the Internet will help you collect the other details needed to find the satellite while it is visible between your horizons. **ANSWER A.**

E2A01 What is the direction of an ascending pass for an amateur satellite?

A. From west to east.
B. From east to west.
C. From south to north
D. From north to south

Some satellites travel over the Equator from north to south (descending) while others pass over the Equator ***from south to north (ascending)***. Direction, height and speed all affect where a satellite is at a specific time. It's easy to remember if you think of the North Pole as being at the top of the earth. **ANSWER C.**

E2A03 What is the orbital period of an Earth satellite?

A. The point of maximum height of a satellite's orbit.
B. The point of minimum height of a satellite's orbit.
C. The time it takes for a satellite to complete one revolution around the Earth.
D. The time it takes for a satellite to travel from perigee to apogee.

The time it takes to make one complete orbit of the Earth is the period of a satellite. Low orbit satellites can take less than an hour to complete one orbit, while those at higher altitudes can takes days. **ANSWER C.**

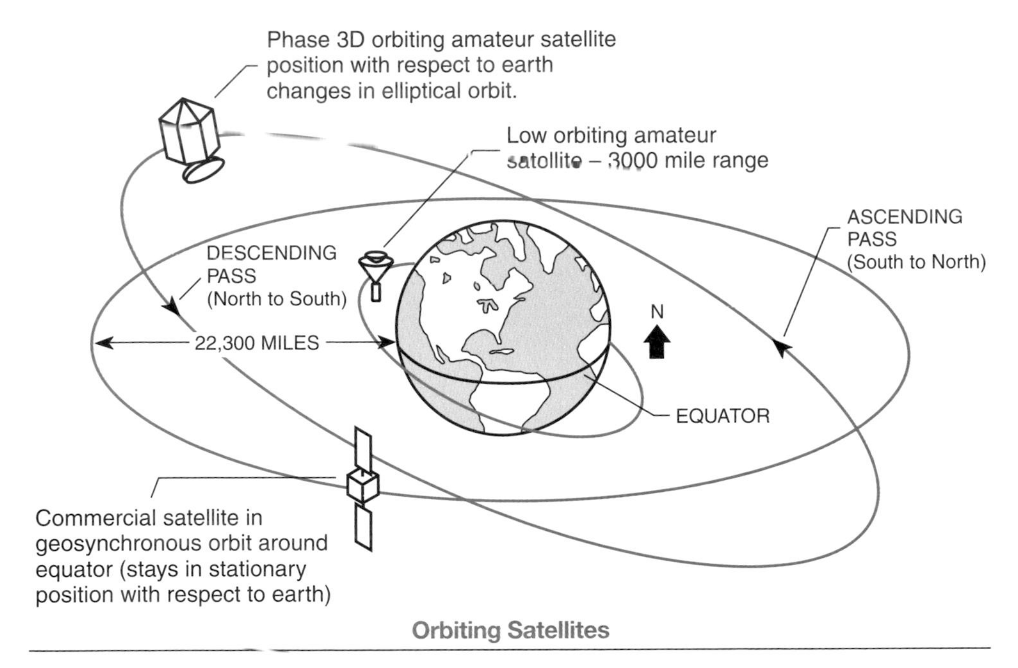

Orbiting Satellites

E1D04 What is an Earth station in the amateur satellite service?

A. An amateur station within 50 km of the Earth's surface intended for communications with amateur stations by means of objects in space.
B. An amateur station that is not able to communicate using amateur satellites.
C. An amateur station that transmits telemetry consisting of measurement of upper atmosphere data.
D. Any amateur station on the surface of the Earth.

Below 50 kilometers, you're considered an ***Earth station*** when you are communicating via the amateur satellite service. [97.3] **ANSWER A.**

E1D11 Which amateur stations are eligible to operate as Earth stations?

A. Any amateur station whose licensee has filed a pre-space notification with the FCC's International Bureau.
B. Only those of General, Advanced or Amateur Extra Class operators.
C. Only those of Amateur Extra Class operators.
D. Any amateur station, subject to the privileges of the class of operator license held by the control operator.

An amateur Earth station is located down here on planet Earth. Operating as an ***Earth station*** for space communications ***is available to all amateurs***. Be sure to use only those frequency privileges allowed by your license class. [97.209] **ANSWER D.**
☞ **www.amsat.org**

Ready to hear a spacecraft data beacon? There are plenty on the air every day all over the world as they fly by.

E1D03 What is a telecommand station in the amateur satellite service?

A. An amateur station located on the Earth's surface for communication with other Earth stations by means of Earth satellites.
B. An amateur station that transmits communications to initiate, modify or terminate functions of a space station.
C. An amateur station located more than 50 km above the Earth's surface.
D. An amateur station that transmits telemetry consisting of measurements of upper atmosphere data.

Telecommand means ***initiating, modifying, or terminating a function of another station***, such as a space station. [97.3] **ANSWER B.**

E1D10 Which amateur stations are eligible to be telecommand stations?

A. Any amateur station designated by NASA.
B. Any amateur station so designated by the space station licensee, subject to the privileges of the class of operator license held by the control operator.
C. Any amateur station so designated by the ITU.
D. All of these choices are correct.

Any amateur station so designated by the space station licensee may be a telecommand station. [97.211] **ANSWER B.**

E1D01 What is the definition of the term telemetry?

A. One-way transmission of measurements at a distance from the measuring instrument.
B. Two-way radiotelephone transmissions in excess of 1000 feet.
C. Two-way single channel transmissions of data.
D. One-way transmission that initiates, modifies, or terminates the functions of a device at a distance.

We hear the word "telemetry" a lot in satellite service. Satellite telemetry may give ground controllers vital information on satellite battery power, power loads, and satellite solar power charging. ***Telemetry is a one-way transmission of measurements at a distance*** from the measuring instrument. [97.3] **ANSWER A.**

E2A04 What is meant by the term mode as applied to an amateur radio satellite?

A. The type of signals that can be relayed through the satellite.
B. The satellite's uplink and downlink frequency bands.
C. The satellite's orientation with respect to the Earth.
D. Whether the satellite is in a polar or equatorial orbit.

Working through amateur satellites is exciting, challenging, and fun! Transmit and receive frequencies for satellite operations are on separate bands. The satellite receives on one frequency band and transmits on another frequency band. Since two bands are involved, we have two *mode designators* to best describe the transmit and receive bands. Each band has its own designator. The Earth transmit band is called the "*uplink*," and the receive band is the "*downlink*." **ANSWER B.**

E2A05 What do the letters in a satellite's mode designator specify?

A. Power limits for uplink and downlink transmissions.
B. The location of the ground control station.
C. The polarization of uplink and downlink signals.
D. The uplink and downlink frequency ranges.

The two *mode designators indicate* the transmit *uplink frequency* and the receive *downlink frequency* bands that the satellite uses. **ANSWER D.**

E2A09 What do the terms L band and S band specify with regard to satellite communications?

A. The 23 centimeter and 13 centimeter bands.
B. The 2 meter and 70 centimeter bands.
C. FM and Digital Store-and-Forward systems.
D. Which sideband to use.

You can work this out by recalling the MHz to meters and meters to MHz formula, but there are lots of conversions because we are working with centimeters and GHz. The L band at 1.26 GHz works out to be 23 centimeters. The S band, at 2.4 GHz, works out to be 13 centimeters. You can rule out Answer B as that would be VHF and UHF, with C and D having nothing to do with band specifics. *23 centimeters is L band, and 13 centimeters is S band.* **ANSWER A.**

Frequency Bands	Frequency Range	Modes
High Frequency	21 - 30 MHz	Mode H
VHF	144 - 146 MHz	Mode V
UHF	435 - 438 MHz	Mode U
L band	1.26 - 1.27 GHz	Mode L
S band	2.4 - 2.45 GHz	Mode S
C band	5.8 GHz	Mode C
X band	10.4 GHz	Mode X
K band	24 GHz	Mode K

E2A06 On what band would a satellite receive signals if it were operating in mode U/V?

A. 435 MHz - 438 MHz.
B. 144 MHz - 146 MHz.
C. 50.0 MHz - 50.2 MHz.
D. 29.5 MHz - 29.7 MHz.

Watch out on this question – we don't want you to miss it even though you thought they might be trying to trick you. The question asks on what band would a ***satellite receive*** signals – this is the same as asking on what band would you transmit to the satellite. Since the ***mode*** designator for the satellite begins with the letter ***U***, we transmit to the satellite on ***UHF 435 - 438 MHz*** where the satellite receives our transmitted signals. **ANSWER A.**

E2A07 Which of the following types of signals can be relayed through a linear transponder?

A. FM and CW.
B. SSB and SSTV.
C. PSK and Packet.
D. All of these choices are correct.

Thanks to AMSAT, many of our OSCAR satellites offer a "bent pipe" re-transmission of many transmission modes. If we uplink FM or CW, it comes back down as FM or CW, respectively, through the satellite linear transponder. Same thing with single sideband, slow scan television, and the digital modes – ***all of these signals are relayed through*** the satellite's on-board ***linear transponder***. **ANSWER D.**

E2A08 Why should effective radiated power to a satellite which uses a linear transponder be limited?

A. To prevent creating errors in the satellite telemetry.
B. To avoid reducing the downlink power to all other users.
C. To prevent the satellite from emitting out-of-band signals.
D. To avoid interfering with terrestrial QSOs.

Satellites are powered by on-board batteries charged by solar panels. On older satellites, ***any strong input signal is re-transmitted at the same strength. This quickly depletes satellite battery power***. Our modern constellation of low Earth orbit satellites re-transmits modest power signals at approximately the same output power to conserve battery life. An overly strong input signal can cause the satellite to reduce its output signal to protect the batteries from a high power drain. Once the satellite cuts power back in response to a strong signal, all of its re-transmissions are done at a reduced power. Everyone suffers even if they are not using high power. So don't run any more power than it takes for you to hear your own signal coming back clearly. **ANSWER B.**

E2A10 Why may the received signal from an amateur satellite exhibit a rapidly repeating fading effect?

A. Because the satellite is spinning.
B. Because of ionospheric absorption.
C. Because of the satellite's low orbital altitude.
D. Because of the Doppler Effect.

Many amateur satellites continuously rotate to keep all sides of the "bird" equally bathed in sunlight. This ***spinning*** distributes the heat build-up and allows all of the solar panels to receive energy from the Sun. This spinning moves not only the solar panels but also the antennas. Because of the rotation, we hear a ***characteristic rapid pulse in and out on the downlink***. Using circular polarization on the receiving end can largely eliminate this fading. **ANSWER A.**

E2A11 What type of antenna can be used to minimize the effects of spin modulation and Faraday rotation?

A. A linearly polarized antenna.
B. A circularly polarized antenna.
C. An isotropic antenna.
D. A log-periodic dipole array.

A circularly polarized antenna will give you the best results when working amateur satellites. Although you lose approximately 3 dB gain over a horizontal antenna, ***circular polarization*** minimizes fading. **ANSWER B.**

A small 3 element beam can easily hear a satellite signal, but continuously rotate the beam from vertical to horizontal to peak the signal. Good exercise!

Cross-polarized, dual-feed beams take the polarization challenge away for smooth copy.

E2A12 What is one way to predict the location of a satellite at a given time?

A. By means of the Doppler data for the specified satellite.
B. By subtracting the mean anomaly from the orbital inclination.
C. By adding the mean anomaly to the orbital inclination.
D. By calculations using the Keplerian elements for the specified satellite.

You can log onto many satellite web pages and learn the ***Keplerian elements*** for a specific satellite or a specific day and time ***for the location of the satellite***. You can find this information on the AMSAT website at ☞ **www.AMSAT.org**. **ANSWER D.**

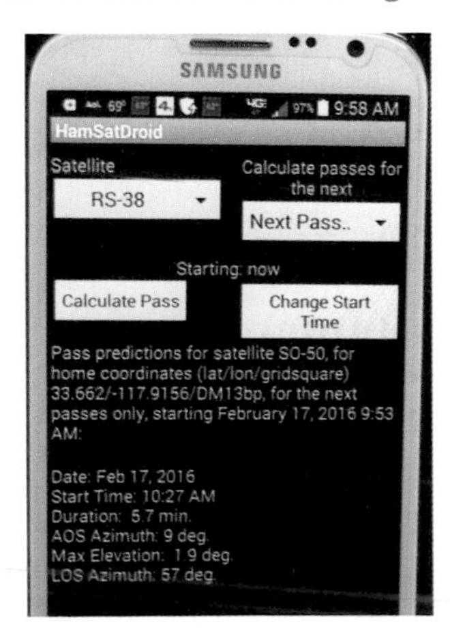

You can use smart phone apps to access computer programs and websites that can show you where and when an amateur satellite or the Space Station will be in range of your ham station

Where's the amateur satellite?

AMSAT is the North American distributor of SatPC32, a tracking program designed for ham satellite applications. It has a host of features that help you locate and track amateur satellites, set your antenna rotator for automatic control, and provides visual mapping of satellite locations as they orbit the Earth. You can download a demo version from www.dk1tb.de/indexeng.htm, or order online at www.amsat.org or by calling 1-888-322-6728. BTW – the author of SatPC32, DK1TB, donated the program to AMSAT and all proceeds support AMSAT.

E2D04 What is the purpose of digital store-and-forward functions on an Amateur Radio satellite?

A. To upload operational software for the transponder.
B. To delay download of telemetry between satellites.
C. To store digital messages in the satellite for later download by other stations.
D. To relay messages between satellites.

Orbiting amateur radio satellites make great store-and-forward devices to ***hold messages that may be downloaded later*** by other stations. **ANSWER C.**

E2D05 Which of the following techniques is normally used by low Earth orbiting digital satellites to relay messages around the world?

A. Digipeating.
B. Store-and-forward.
C. Multi-satellite relaying.
D. Node hopping.

Store-and-forward is a popular way of sending a wireless e-mail that may be received halfway around the world when the satellite passes over the distant station. **ANSWER B.**

E1D06 Which of the following is a requirement of a space station?

A. The space station must be capable of terminating transmissions by telecommand when directed by the FCC.
B. The space station must cease all transmissions after 5 years.
C. The space station must be capable of changing its orbit whenever such a change is ordered by NASA.
D. All of these choices are correct.

All space stations must be capable of receiving a ***telecommand to shut down*** if ordered to do so by the FCC. [97.207] **ANSWER A.**

E1D07 Which amateur service HF bands have frequencies authorized for space stations?

A. Only the 40 m, 20 m, 17 m, 15 m, 12 m and 10 m bands.
B. Only the 40 m, 20 m, 17 m, 15 m and 10 m bands.
C. Only the 40 m, 30 m, 20 m, 15 m, 12 m and 10 m bands.
D. All HF bands.

By international agreement, all high-frequency ITU voice bands plus most of our regular high-frequency ham bands are available for space operation. This includes ***40 meters, 20 meters, 17 meters, 15 meters, 12 meters, and 10 meters***. To help you remember the correct bands, think of "voice bands." This eliminates the answers that contains 30 meters since this band does not include voice privileges. [97.207] **ANSWER A.**

E2A13 What type of satellite appears to stay in one position in the sky?

A. HEO.
B. Geostationary.
C. Geomagnetic.
D. LEO.

A satellite in a ***geostationary*** orbit remains ***in one position*** above the Earth. No tracking necessary. You always know where it will be. **ANSWER B.** To learn more, ☞ **www.amsat.org**

Good To Know:

Geostationary or Geosynchronous?

Great news from AMSAT! They are working with the U.S. Air Force for a Geosynchronous digital satellite scheduled to lift off in 2018 that will "park" at 74 degrees west over the Equator. As designed, this satellite will have a 12 foot wide "Field of View" seeing all of the U.S., Canada, Hawaii, and Mexico. In addition, it will also see a portion of lower Alaska, northern South America, the western edge of Europe and part of western Africa! The satellite will receive digital uplinks on 5.6 GHz, and downlink on 10 GHz. The on-board linear transponder may offer up to 100 channels of data capability! If this comes to pass, it will greatly simplify and expand ham satellite operation.

If you have worked any amateur satellites, you know that you have to track the satellite as it moves from horizon to horizon. The Keplerian Elements, or "Keps" are the positional data you need to "find" the satellite in its overhead pass. The ARRL and other amateur satellite entities faithfully broadcast the Keps or otherwise disseminate the data on a daily basis, which you can either insert into your automatic antenna steering program, or use as a rough guideline if you're using the "Armstrong Method" of manually tracking a bird.

Now, there is a bit of difference between geostationary orbit and geosynchronous orbit. A true geostationary orbit must be perfectly circular (or very nearly so) and directly above the equator. This is the orbit used for broadcast satellites, and is generally a very expensive proposition, initially, as the satellite needs to be launched in a very high orbit, around 26,199 miles. However, a true geostationary orbit allows you to dispense entirely with tracking, since the satellite is in a relative fixed position over the earth.

A geosynchronous orbit, on the other hand, can be highly elliptical. It can appear to "wobble" in its orbit around the earth as it moves closer or farther away from the equator. At least once a day, a geosynchronous orbit will appear over the same location. So, even with a fixed antenna, you will be able to regularly use the satellite.

The distinction between the two orbital types is actually not too important for amateurs in the long run, because geosynchronous orbits are almost always an intermediate stage on the way to geostationary orbit. Space agencies generally launch into a geosynchronous orbit first, and then begin the rather long process of "rounding out" to a true geostationary orbit.

E1D05 What class of licensee is authorized to be the control operator of a space station?

A. All except Technician Class.
B. Only General, Advanced or Amateur Extra Class.
C. Any class with appropriate operator privileges.
D. Only Amateur Extra Class.

Originally, only an Extra Class operator could transmit from space. Today, there is no longer a specific license class requirement for space operation. ***Any grade of license*** now qualifies to be the control operator as long as they are sure to operate within the privileges of their class. [97.207] **ANSWER C.**

E1D08 Which VHF amateur service bands have frequencies available for space stations?

A. 6 meters and 2 meters.
B. 6 meters, 2 meters, and 1.25 meters.
C. 2 meters and 1.25 meters.
D. 2 meters.

On VHF (30-300 MHz), ***only the 2-meter band allows space station operation***. Space station operation is prohibited on 1.25 meters (the 222 MHz band) and 6 meters. Notice that the wrong answers includes the other bands. [97.207] **ANSWER D.**

E1D09 Which UHF amateur service bands have frequencies available for a space station?

A. 70 cm only.
B. 70 cm and 13 cm.
C. 70 cm and 33 cm.
D. 33 cm and 13 cm.

On UHF (300-3000 MHz), the 33 cm (900 MHz) band is NOT allowed for space station operations, but all other bands are allowed. This includes ***70 cm (435 MHz) and 13 cm (2300 MHz)***. [97.207] **ANSWER B.**

E3A01 What is the approximate maximum separation measured along the surface of the Earth between two stations communicating by Moon bounce?

A. 500 miles, if the Moon is at perigee.
B. 2000 miles, if the Moon is at apogee.
C. 5000 miles, if the Moon is at perigee.
D. 12,000 miles, if the Moon is visible by both stations.

It takes a minimum of a pair of long-boom Yagis, plus the maximum legal power output, to establish a Moon bounce QSO on CW or SSB phone. If you are working a station with a huge dish antenna, you might get by with a single long-boom Yagi and a good VHF or UHF kilowatt amplifier. Moon bounce is most common on the bottom edge of the 2 meter band. Communications over a great distance and even beyond the horizon are possible when ***both stations can see the Moon.*** Monthly ham magazines publish the best EME (Earth-Moon-Earth) days. **ANSWER D.**

EME communications require specialized equipment and patience.

E3A02 What characterizes libration fading of an EME signal?

A. A slow change in the pitch of the CW signal.
B. A fluttery irregular fading.
C. A gradual loss of signal as the Sun rises.
D. The returning echo is several hertz lower in frequency than the transmitted signal.

Remember, EME means Earth-Moon-Earth and that the Moon is not a perfect reflector of radio signals. All those craters on the Moon will cause your ***return signal to sound fluttery, with rapid, irregular fading***. **ANSWER B.**

E3A03 When scheduling EME contacts, which of these conditions will generally result in the least path loss?

A. When the Moon is at perigee.
B. When the Moon is full.
C. When the Moon is at apogee.
D. When the MUF is above 30 MHz.

PERigee is the time when the Moon is closest to the Earth. This is the PERfect condition for EME to occur, and a PERfect day to schedule an EME contact. When the ***Moon is at perigee*** and just appearing above the horizon, your will get the ***best opportunity for Moon-bounce*** success. **ANSWER A.**

E2D06 Which of the following describes a method of establishing EME contacts?

A. Time synchronous transmissions alternately from each station.
B. Storing and forwarding digital messages.
C. Judging optimum transmission times by monitoring beacons reflected from the Moon.
D. High speed CW identification to avoid fading.

Although "in the wild" EME contacts have occurred, your chances of making an EME contact are much, much better if you have some means of scheduling ahead of time. The "JT" digital modes are ***scheduled modes*** especially suitable for EME contacts. **ANSWER A.**

Gordo hears signals from the Moon on a small beam. The Arecibo signal certainly helps overcome the path loss.

E2D03 Which of the following digital modes is especially useful for EME communications?

A. FSK441.
B. PACTOR III.
C. Olivia.
D. JT65.

It takes a mighty big long-boom antenna to bounce signals off the Moon in an Earth-Moon-Earth contact. With voice, it is a huge effort requiring at a minimum a pair of long-boom Yagi antennas with the maximum legal power output. But going ***digital with JT65*** can make it possible for you to begin receiving Moon echoes with just a modest antenna system. You can hear echoes from the Moon of digital signals on Gordo's exclusive audio CD course. Take a listen! **ANSWER D.**

E2D13 What type of modulation is used for JT65 contacts?

A. Multi-tone AFSK.
B. PSK.
C. RTTY.
D. IEEE 802.11.

Multi-tone modulation methods allow the use of a much narrower bandwidth (for a given data rate) than would ordinarily be possible with a single frequency shifted tone. JT65 uses ***multi-tone Audio Frequency Shift Keying (AFSK)***, which is converted to multiple radio frequency shift keying when applied to a single sideband transmitter. When this FSK signal is received by a single sideband receiver, it is converted back to AFSK. You can see all the different audio tones on a standard waterfall display. **ANSWER A.**

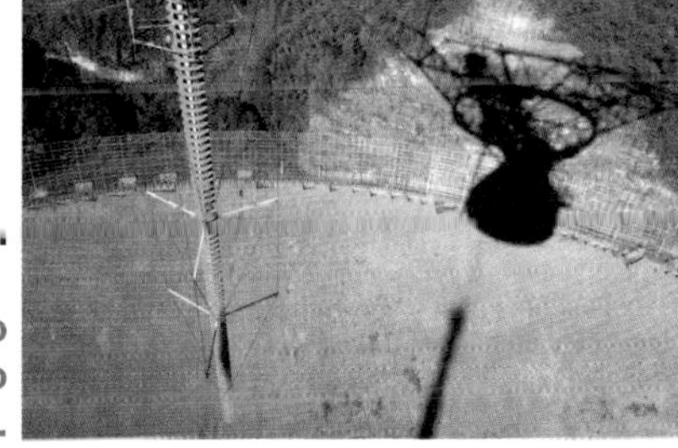

Licensed local hams in Puerto Rico sometimes get to work the Arecibo radio telescope on ham bands.

E2E03 How is the timing of JT65 contacts organized?

A. By exchanging ACK/NAK packets.
B. Stations take turns on alternate days.
C. Alternating transmissions at 1 minute intervals.
D. It depends on the lunar phase.

JT65 certainly seems like an oddball mode if you're used to more conventional digital modes. Its ability to pull very weak signals out of the noise is largely a result of its ***precise timing and regimented transmission schedule***. You will want your computer's clock accurately synchronized to GPS, Internet network time, or WWV for reliable operation. If you find you cannot decode any JT65 signals, check your computer's clock. **ANSWER C.**

Good To Know:

Weak Signal Band Plan Recommendations

Over the past few years, tremendous advances have been made in weak-signal communications on the VHF and UHF bands using computer-generated data modes. These modes enable communications at or below the noise floor. Amateurs with modest stations are now making contacts over thousands of kilometers via EME and Meteor Scatter using these new modes. However, these very-weak data signals often cannot be detected by ear, which makes them incompatible with normal analog SSB or CW signals. There have been instances of interference caused by SSB operators who obviously can't hear the data signals. The recent EME Conference held in Germany recommended the creation of worldwide VHF and UHF sub-bands for JT-65, the weak signal data protocol. The suggested segments are 50.185 to 50.195 MHz on 6 meters; 144.115 to 144.135 MHz on 2 meters; 432.060 to 432.070 MHz on 70 centimeters; and 1296.060 to 1296.070 MHz.

E2D14 What is one advantage of using JT65 coding?

A. Uses only a 65 Hz bandwidth.

B. The ability to decode signals which have a very low signal to noise ratio.

C. Easily copied by ear if necessary.

D. Permits fast-scan TV transmissions over narrow bandwidth.

Weak signal VHF/UHF and microwave operators know well the popular WSJT software that enables miraculous ***reception of digital signals well below a normal noise floor. JT65*** coding allows your radio to capture everything from meteor scatter CQs to conversations off the surface of the Moon! Just because you can't hear anything coming out of the speaker, many times your computer's sound card can deliver enough meaningful audio to the decoding programs to utterly amaze you! **ANSWER B.**

E2D12 How does JT65 improve EME communications?

A. It can decode signals many dB below the noise floor using FEC.

B. It controls the receiver to track Doppler shift.

C. It supplies signals to guide the antenna to track the Moon.

D. All of these choices are correct.

The best digital emission for Earth-Moon-Earth (EME) communications is ***JT65***, a data mode that ***can detect signals down to 30 dB lower than your VHF/UHF noise floor by using forward error correction (FEC)***, well-structured formats, a one minute transmit period, and message frames repeated up to five times. Let your computer and transceiver work all night banging signals off the Moon. When you wake up the next morning, Voila! Here are the contacts that took place!
ANSWER A.

HERE ARE SOME WEB ADDRESSES OF USEFUL SITES RELATED TO OUTER SPACE COMMS

http://spaceweather.com/
http://prop.hfradio.org/
www.amsat.org/amsat-new/news/
http://hamwaves.com/propagation/
http://www.spacew.com/www/realtime.php
http://www.arrl.org/news/arrl-committee-seeks-microwave-band-plan-input
http://www.arrl.org/news/space-station-active-on-70-cm-packet
http://www.vhfdx.de/wsjt/
http://www.qsl.net/w8wn/hscw/hscwopen.html

CD 2 TRACK 3

Visuals & Video Modes

E2B01 How many times per second is a new frame transmitted in a fast-scan (NTSC) television system?

A. 30.
B. 60.
C. 90.
D. 120.

Operating fast-scan television is fun on the 420 MHz and higher bands. ***A new frame is created 30 times per second*** by two complete fields for each frame. This gives you a picture just like home television. In fact, older analog, cable-ready TVs can still receive local ATV signals! **ANSWER A.**

These are the KH6HME beams in Hawaii which continuously send out beacon signals for tropo ducting experiments to the West Coast of the U.S. – a 2500 mile path!

A 6 MHz wide ATV signal received by Gordo over a 2500 mile tropo path from Hawaii to the West Coast.

E2B02 How many horizontal lines make up a fast-scan (NTSC) television frame?

A. 30. C. 525.
B. 60. D. 1080.

A total of ***525 lines from two fields make up a complete frame***. **ANSWER C.**

E2B03 How is an interlaced scanning pattern generated in a fast-scan (NTSC) television system?

A. By scanning two fields simultaneously.
B. By scanning each field from bottom to top.
C. By scanning lines from left to right in one field and right to left in the next.
D. By scanning odd numbered lines in one field and even numbered lines in the next.

Each field is 262.5 lines, and ***the frame is created by scanning even numbered lines in one field and odd numbered lines in the second field***. **ANSWER D.**

E2B04 What is blanking in a video signal?

A. Synchronization of the horizontal and vertical sync pulses.
B. Turning off the scanning beam while it is traveling from right to left or from bottom to top.
C. Turning off the scanning beam at the conclusion of a transmission.
D. Transmitting a black and white test pattern.

The scanning beam is turned off when it completes its travel from left to right and is returning to the left to start another line. The scanning beam is also blanked when the beam reaches the bottom and is returning to the top to begin the next field. **ANSWER B.**

E2B08 Which of the following is a common method of transmitting accompanying audio with amateur fast-scan television?

A. Frequency-modulated sub-carrier.
B. A separate VHF or UHF audio link.
C. Frequency modulation of the video carrier.
D. All of these choices are correct.

Besides the standard 4.5 MHz sound sub-carrier used by most ATV and analog broadcast stations, there are a variety of ways to pass audio while transmitting an amateur radio fast-scan "live action" video signal. We could ***frequency modulate the video carrier itself***, known as "on carrier" sound, which was used in the early days when hams added video modulators to FM voice transceivers. Audio can be sent simultaneously using FM audio on a ***dedicated VHF or UHF audio link*** in another band. This link can also be used as an ATV calling, coordination and talkback channel for the receiving stations. On the 2 meter band, most ATV stations use 144.340 MHz or 146.430 MHz for this link. **ANSWER D.**

☞ **www.hamtv.com**

E2B07 What is the name of the signal component that carries color information in NTSC video?

A. Luminance. C. Hue.
B. Chroma. D. Spectral Intensity.

Chroma is the video component contained in a phase modulated 3.58 MHz sub-carrier that varies with the instantaneous color in the scene being scanned by the video camera. **ANSWER B.**

E2B06 What is vestigial sideband modulation?

A. Amplitude modulation in which one complete sideband and a portion of the other are transmitted.
B. A type of modulation in which one sideband is inverted.
C. Narrow-band FM modulation achieved by filtering one sideband from the audio before frequency modulating the carrier.
D. Spread spectrum modulation achieved by applying FM modulation following single sideband amplitude modulation.

Vestigial sideband modulation ***(VSB) is amplitude modulation of a carrier in which one complete sideband is transmitted*** (upper VSB is standard) ***and a vestigial sideband filter is used to attenuate the other sideband***. This helps reduce occupied bandwidth to 6 MHz and provides better spectrum efficiency than full double sideband amplitude modulation. **ANSWER A.**

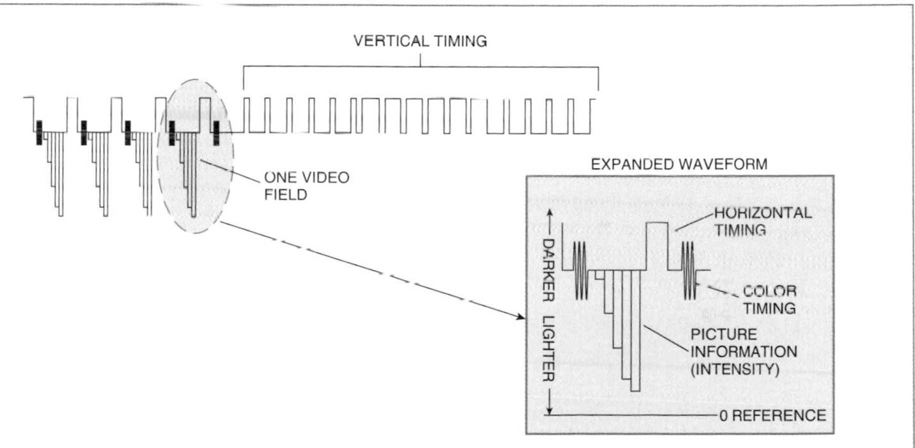

a. A TV Composite Video Signal
Source: *Using Video in Your Home,* G. McComb. © 1989, Master Publishing, Inc., Niles, IL.

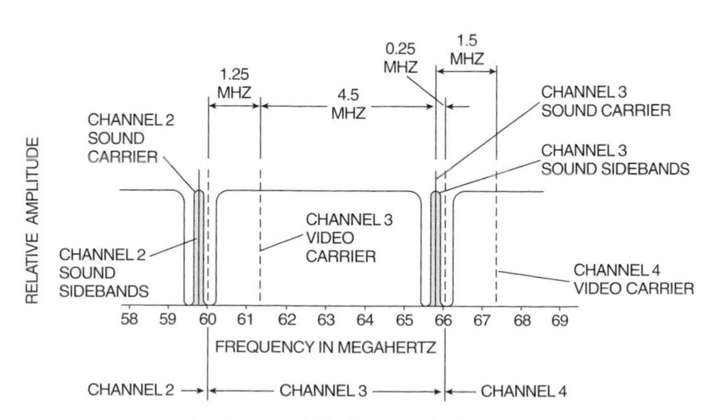

b. Bandwidth for a TV Channel
Source: *Antennas,* A. Evans, K. Britain. © 1988, Master Publishing, Inc., Niles, IL.

The video signal is composed of three major components: picture, timing, and color. Television sets "know" how to decode this stream of information into pictures that reproduce the changing scenes captured by the video camera.

E2B05 Which of the following is an advantage of using vestigial sideband for standard fast-scan TV transmissions?

A. The vestigial sideband carries the audio information.
B. The vestigial sideband contains chroma information.
C. Vestigial sideband reduces bandwidth while allowing for simple video detector circuitry.
D. Vestigial sideband provides high frequency emphasis to sharpen the picture.

Vestigial sideband (VSB) cuts off the lower color and sound sub-carriers at -3.58 and -4.50 MHz from the video carrier, respectively, as they are unnecessary. The frequencies below 1.25 MHz from the video carrier in a VSB transmitter will have very low energy radiated. This frees up room for more users of the band. Total VSB TV channel ***bandwidth*** is (14.75 ÷ 2) - 1.25 = 6 MHz. ***Simple analog video detectors are used*** in the TV receiver with most of the work being done by the VSB filter in the IF ahead of the detector. Even though commercial TV signals have switched from analog to digital, many ham fast-scan TV signals remain analog. New TVs still employ an analog tuner and vestigial sideband receiver with a simple video detector in addition to the digital system to allow compatibility with older VCRs and DVD players. **ANSWER C.** ☞ **http://www.hampubs.com/**

E2B18 On which of the following frequencies is one likely to find FM ATV transmissions?

A. 14.230 MHz.
B. 29.6 MHz.
C. 52.525 MHz.
D. 1255 MHz.

FM fast-scan television is found on the 902 MHz and higher bands because it occupies a whopping 17 to 21 MHz of bandwidth depending on the sound subcarrier frequency used. U.S. standard in the 902 and 1240 MHz bands is 4 MHz deviation and 5.5 MHz sound, resulting in an occupied bandwidth of 19 MHz. The center frequency must be at least 9.5 MHz from the band edge so the sideband energy does not fall outside the band limits. ***1255 MHz*** is a reasonable frequency to use, given these parameters. The beauty of FM ATV, thanks to signal-to-noise limiter multiplication, is a snow-free picture that can be received up to 12 dB weaker in signal strength than an AM ATV picture. **ANSWER D.**

E2B16 Which is a video standard used by North American Fast Scan ATV stations?

A. PAL.
B. DRM.
C. Scottie.
D. NTSC.

North American fast-scan analog ATV stations transmit ***NTSC, National Television System Committee***, the same identifier as the old analog broadcast TV signals. Broadcast is now all digital. (PAL is the analog broadcast TV system used primarily in Europe and other countries. DRM and Scottie are SSTV modes.) **ANSWER D.**

E2B12 How are analog SSTV images typically transmitted on the HF bands?

A. Video is converted to equivalent Baudot representation.
B. Video is converted to equivalent ASCII representation.
C. Varying tone frequencies representing the video are transmitted using PSK.
D. Varying tone frequencies representing the video are transmitted using single sideband.

Transmitting a slow scan picture requires not much more than your HF transceiver, a sound card, SSTV software, and your favorite camcorder to freeze frame your smiling mug. The *varying tone frequencies* you hear on a transmitted SSB signal *represent the video* going out over the airwaves. To receive a color SSTV signal, simply connect your radio to your computer's sound card mic input and the rest is done mostly by the SSTV software! **ANSWER D.**

E2B17 What is the approximate bandwidth of a slow-scan TV signal?

A. 600 Hz.
B. 3 kHz.
C. 2 MHz.
D. 6 MHz.

We can find slow scan television on HF as well as on VHF/UHF. On HF, look for *slow scan* at 14.230 MHz and notice that it takes only *3 kHz of bandwidth*. Be patient! It can take up to 45 seconds to complete a single picture. **ANSWER B.**

E2B19 What special operating frequency restrictions are imposed on slow scan TV transmissions?

A. None; they are allowed on all amateur frequencies.
B. They are restricted to 7.245 MHz, 14.245 MHz, 21.345 MHz, and 28.945 MHz.
C. They are restricted to phone band segments and their bandwidth can be no greater than that of a voice signal of the same modulation type.
D. They are not permitted above 54 MHz.

Slow-scan TV transmissions are allowed in phone band segments of any band as long as their *bandwidth is no greater than that of a voice signal* of the same modulation type. You can easily hear 20 meter slow-scan signals by tuning in 14.230 and 14.233 MHz, a hot-bed of SSTV activity that is easily seen with a computer program or on your little handheld SSTV transmit/receive equipment. **ANSWER C.**

While it takes about 30 to 45 seconds to download a photo, slow scan TV is popular on the HF bands, plus from the International Space Station on 2 meters!

E2B13 How many lines are commonly used in each frame of an amateur slow-scan color television picture?

A. 30 or 60.
B. 60 or 100.
C. 128 or 256.
D. 180 or 360.

It takes *128 lines to make up a complete frame. Using 256 lines increases the resolution* of the color. **ANSWER C.**

E2B15 What signals SSTV receiving equipment to begin a new picture line?

A. Specific tone frequencies.
C. Specific tone amplitudes.
B. Elapsed time.
D. A two-tone signal.

In most SSTV receiving equipment, ***new picture lines are triggered when specific tone frequencies are received***. The software within the equipment knows what to listen for and exactly what to do when that tone is heard. **ANSWER A.**

E2B14 What aspect of an amateur slow-scan television signal encodes the brightness of the picture?

A. Tone frequency.
B. Tone amplitude.
C. Sync amplitude.
D. Sync frequency.

Brightness of an SSTV picture depends on the ***tone frequencies*** received. The higher the frequency the brighter the picture. **ANSWER A.**

Don't throw out your old tube TV set – it can do double duty displaying both ATV as well as SSTV pictures in color!

E2B09 What hardware, other than a receiver with SSB capability and a suitable computer, is needed to decode SSTV using Digital Radio Mondiale (DRM)?

A. A special IF converter.
B. A special front end limiter.
C. A special notch filter to remove synchronization pulses.
D. No other hardware is needed.

The term "Digital Radio Mondiale" ***(DRM) is not a special kind of circuit*** with a big chip in the middle. Rather, DRM is a panel of commercial and non-commercial radio users who have developed the sole standard for sending broadcasts over worldwide low-frequency and high-frequency bands. SSTV is easy to hear at 14.230 MHz, and it's easy to capture a full-color, high resolution picture with just your HF transceiver and a suitably-equipped computer with a sound card running SSTV software. **ANSWER D.** ☞ **http://tvcomm.co.uk/g7izu/**

E2B10 Which of the following is an acceptable bandwidth for Digital Radio Mondiale (DRM) based voice or SSTV digital transmissions made on the HF amateur bands?

A. 3 KHz.
B. 10 KHz.
C. 15 KHz.
D. 20 KHz.

The restrictions for SSTV specify that the typical high-frequency, *slow-scan signal* can occupy no more room than a conventional single sideband voice signal, *3 kHz or less*. **ANSWER A.**

E2B11 What is the function of the Vertical Interval Signaling (VIS) code sent as part of an SSTV transmission?

A. To lock the color burst oscillator in color SSTV images.
B. To identify the SSTV mode being used.
C. To provide vertical synchronization.
D. To identify the call sign of the station transmitting.

There are over a dozen individual methods for sending a color slow scan television signal over HF and VHF airwaves. The most common are Martin and Scottie. These popular modes require the initial reception of *the 300 ms VIS code, which is necessary for your computer to know which SSTV mode is being sent*. **ANSWER B.**

E6F13 What is a liquid crystal display (LCD)?

A. A modern replacement for a quartz crystal oscillator which displays its fundamental frequency.
B. A display utilizing a crystalline liquid and polarizing filters which becomes opaque when voltage is applied.
C. A frequency-determining unit for a transmitter or receiver.
D. A display that uses a glowing liquid to remain brightly lit in dim light.

A *liquid-crystal display*, found on most amateur radio handheld units, *uses a crystalline liquid in conjunction with polarizing filters* to change the way light is refracted from a rear mirror through the liquid. **ANSWER B.**

☞ **www.hamtv.com**

The trans-reflective color LCD is popular on HF and VHF/UHF equipment.

E6F14 Which of the following is true of LCD displays?

A. They are hard to view in high ambient light conditions.
B. They may be hard to view through polarized lenses.
C. They only display alphanumeric symbols.
D. All of these choices are correct.

It seems like every year, some new, exotic, and exciting display technology arises. The LCD has been around a long time now and is still an efficient and convenient way of displaying small amounts of data and some very rudimentary graphics. But since they emit no light of their own, they must be used in a carefully controlled lighting environment. They are polarized themselves and may be ***difficult to view if the operator is using polarized sunglasses*** of the opposite polarization. **ANSWER B.**

In direct sunlight the amber and black LCD display is easy to see on your hand held HT or mobile rig.

E6E01 Which of the following is true of a charge-coupled device (CCD)?

A. Its phase shift changes rapidly with frequency.
B. It is a CMOS analog-to-digital converter.
C. It samples an analog signal and passes it in stages from the input to the output.
D. It is used in a battery charger circuit.

While most commonly associated with video devices such as CCD cameras, CCD technology is not limited to just imaging. Another type of charge-coupled device, sometimes known as a "bucket brigade" device, is often used to create a time delay of an analog signal. The familiar switched capacitor filter is another example of a ***CCD where a sampled analog voltage is stored in a capacitor and then transferred to a subsequent stage*** after a precise length of time. **ANSWER C.**

Gordo is ready to operate video for his Wednesday evening Ham Nation pod-cast. Watch the show at www.live.twit.tv.

Digital Excitement with Your Computers & Radios

E2E01 Which type of modulation is common for data emissions below 30 MHz?

A. DTMF tones modulating an FM signal.
B. FSK.
C. Pulse modulation.
D. Spread spectrum.

Frequency shift keying (FSK) is created by data pulses from your computer that shift the master oscillator back and forth to create two distinct tones in the other operator's receiver. Audio Frequency Shift Keying (AFSK) generates audio tones from your computer's sound card or multi-mode controller and is used for more complex digital signaling like PSK 31, MFSK, and CLOVER. These modes have more than two distinct frequency signals. Both ***Frequency Shift Keying (FSK)*** and Audio Frequency Shift Keying (AFSK) can be found on HF, ***below 30 MHz***. Listen to Gordo's audio CDs to hear the difference between these two modes. **ANSWER B.**

E2E11 What is the difference between direct FSK and audio FSK?

A. Direct FSK applies the data signal to the transmitter VFO.
B. Audio FSK has a superior frequency response.
C. Direct FSK uses a DC-coupled data connection.
D. Audio FSK can be performed anywhere in the transmit chain.

With ***direct frequency shift keying*** (FSK) the ***variable frequency oscillator*** in your transceiver ***changes the actual frequency*** in sync with the 1 and 0 data from your computer. With audio frequency shift keying (AFSK), your master oscillator stays on frequency and audio tones are generated by a sound card or modem to correspond precisely with the data being sent. **ANSWER A.**

E4A09 When using a computer's sound card input to digitize signals, what is the highest frequency signal that can be digitized without aliasing?

A. The same as the sample rate.
B. One-half the sample rate.
C. One-tenth the sample rate.
D. It depends on how the data is stored internally.

Don't you just hate laws? Nyquist's sampling theorem is one of those laws you just can't violate. The sampling rate of your sound card must be greater than or equal to twice the highest frequency signal you're measuring. In terms of the question, ***the frequency can be no more than half the sampling rate of the sound card***. **ANSWER B.**

E8D07 What is a common cause of overmodulation of AFSK signals?

A. Excessive numbers of retries.
B. Ground loops.
C. Bit errors in the modem.
D. Excessive transmit audio levels.

It is important to know that many digital modes referred to as AFSK also include amplitude modulation components. If a mode is truly AFSK, transmitter linearity is not an issue; you can run "plain vanilla" RTTY through a class C amplifier with no problem whatsoever (whether using a sound card mode, or direct FSK). However, if you are using a digital mode that has amplitude modulation as well, you MUST have a linear system from microphone to antenna. ***Overdriving the audio input*** of a receiving SSB transmitter is a sure way to create non-linearity; a signal that is difficult if not impossible to copy. **ANSWER D.**

E8D06 Which of the following indicates likely overmodulation of an AFSK signal such as PSK or MFSK?

A. High reflected power.
B. Strong ALC action.
C. Harmonics on higher bands.
D. Rapid signal fading.

Sound card digital modes seldom require more than about 25% of your transmitter's available power. If you keep the audio drive below the level that causes ***strong ALC action***, you will probably not overmodulate. **ANSWER B.**

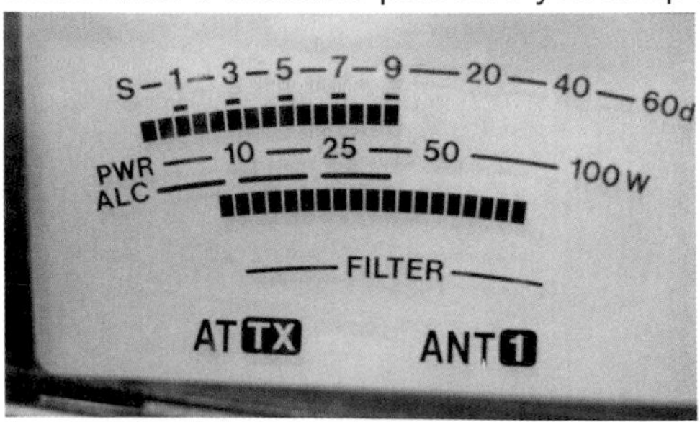

When you see an ALC reading that shoots to the right, turn down the digital drive level or your signal may be distorted.

E8D08 What parameter might indicate that excessively high input levels are causing distortion in an AFSK signal?

A. Signal to noise ratio.
B. Baud rate.
C. Repeat Request Rate (RRR).
D. Intermodulation Distortion (IMD).

Intermodulation distortion is the primary reason why many digital modes do not decode properly. IMD doesn't always show itself on a waterfall display, although a signal with lots of IMD may appear wider than normal. Always be safe and run your audio well below the level that creates any ALC action on your transmitter. Intermodulation is not as much an issue when using sound card digital modes since they work with greatly reduced power. **ANSWER D.**

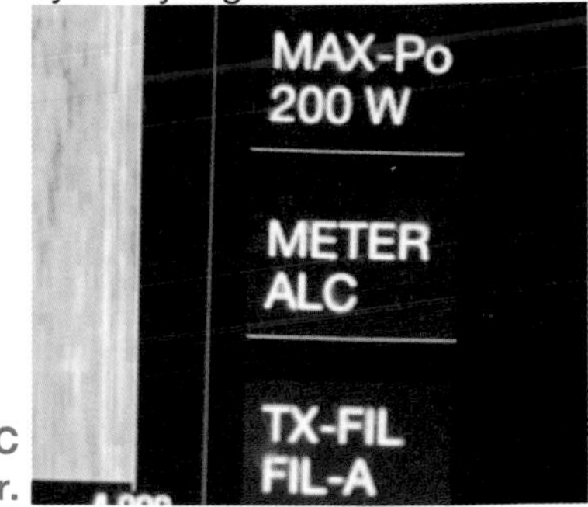

When operating digital, keep your ALC settings always in view on the meter.

E8D09 What is considered a good minimum IMD level for an idling PSK signal?

A. +10 dB.
B. +15 dB.
C. -20 dB.
D. -30 dB.

For IMD, ***the bigger the negative number the better. -30dB*** means that the sum of all intermodulation products is less than 1/1000th the power level of your desired signal. **ANSWER D.**

E8C03 When performing phase shift keying, why is it advantageous to shift phase precisely at the zero crossing of the RF carrier?

A. This results in the least possible transmitted bandwidth for the particular mode.
B. It is easier to demodulate with a conventional, non-synchronous detector.
C. It improves carrier suppression.
D. All of these choices are correct.

Zero crossing not only allows the ***least amount of spectrum*** to be occupied, but it also results in the least amount of confusion or "blurring" between symbols for a given rate. While commonly implemented in most digital modes, experimental modes such as coherent CW take advantage of this zero crossing benefit. **ANSWER A.**

E2E07 What is the typical bandwidth of a properly modulated MFSK16 signal?

A. 31 Hz.
B. 316 Hz.
C. 550 Hz.
D. 2.16 kHz.

MFSK refers to a super-charged RTTY signal, Millennium Frequency Shift Keying. While simple RTTY uses two tones, ***MFSK16 uses*** 16 tones, one at a time at 15.625 baud, spaced at 15.625 Hertz apart. This works out to a relatively narrow ***316 Hertz bandwidth***, just like RTTY, and offers higher non-error rates with improved reception under noisy conditions. It is an ideal mode for 40, 80, and 160 meters. **ANSWER B.**

E2E06 What is the most common data rate used for HF packet?

A. 48 baud.
B. 110 baud.
C. 300 baud.
D. 1200 baud.

High frequency packet moves along at a relatively slow ***300 baud*** rate so these signals do not exceed bandwidth limits on frequencies below 30 MHz. 300 baud is relatively slow compared to higher speeds on VHF and UHF. HF packet is not necessarily a "right now" transmission method, however, it is highly accurate, and even binary files may be sent around the world via HF packet! **ANSWER C.**

E8C11 What is the relationship between symbol rate and baud?

A. They are the same.
B. Baud is twice the symbol rate.
C. Symbol rate is only used for packet-based modes.
D. Baud is only used for RTTY.

Symbol rate and baud rate are the same, and the best indication of the bandwidth necessary for transmission. The number of symbols per character may determine the ultimate "data rate" but not the bandwidth. **ANSWER A.**

E8C02 What is the definition of symbol rate in a digital transmission?

A. The number of control characters in a message packet.
B. The duration of each bit in a message sent over the air.
C. The rate at which the waveform of a transmitted signal changes to convey information.
D. The number of characters carried per second by the station-to-station link.

A character can consist of several symbols. The simplest application of this is Morse code, where we have two basic symbols, a dot and a dash, that are used in combinations to form characters. ***Symbol rate is the rate at which the waveform of a transmitted signal changes to convey information.*** Depending on the complexity of the code, symbol rate and character rate can be two very different things! **ANSWER C.**

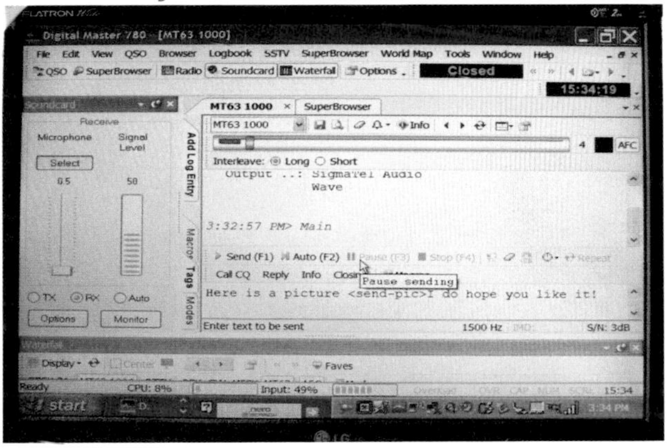

Going digital with your ham radio and computer will let you work stations you can't even hear above local noise!

E2D09 Which of these digital modes has the fastest data throughput under clear communication conditions?

A. AMTOR.
B. 170 Hz shift, 45 baud RTTY.
C. PSK31.
D. 300 baud packet.

Although ***300 baud packet*** sounds relatively slow, on the worldwide bands it's faster than most other modes. **ANSWER D.**

E2E08 Which of the following HF digital modes can be used to transfer binary files?

A. Hellschreiber.
B. PACTOR.
C. RTTY.
D. AMTOR.

PACTOR transmissions sound like two birds chirping. Data is sent in blocks and the receiving station sends ACK for acknowledging everything sent or NAK indicating the request to re-send the block. ***PACTOR may be used to transfer binary files***, allowing you to download programs from a BBS and then work with the program on your computer. **ANSWER B.**

E8D10 What are some of the differences between the Baudot digital code and ASCII?

A. Baudot uses 4 data bits per character, ASCII uses 7 or 8; Baudot uses 1 character as a letters/figures shift code, ASCII has no letters/figures code.
B. Baudot uses 5 data bits per character, ASCII uses 7 or 8; Baudot uses 2 characters as letters/figures shift codes, ASCII has no letters/figures shift code.
C. Baudot uses 6 data bits per character, ASCII uses 7 or 8; Baudot has no letters/figures shift code, ASCII uses 2 letters/figures shift codes.
D. Baudot uses 7 data bits per character, ASCII uses 8; Baudot has no letters/figures shift code, ASCII uses 2 letters/figures shift codes.

Baudot is a 5-bit code; ASCII is a 7-bit code. In actuality, ASCII is an 8-bit code with only 7 bits uniformly defined and no shift code. There are ***2 shift codes in Baudot***: an up-shift and a down-shift. **ANSWER B.**

E2E02 What do the letters FEC mean as they relate to digital operation?

A. Forward Error Correction.
B. First Error Correction.
C. Fatal Error Correction.
D. Final Error Correction.

AMTOR is a digital mode with error correction leading to perfect copy when HF band conditions are favorable. ***"FEC" (forward error correction)*** is done in AMTOR Mode B transmissions by sending each character twice. **ANSWER A.**

E8C01 How is Forward Error Correction implemented?

A. By the receiving station repeating each block of three data characters.
B. By transmitting a special algorithm to the receiving station along with the data characters.
C. By transmitting extra data that may be used to detect and correct transmission errors.
D. By varying the frequency shift of the transmitted signal according to a predefined algorithm.

During AMTOR FEC, ***each data character is sent twice*** with no acknowledgment by the receiving station. The software used by the receiving station is expecting this extra data and uses it for error free reception. **ANSWER C.**

E8C08 How does ARQ accomplish error correction?

A. Special binary codes provide automatic correction.
B. Special polynomial codes provide automatic correction.
C. If errors are detected, redundant data is substituted.
D. If errors are detected, a retransmission is requested.

ARQ (Automatic Repeat reQuest) is a digital protocol that acknowledges individual data blocks and ***requests a repeat of any missed block***. ARQ is part of the PACTOR and CLOVER digital signaling systems. **ANSWER D.**

E8C09 Which is the name of a digital code where each preceding or following character changes by only one bit?

A. Binary Coded Decimal Code.
B. Extended Binary Coded Decimal Interchange Code.
C. Excess 3 code.
D. Gray code.

In so-called coding theory, the ***Gray code*** has the shortest "distance" between any two symbols. This coding scheme reduces bandwidth and provides other benefits such as simpler error correction. **ANSWER D.**

E8C10 What is an advantage of Gray code in digital communications where symbols are transmitted as multiple bits?

A. It increases security.
B. It has more possible states than simple binary.
C. It has more resolution than simple binary.
D. It facilitates error detection.

Since ***Gray code*** only changes the last bit, it is ***easy*** to use that bit as a "flag" for ***error detection***. If more than one character changes at a time, it is a sure indication that an error has been transmitted, and steps can be taken to correct this, such as an Automatic Repeat Request (ARQ). **ANSWER D.**

E8B07 Orthogonal Frequency Division Multiplexing is a technique used for which type of amateur communication?

A. High speed digital modes.
B. Extremely low-power contacts.
C. EME.
D. OFDM signals are not allowed on amateur bands.

ODFM uses the clever principle of frequency interleaving to allow ***much higher data rates*** than "should" be possible on a given bandwidth channel. **ANSWER A.**

E8B08 What describes Orthogonal Frequency Division Multiplexing?

A. A frequency modulation technique which uses non-harmonically related frequencies.
B. A bandwidth compression technique using Fourier transforms.
C. A digital mode for narrow band, slow speed transmissions.
D. A digital modulation technique using subcarriers at frequencies chosen to avoid intersymbol interference.

ODFM requires precise timing and phasing of the component signals. By carefully choosing the spectrum generated by each of the modulated subcarriers, the ***signals can be interleaved without any mutual interference.*** **ANSWER D.**

E2E04 What is indicated when one of the ellipses in an FSK crossed-ellipse display suddenly disappears?

A. Selective fading has occurred.
B. One of the signal filters is saturated.
C. The receiver has drifted 5 kHz from the desired receive frequency.
D. The mark and space signal have been inverted.

Most high-frequency stations feature a companion station monitor oscilloscope that will display two crossed ellipses, indicating a properly tuned-in digital signal. ***If one of the ellipses disappears***, the phenomenon known as ***selective fading has occurred***. This will cause your received copy to stall until the signal comes out of the fade. Don't touch the receiver during selective fading – you are on frequency; it's just that propagation is causing the signal to fade in and out. **ANSWER A.**

E8D11 What is one advantage of using ASCII code for data communications?

A. It includes built in error correction features.
B. It contains fewer information bits per character than any other code.
C. It is possible to transmit both upper and lower case text.
D. It uses one character as a shift code to send numeric and special characters.

ASCII code allows upper and lower case text to be transmitted; Baudot does not. **ANSWER C.**

E8D12 What is the advantage of including a parity bit with an ASCII character stream?

A. Faster transmission rate.
B. The signal can overpower interfering signals.
C. Foreign language characters can be sent.
D. Some types of errors can be detected.

The parity bit is used for error detection in a data transmission. A parity bit 1 or 0 might indicate the odd or even number of 1s or 0s in the data block. It is not foolproof – if a pair of 1s or a pair of 0s is missed, the error detection circuit might consider no errors. Yet this simple error detection method is commonly used and works surprisingly well to confirm that the other station received the character intact or needs a repeat. **ANSWER D.**

E8C06 What is the necessary bandwidth of a 170-hertz shift, 300-baud ASCII transmission?

A. 0.1 Hz.
B. 0.3 kHz.
C. 0.5 kHz.
D. 1.0 kHz.

Digital bandwidth is:

BW = baud rate + (1.2 × frequency shift)

BW = 300 + (1.2 × 170) = 504 Hz = ***0.504 kHz***.

Remember, to change Hz to kHz, simply move the decimal three places to the left. **ANSWER C.**

E8C07 What is the necessary bandwidth of a 4800-Hz frequency shift, 9600-baud ASCII FM transmission?

A. 15.36 kHz.
B. 9.6 kHz.
C. 4.8 kHz.
D. 5.76 kHz.

Digital bandwidth again – use the formula at question E8C06. In this problem, the baud rate is 9600 and the frequency shift is 4800. ***Bandwidth = 9600 + (1.2 × 4800) = 15360***. The answer is 15,360 Hz, and when the decimal point is moved three places to the left, the bandwidth works out to be ***15.36 kHz***. **ANSWER A.**

E2E09 Which of the following HF digital modes uses variable-length coding for bandwidth efficiency?

A. RTTY.
B. PACTOR.
C. MT63.
D. PSK31.

PSK31 stands for Phase Shift Keying with a bit rate slightly over 31. It is a little like Morse code, with the more common letters represented by a short count of bits: E (11), A (1011), T (101), or O (111), while less used characters such as Q and X, might contain up to 10 bits. Using PSK31, this ***Varicode is transmitted*** with audio phase shifts which occupy a mere 31 Hz of bandwidth, leading to bandwidth efficiency. **ANSWER D.**

E8C04 What technique is used to minimize the bandwidth requirements of a PSK31 signal?

A. Zero-sum character encoding.
B. Reed-Solomon character encoding.
C. Use of sinusoidal data pulses.
D. Use of trapezoidal data pulses.

Any digital mode, such as PSK31, that causes a sudden change of frequency, phase, or amplitude during data transmission will result in the generation of sidebands which increase the occupied bandwidth of the signal, sometimes dramatically. If the transitions between "mark" and "space" are made smoothly ***using sinusoidal data pulses*** the ***bandwidth can be greatly reduced***. **ANSWER C.**

E2E10 Which of these digital modes has the narrowest bandwidth?

A. MFSK16.
B. 170 Hz shift, 45 baud RTTY.
C. PSK31.
D. 300-baud packet.

Of those listed, ***PSK31 has the narrowest bandwidth***. Phase shift keying using sinusoidal data pulses and efficient Varicode characters generates a PSK31 signal with a very small bandwidth. **ANSWER C.**

E2E13 Which of the following is a possible reason that attempts to initiate contact with a digital station on a clear frequency are unsuccessful?

A. Your transmit frequency is incorrect.
B. The protocol version you are using is not supported by the digital station.
C. Another station you are unable to hear is using the frequency.
D. All of these choices are correct.

There are a lot of things that can go wrong in digital communications. Since there are so many new digital modes, it's difficult to provide a "silver bullet" that will fix any problem you could possibly have with digital communications. But, it's always best to clean your own house first! Be sure you are not overmodulating your transmitter – to do so can wreak havoc with just about any digital mode. Then check for other obvious problems, such as trying to use a different mode than the other station – again a common problem given the large number of new modes. Be sure your transmitter is on the right sideband, nearly always USB. And also realize that the frequency may sound clear to you but the guy at the other end may be buried under a pile of QRM! ***All of these choices are correct***. **ANSWER D.**

E2A14 What technology is used to track, in real time, balloons carrying amateur radio transmitters?

A. Radar.
B. Bandwidth compressed LORAN.
C. APRS.
D. Doppler shift of beacon signals.

The Automatic Packet Reporting System (APRS) was developed in 1992 as an innovative way of tracking mobile amateur radio stations. It turns out to be a very practical way of tracking objects in the sky too! ***APRS*** has added a new dimension to ham radio science, allowing the ***precise tracking of high altitude balloons*** carrying amateur radio and scientific equipment. **ANSWER C.**

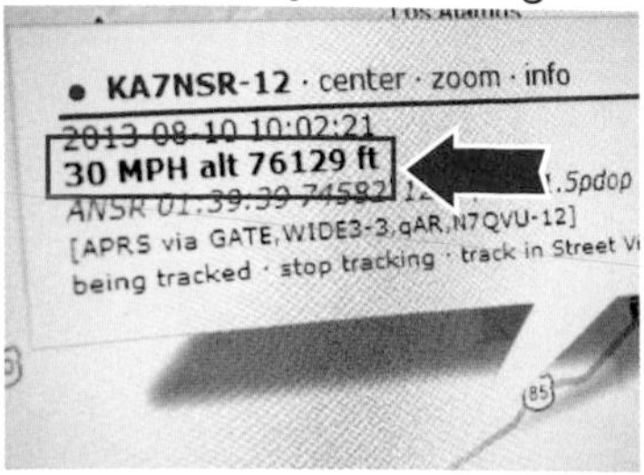

Look carefully at the balloon's altitude, 76,129 feet and drifting at 30 MPH, all reported via APRS!

E2D10 How can an APRS station be used to help support a public service communications activity?

A. An APRS station with an emergency medical technician can automatically transmit medical data to the nearest hospital.
B. APRS stations with General Personnel Scanners can automatically relay the participant numbers and time as they pass the check points.
C. An APRS station with a GPS unit can automatically transmit information to show a mobile station's position during the event.
D. All of these choices are correct.

Many marathons are supported by ham operators on foot, bicycles, and motorcycles with an ***APRS station tied into a GPS unit to automatically transmit position information*** along the course route. All this can be done without the operator needing to do a thing – simply set the time interval you want your APRS packets transmitted and, presto!, everyone with the right equipment can easily see the APRS position information. **ANSWER C.** ☞ **www.findu.com**

Emergency communicators rely on APRS for both "walking" hams, as well as those in automobiles and at base stations.

E2D07 What digital protocol is used by APRS?

A. PACTOR.
B. 802.11.
C. AX.25.
D. AMTOR.

The digital packet protocol in use for ***APRS is AX.25***, the ham version of commercial packet networks operating X.25. The AX.25 protocol is recognized and authorized by the FCC with specific speed limitations depending on the frequency being used. **ANSWER C.**

E2D08 What type of packet frame is used to transmit APRS beacon data?

A. Unnumbered Information.
B. Disconnect.
C. Acknowledgement.
D. Connect.

The ***packet frames*** used to transmit APRS beacon data ***are called UIF – Unnumbered Information Frames***. Networking systems can allow an APRS position and short message to be transmitted and relayed throughout the country. Several manufacturers produce handheld and mobile radios with the APRS terminal node controller already built-in. Handheld GPS equipment that outputs NMEA data is available for under $99. A mapping GPS can show ham call signs and locations. **ANSWER A.**

E2D11 Which of the following data are used by the APRS network to communicate your location?

A. Polar coordinates.
B. Time and frequency.
C. Radio direction finding spectrum analysis.
D. Latitude and longitude.

It is ***Latitude and longitude*** in degrees, minutes, and fractions of a minute, that ***are transmitted by your APRS*** system, usually on 144.390 MHz. The data transmitted is usually in the NMEA 0183 format. I-Gate stations tied into the internet may take your GPS position and have it displayed at ☞ **http://www.aprs.fi**. **ANSWER D.**
☞ **www.aprs.fi**

E1F01 On what frequencies are spread spectrum transmissions permitted?

A. Only on amateur frequencies above 50 MHz.
B. Only on amateur frequencies above 222 MHz.
C. Only on amateur frequencies above 420 MHz.
D. Only on amateur frequencies above 144 MHz.

Spread spectrum is still new to ham radio operators and ***is confined to the 222 MHz band and above***. For the time being, spread spectrum is not allowed on 2 or 6 meters. And while spread spectrum is allowed on frequencies above 420 MHz, that answer leaves out the 220 MHz band. [97.305] **ANSWER B.**

E8D03 How does the spread spectrum technique of frequency hopping work?

A. If interference is detected by the receiver it will signal the transmitter to change frequencies.
B. If interference is detected by the receiver it will signal the transmitter to wait until the frequency is clear.
C. A pseudo-random binary bit stream is used to shift the phase of an RF carrier very rapidly in a particular sequence.
D. The frequency of the transmitted signal is changed very rapidly according to a particular sequence also used by the receiving station.

Frequency hopping depends on a particular signaling sequence, and the frequency of the RF carrier is changed so quickly that it is almost indistinguishable except to other spread spectrum receivers locked on to the same signaling sequence. **ANSWER D.**

Spread spectrum phones are common!

E8D02 What spread spectrum communications technique uses a high speed binary bit stream to shift the phase of an RF carrier?

A. Frequency hopping.
B. Direct sequence.
C. Binary phase-shift keying.
D. Phase compandored spread spectrum.

The term "***direct sequence***" is used to describe a system where a ***fast binary stream shifts*** the phase of an RF carrier. This is another spread spectrum communications technique. Using only a single carrier frequency, and then modulating the signal by shifting the phase in accordance with a high-speed binary bit stream, results in a signal whose pattern on a spectrum analyzer spreads out into a pattern of hills and valleys. This is called direct sequence spread spectrum. **ANSWER B.**

E8D01 Why are received spread spectrum signals resistant to interference?

A. Signals not using the spread spectrum algorithm are suppressed in the receiver.
B. The high power used by a spread spectrum transmitter keeps its signal from being easily overpowered.
C. The receiver is always equipped with a digital blanker.
D. If interference is detected by the receiver it will signal the transmitter to change frequencies.

Amateur operators transmitting and receiving spread spectrum signals are just pioneering one more way to conserve our valuable frequency resources. Although spread spectrum may utilize an entire band, the hopping technique won't disturb other signals on the band, and all other ***signals not using the same spread spectrum algorithm are suppressed in the receiver***. So just when you thought there was no more room for any more signals on a particular band, spread spectrum gets through! **ANSWER A.**

E1F10 What is the maximum permitted transmitter peak envelope power for an amateur station transmitting spread spectrum communications?

A. 1 W.
B. 1.5 W.
C. 10 W.
D. 1.5 kW.

Recent FCC rule changes now limit 222 MHz and up ***spread spectrum power output to no more than 10 watts***. [97.313] **ANSWER C.**

E1F09 Which of the following conditions apply when transmitting spread spectrum emission?

A. A station transmitting SS emission must not cause harmful interference to other stations employing other authorized emissions.
B. The transmitting station must be in an area regulated by the FCC or in a country that permits SS emissions.
C. The transmission must not be used to obscure the meaning of any communication.
D. All of these choices are correct.

Spread spectrum communications are ***allowed to any other station under FCC regulation and*** to ***stations in foreign countries*** that have specifically permitted spread spectrum communications to their country from ours. Spread spectrum ***may not be used to obscure the meaning of any communication***. If these three requirements are met, and as long as you stay at no more than 10 watts, you are entitled to use spread spectrum on 222 MHz and up. [97.311] **ANSWER D.**

E2C04 What type of transmission is most often used for a ham radio mesh network?

A. Spread spectrum in the 2.4 GHz band.
B. Multiple Frequency Shift Keying in the 10 GHz band.
C. Store and forward on the 440 MHz band.
D. Frequency division multiplex in the 24 GHz band.

Wireless commercial networks generally require more bandwidth than is normally available on the ham bands – unless you move up into the microwave frequencies. ***2.4 GHz*** has plenty of "elbow room" available for experimentation with wideband signals such as ***spread spectrum***. **ANSWER A.**

Take time to familiarize yourself with the Amateur Radio microwave band allocations listed below. All modes and licensees (except Novices) are authorized on the following bands [FCC Rules, Part 97.301(a)]:

2300-2310 MHz	*47.0-47.2 GHz*
2390-2450 MHz	*76.0-81.0 GHz**
3300-3500 MHz	*122.25 -123.00 GHz*
5650-5925 MHz	*134-141 GHz*
10.0-10.5 GHz	*241-250 GHz*
24.0-24.25 GHz	*All above 300 GHz*

**Amateur operation at 76-77 GHz has been suspended till the FCC can determine that interference will not be caused to vehicle radar systems.*

Commercial Wi-Fi (mesh) equipment is available in the shared 2390-2450 MHz band and less commonly in the 5650-5925 MHz band. Every ham should be aware that most of our microwave bands are unused and are wide open for experimentation and development!

E2C09 What type of equipment is commonly used to implement a ham radio mesh network?

A. A 2 meter VHF transceiver with a 1200 baud modem.
B. An optical cable connection between the USB ports of 2 separate computers.
C. A standard wireless router running custom software.
D. A 440 MHz transceiver with a 9600 baud modem.

The commercial frequencies in the 2.4 GHz band actually overlap amateur radio frequencies in that band. However, hams can use more power than the unlicensed wireless equipment, so you can do a few things the commercial folks can't when it comes to wireless networking. ***Off the shelf wireless routers*** can easily be configured to accommodate a wide area ham radio mesh network. **ANSWER C.**

E2E05 Which type of digital mode does not support keyboard-to-keyboard operation?

A. Winlink.
B. RTTY.
C. PSK31.
D. MFSK.

It's fun to exchange messages with other hams using your keyboard in real time. You can hear the sounds of RTTY, PSK 31 and MFSK on Gordo's lively audio CD course. This is keyboard-to-keyboard at its best! Winlink is a great way to send digital traffic into the e-mail system using PACTOR protocols, and can also be used to send graphics and weather picture charts. But ***Winlink is not designed for keyboard-to-keyboard communications***. **ANSWER A.**

E8C05 What is the necessary bandwidth of a 13-WPM International Morse code transmission?

A. Approximately 13 Hz.
B. Approximately 26 Hz.
C. Approximately 52 Hz.
D. Approximately 104 Hz.

To determine the bandwidth of a Morse code signal, use this formula:

$$BW_{CW} = \text{baud rate} \times \text{wpm} \times \text{fading factor}$$

For CW, the baud rate is 0.8 and the fading factor is 5, therefore, ***multiply 0.8 times 13 wpm times 5. The answer works out to be 52 Hz***. **ANSWER C.**

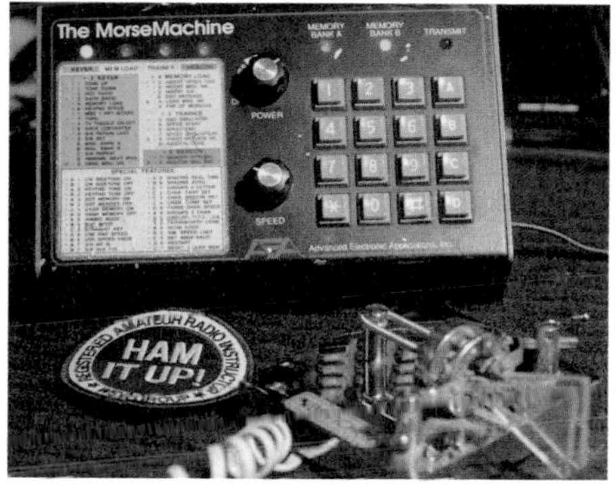

Morse code is more popular than ever these days on ham radio!

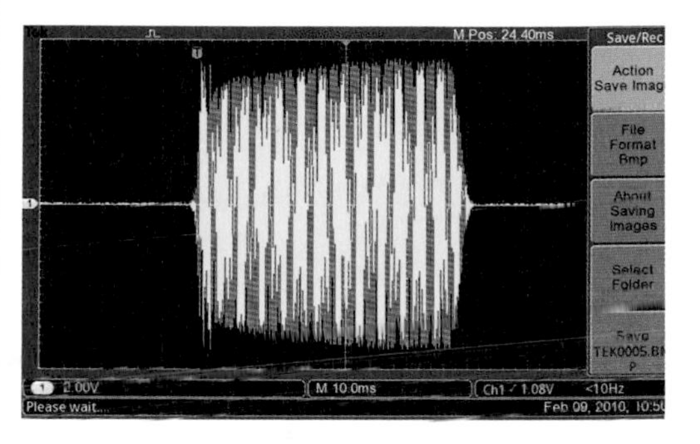

Waveform of a single CW dit.

E8D04 What is the primary effect of extremely short rise or fall time on a CW signal?

A. More difficult to copy.
B. The generation of RF harmonics.
C. The generation of key clicks.
D. Limits data speed.

The process of keying a carrier on or off very quickly is identical to AM modulating it at a very high frequency; this generates sidebands resulting in ***key clicks***, which can be far removed from the carrier frequency. These sidebands can splatter across a large swath of an amateur band. Care should be applied to your keying wave shape for good CW operations. **ANSWER C.**

Morse code for low power (QRP) is just the ticket on a small, battery operated rig with built in paddles.

E8D05 What is the most common method of reducing key clicks?

A. Increase keying waveform rise and fall times.
B. Low-pass filters at the transmitter output.
C. Reduce keying waveform rise and fall times.
D. High-pass filters at the transmitter output.

The CW envelope shape will greatly affect the emitted spectrum of a CW signal, as well as its sound. If the rise and fall times are too abrupt, it will emit key clicks, which are sidebands that can splatter across a large swath of an amateur band. If the keying is too soft, the tone can sound a bit mushy on the receiving end and high speed characters may overlap with succeeding characters. Linear amplifiers can alter the keying envelope too, so it's a great idea to be able to ***adjust both the rise and fall times of you CW "note."*** You might want to check out the December, 1984 QST article, "Try this Versatile CW Shaper" by Eric for ideas on how to do this on a variety of radios. **ANSWER A.**

Good To Know:

A New Mission for CW

For a few years now, knowing the International Radiotelegraph Code (usually synonymous with CW) has not been required for any class of amateur radio license. However, for the weak signal operator and the experimenter, CW still has a prominent role to play. Not only is a CW transmitter the simplest transmitter to build for any frequency from the lowest frequency medium wave bands clear up through our microwave bands, but CW has the best absolute weak signal performance of any mode. This includes exotic digital modes. Most of the activity on the experimental MF and lower frequency bands is by means of very slow CW, in ham vernacular QRSs... or QRSss...or QRSsss...depending on just how slow the CW is transmitted...sometimes as slow as a few characters per hour! By means of the lock-in amplifier, *a special type of Direct Conversion receiver, signals as weak as 60 dB below the noise floor can be detected and decoded! It just takes a little time.*

Amateur Radio still has exciting frontiers to explore, not only at the very highest frequencies of radio, but also at the very lowest frequencies of radio. We are currently poised to obtain a new amateur radio allocation at 2200 meters! One can rest assured that any successful communications achieved here will be by means of CW!

For the somewhat less-patient experimenter, high speed CW (QRR, QRRr, QRRrr, and QRRrrr) is still the preferred mode for fleeting phenomena like meteor burst, as well.

Until fairly recently, better-than-average CW proficiency was the hallmark of the Extra Class radio amateur. While this is no longer a compelling reason for most hams to become Extras, many many Extra Class radio amateurs still take great pride and joy in effortless CW operation...at any speed.

At one time or another, Eric has worked just about every mode of amateur radio communications there is, from CW, to SSB, to RTTY, to AMTOR, to PACKET, to SSTV, to exotic sound card digital modes that didn't even have a name at the time...not counting several military modes that you will never hear on the ham bands. But after 44 years on the air, CW remains his default mode.

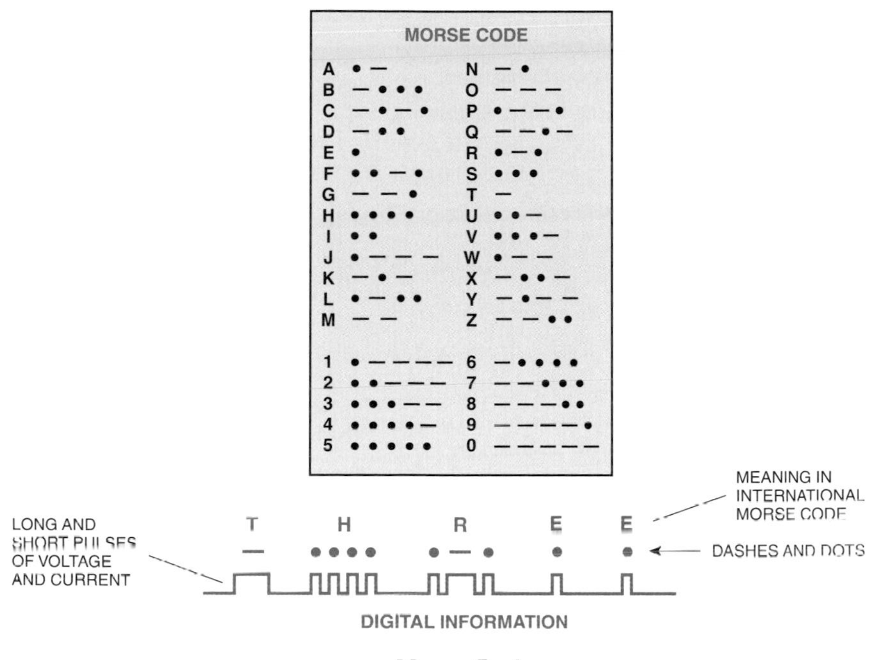

MORSE CODE

A	•—	N	—•
B	—•••	O	———
C	—•—•	P	•——•
D	—••	Q	——•—
E	•	R	•—•
F	••—•	S	•••
G	——•	T	—
H	••••	U	••—
I	••	V	•••—
J	•———	W	•——
K	—•—	X	—••—
L	•—••	Y	—•——
M	——	Z	——••
1	•————	6	—••••
2	••———	7	——•••
3	•••——	8	———••
4	••••—	9	————•
5	•••••	0	—————

Morse Code

SITES RELATED TO DIGITAL EXCITEMENT FOR YOUR COMPUTERS & RADIOS:

www.dstarinfo.com/
www.dstarinfo.com/repeater-lists.aspx
www.dstarusers.org/repeaters.php
www.cq-amateur-radio.com
www.dvdongle.com
www.rsgbiota.org/
www.dailydx.com
http://cluster.sdr-radio.com

FM versus AM

Greater Number of Sidebands

In FM modulation, one sine wave is used to modify another sine wave and multiple sums and differences result, depending on the deviation ratio. This is illustrated in Figure a. *Sidebands appear on either side of* f_c *at* f_m, $2f_m$, $3f_m$, *etc., depending upon the deviation ratio.* Figure b *shows that there are five significant sidebands for a deviation ratio of 3, and* Figure c *shows three significant sidebands for a deviation ratio of 1.67. The multiple sidebands are formed in both the upper and lower sideband positions, separated by the modulating frequency.*

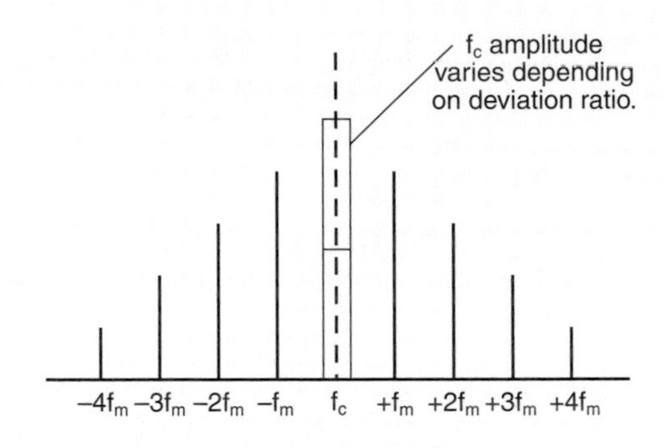

a. General Spectrum

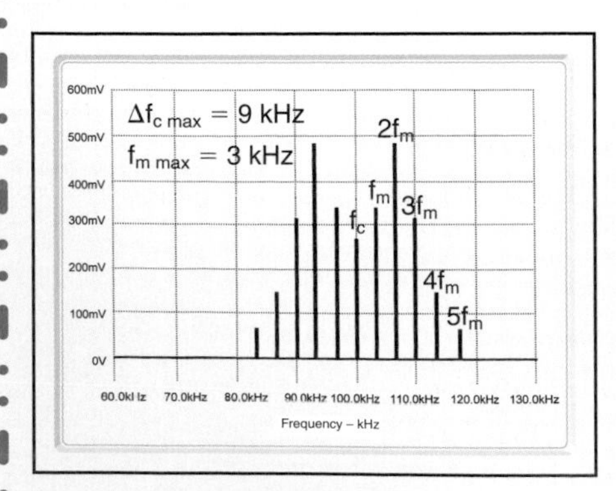

b. Deviation Ratio of 3

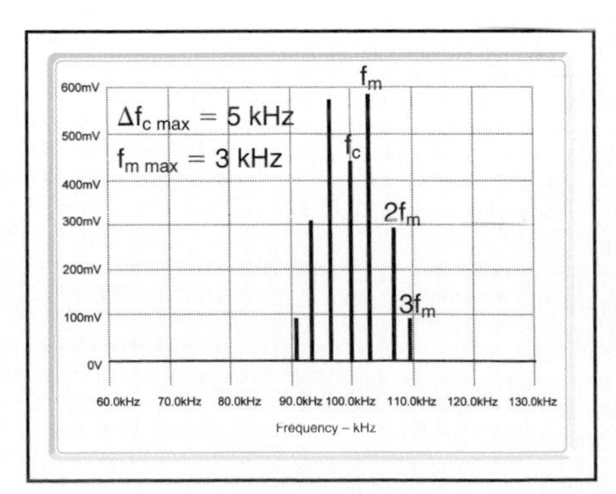

c. Deviation Ratio of 1.67

Frequency spectrum of FM signals.

Source: *Basic Communications Electronics*, Hudson & Luecke, © 1999 Master Publishing, Inc., Niles, IL

CD 2 TRACK 5

Modulate Your Transmitters

E8B01 What is the term for the ratio between the frequency deviation of an RF carrier wave and the modulating frequency of its corresponding FM-phone signal?

A. FM compressibility.
B. Quieting index.
C. Percentage of modulation.
D. Modulation index.

Be careful with this one. Since they don't say the highest modulating frequency, but say only "modulating frequency," they are looking for the ***modulation index***. Notice that there is no answer that says "deviation ratio" to get you in trouble. Modulation index is a continually changing instantaneous value, while deviation ratio is a legally mandated maximum for a given radio service. **ANSWER D.**

E8B02 How does the modulation index of a phase-modulated emission vary with RF carrier frequency (the modulated frequency)?

A. It increases as the RF carrier frequency increases.
B. It decreases as the RF carrier frequency increases.
C. It varies with the square root of the RF carrier frequency.
D. It does not depend on the RF carrier frequency.

This is one of those answers that has a "not" in it that makes it correct. When you calculate the ***modulation index***, you ***do NOT look at the actual RF carrier frequency*** as part of your calculations. **ANSWER D.**

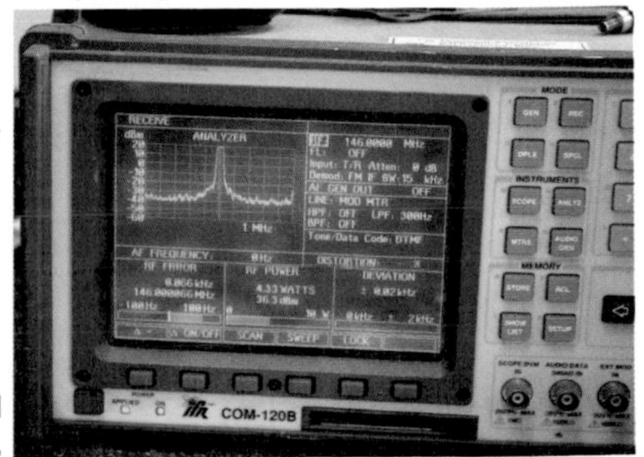

Close look at a clean FM signal on the scope.

E8B03 What is the modulation index of an FM-phone signal having a maximum frequency deviation of 3000 Hz either side of the carrier frequency when the modulating frequency is 1000 Hz?

A. 3. C. 3000.
B. 0.3. D. 1000.

This one is easy since both frequencies are given in Hz. Be careful of your units as you do this type of calculation. ***Simply divide 3000 by 1000 to get a modulation index of 3.*** **ANSWER A.**

$$\text{Modulation Index } (\chi) = \frac{\text{Peak Deviation (D)}}{\text{Modulation Frequency } (m)} \text{ or } \chi = \frac{D}{m}$$

E8B04 What is the modulation index of an FM-phone signal having a maximum carrier deviation of plus or minus 6 kHz when modulated with a 2 kHz modulating frequency?

A. 6000. C. 2000.
B. 3. D. 1/3.

Again, easy because both frequencies are in kHz. Simply divide 6 by 2 to get a ***modulation index of 3.*** **ANSWER B.**

E1B07 What is the highest modulation index permitted at the highest modulation frequency for angle modulation below 29.0 MHz?

A. 0.5. C. 2.0.
B. 1.0. D. 3.0.

With angle modulation, a sine wave carrier is varied by various angles to replicate the incoming modulation source. The ***highest modulation index permitted for angle modulation is exactly 1.0.*** [97.307] **ANSWER B.**

E8B09 What is meant by deviation ratio?

A. The ratio of the audio modulating frequency to the center carrier frequency.
B. The ratio of the maximum carrier frequency deviation to the highest audio modulating frequency.
C. The ratio of the carrier center frequency to the audio modulating frequency.
D. The ratio of the highest audio modulating frequency to the average audio modulating frequency.

It's important to set the deviation ratio properly on your FM transceiver to prevent splatter and a signal bandwidth that is too wide. ***Deviation ratio*** is the ratio of the ***maximum carrier swing to the highest audio modulating frequency*** when you speak into the microphone. **ANSWER B.**

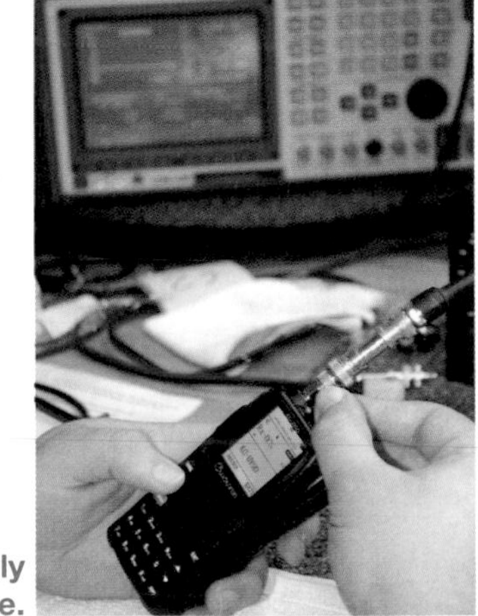

Check your deviation ratio by speaking loudly into the microphone while watching the scope.

E8B05 What is the deviation ratio of an FM-phone signal having a maximum frequency swing of plus-or-minus 5 kHz when the maximum modulation frequency is 3 kHz?

A. 60.
B. 0.167.
C. 0.6.
D. 1.67.

Deviation ratio is the ratio of the maximum carrier swing to the highest audio modulating frequency when you speak into the microphone. To calculate the deviation ratio, divide the maximum carrier frequency swing by the maximum modulation frequency. 5 kHz frequency swing divided by 3 kHz modulation frequency yields a ***deviation ratio of 1.67***. **ANSWER D.**

$$\text{Deviation Ratio} = \frac{\text{Maximum Carrier Frequency Deviation in kHz}}{\text{Maximum Modulation Frequency in kHz}}$$

E8B06 What is the deviation ratio of an FM-phone signal having a maximum frequency swing of plus or minus 7.5 kHz when the maximum modulation frequency is 3.5 kHz?

A. 2.14.
B. 0.214.
C. 0.47.
D. 47.

To calculate deviation ratio of an FM signal, simply divide the maximum carrier frequency swing by the maximum modulation frequency. 7.5 kHz carrier frequency divided by 3.5 kHz modulation frequency yields a ***deviation ratio of 2.14***. Simple, huh? **ANSWER A.**

Check how many kHz of deviation you have on FM with this handy deviation meter.

E8B10 What describes frequency division multiplexing?

A. The transmitted signal jumps from band to band at a predetermined rate.
B. Two or more information streams are merged into a baseband, which then modulates the transmitter.
C. The transmitted signal is divided into packets of information.
D. Two or more information streams are merged into a digital combiner, which then pulse position modulates the transmitter.

The ***"information streams" are the baseband***. Each information stream is modulated onto a sub-carrier (up-converting it), which is then used to modulate the carrier. In the receiver, the sub-carriers are removed by pass band filters, then each information stream is recovered by down-converting the sub-carrier to baseband. **ANSWER B.**

E8B11 What is digital time division multiplexing?

A. Two or more data streams are assigned to discrete sub-carriers on an FM transmitter.

B. Two or more signals are arranged to share discrete time slots of a data transmission.

C. Two or more data streams share the same channel by transmitting time of transmission as the sub-carrier.

D. Two or more signals are quadrature modulated to increase bandwidth efficiency.

Time division multiplexing (TDM) allows multiple users to share the same transmit and receive channel, divided into ***discreet time slots***, for sending data. This modern form of signaling is used most by commercial and telecommunications point-to-point broadcasting, but hams with the right equipment can take advantage of this frequency conserving technology. The two signals don't actually share discrete time slots – rather, they are separated by each using a different time slot. **ANSWER B.**

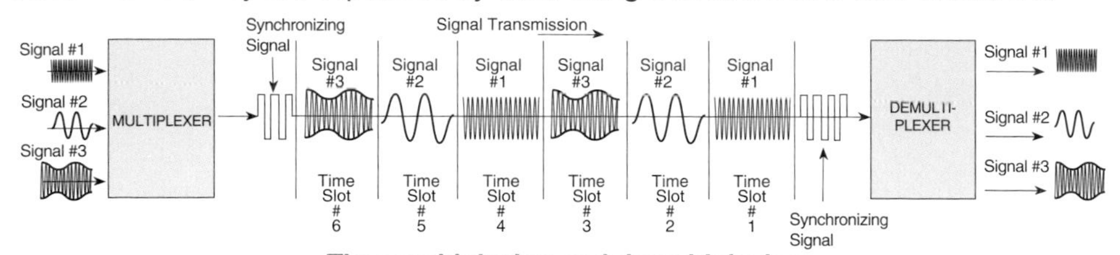

Time multiplexing and demultiplexing.

Source: , Hudson & Luecke, © 1999 Master Publishing, Inc., Niles, IL

E8A02 What type of wave has a rise time significantly faster than its fall time (or vice versa)?

A. A cosine wave.

B. A square wave.

C. A sawtooth wave.

D. A sine wave.

The ***sawtooth wave*** has an immediate rise time significantly faster than the fall time resulting in a wave form that looks like a saw blade. **ANSWER C.**

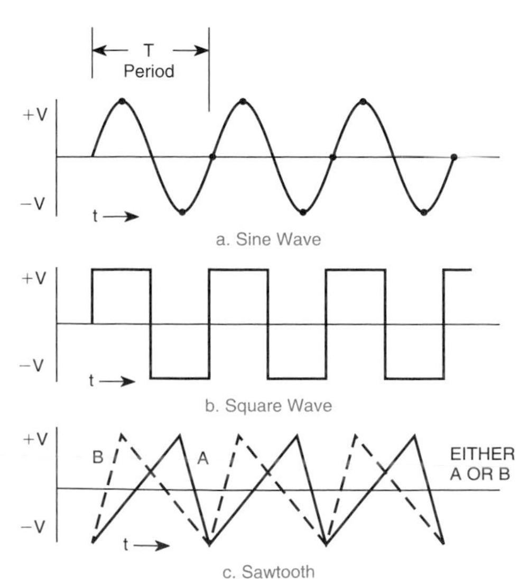

Sine, Square and Sawtooth Waveforms

E8A01 What is the name of the process that shows that a square wave is made up of a sine wave plus all of its odd harmonics?

A. Fourier analysis.
B. Vector analysis.
C. Numerical analysis.
D. Differential analysis.

Fourier analysis was one of the greatest mathematical triumphs of the past millennium. It was so profound that it is the basis of all we do in digital signal processing. Joseph Fourier would be so proud. But alas, poor Joseph didn't have very good computers to work with in the 17th century, so he had to use a quill pen to do his calculations, which is why he had to settle for Really Slow Fourier Transforms (RSFT) instead of Fast Fourier Transforms (FFT). But somehow he still got the job done. **ANSWER A.**

E8A03 What type of wave does a Fourier analysis show to be made up of sine waves of a given fundamental frequency plus all its harmonics?

A. A sawtooth wave.
B. A square wave.
C. A sine wave.
D. A cosine wave.

If it has ***all the harmonics*** it is a ***sawtooth*** wave. A triangle wave, which is simply a symmetrical sawtooth wave, fills the bill here, too. **ANSWER A.**

E8A05 What would be the most accurate way of measuring the RMS voltage of a complex waveform?

A. By using a grid dip meter.
B. By measuring the voltage with a D'Arsonval meter.
C. By using an absorption wave meter.
D. By measuring the heating effect in a known resistor.

In a very complex wave form, one way to determine its effective value is to ***measure the heating effect in a known resistor***, compared to that of a known DC voltage. **ANSWER D.**

E8A07 What determines the PEP-to-average power ratio of a single-sideband phone signal?

A. The frequency of the modulating signal.
B. The characteristics of the modulating signal.
C. The degree of carrier suppression.
D. The amplifier gain.

It's not all that easy to determine the ratio of peak envelope power to average power in your worldwide SSB set because your speech characteristics (the ***characteristics of the modulating signal***) may vary from those of another operator when using the same equipment. Speech compression is specifically designed to reduce the peak-to-average power of an SSB signal. **ANSWER B.**

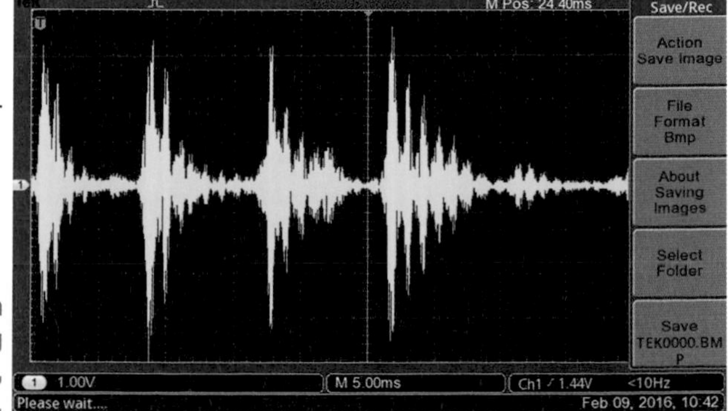

Average modulation can be increased by turning on speech processing, as seen on the right.

E8A06 What is the approximate ratio of PEP-to-average power in a typical single-sideband phone signal?

A. 2.5 to 1.
B. 25 to 1.
C. 1 to 1.
D. 100 to 1.

When you talk about power output from your SSB worldwide set, we normally refer to peak envelope power. Depending on your modulation, ***the ratio of PEP to average power is about 2.5 to 1***, assuming no artificial speech compression is employed. You can remember 2.5 because that's very close to 2.5 kHz of bandwidth occupied by a properly-modulated SSB signal. **ANSWER A.**

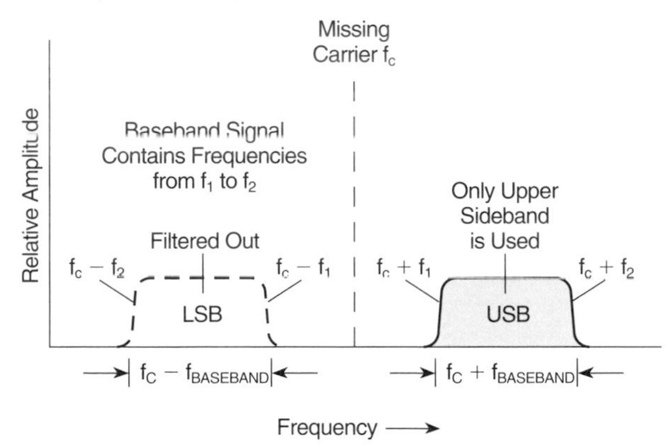

Spectral plot of an SSB signal.

Source: , Hudson & Luecke, © 1999 Master Publishing, Inc., Niles, IL

E8A12 What is an advantage of using digital signals instead of analog signals to convey the same information?

A. Less complex circuitry is required for digital signal generation and detection.
B. Digital signals always occupy a narrower bandwidth.
C. Digital signals can be regenerated multiple times without error.
D. All of these choices are correct.

Digital signals can be ***relayed multiple times without error***. Digital signals in a noisy environment may be forward-error-corrected, giving your equipment and computer a second chance at decoding. Many times you can't even hear the data signal, yet your equipment faithfully decodes that signal. Noise isn't absent in digital signals, but the fact that the signal is either in one state or another (sometimes on or off, or other times at one discrete frequency or another) means that it is easier for the computer to detect its exact state. In comparison, analog signals vary over a wide range of amplitudes and frequencies, and thus are more prone to becoming lost in the noise and interference. Because a digital signal is recovered as only the discrete states, the original signal can be faithfully reproduced without distortion. **ANSWER C.**

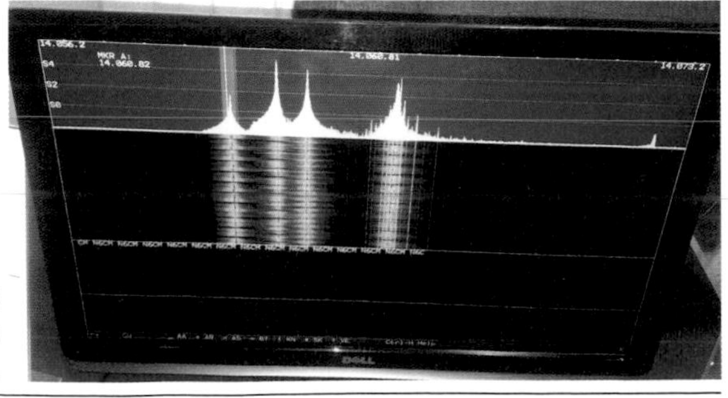

Digital signals take up less bandwidth allowing more signals in a tighter area without interference.

E8A11 What type of information can be conveyed using digital waveforms?

A. Human speech.
B. Video signals.
C. Data.
D. All of these choices are correct.

Even analog voice may be converted into data digital signals, just like what happens inside your cell phone. Over-the-air television video signals are now digital and, of course, data is digital. So *all of the answer choices are correct*. **ANSWER D.**

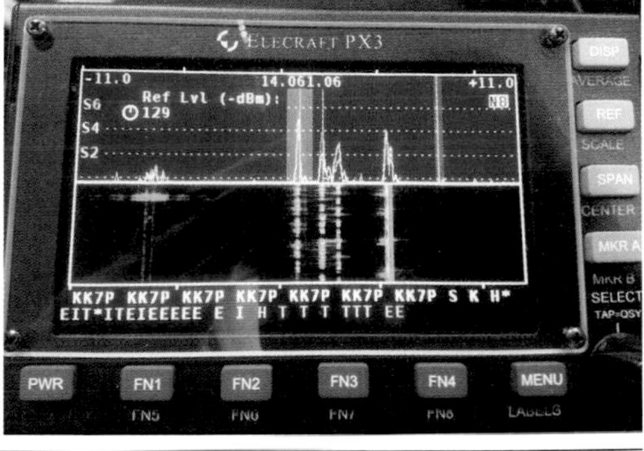

This modern spectrum scope lets you see the signal peaks in real time, plus the waterfall memory traces let you see signals that you might have missed. The waterfall memory won't let you miss any signal!

E8A13 Which of these methods is commonly used to convert analog signals to digital signals?

A. Sequential sampling.
B. Harmonic regeneration.
C. Level shifting.
D. Phase reversal.

Changing an analog signal into a digital signal takes place in the analog-to-digital converter through ***sequential sampling*** of the waveform. Thousands of tiny audio slices are held in a sequential sample-and-hold circuit where they are converted into binary numbers. Some ADCs count with a staircase generator. Others convert analog voltage into digital values with multiple comparators, and others in DDS (Direct Digital Synthesis) use a phase accumulator to identify points on a specific waveform cycle. Sequential sampling of the incoming waveform is the heart of converting analog signals to digital. **ANSWER A.**

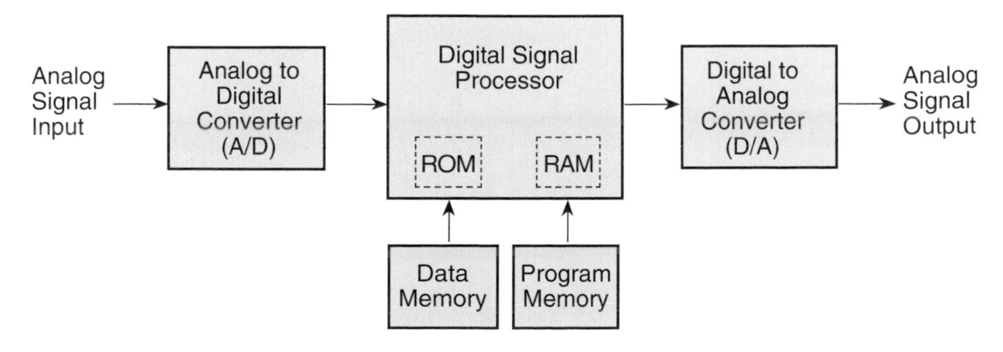

Block diagram of a basic digital signal processing (DSP) system

---

E7E01 Which of the following can be used to generate FM phone emissions?

A. A balanced modulator on the audio amplifier.
B. A reactance modulator on the oscillator.
C. A reactance modulator on the final amplifier.
D. A balanced modulator on the oscillator.

The ***reactance modulator*** causes the ***oscillator*** to vary frequency in accordance with the modulation. It's the oscillator that is *modulated*. **ANSWER B.**

E7E02 What is the function of a reactance modulator?

A. To produce PM signals by using an electrically variable resistance.
B. To produce AM signals by using an electrically variable inductance or capacitance.
C. To produce AM signals by using an electrically variable resistance.
D. To produce PM signals by using an electrically variable inductance or capacitance.

The ***reactance modulator*** is found in a frequency modulation transceiver, ***varying inductance or capacitance to produce*** the ***phase modulation (PM)*** signal where the phase of the carrier current is varied with the frequency of the modulating voltage. **ANSWER D.**

E7E03 How does an analog phase modulator function?

A. By varying the tuning of a microphone preamplifier to produce PM signals.
B. By varying the tuning of an amplifier tank circuit to produce AM signals.
C. By varying the tuning of an amplifier tank circuit to produce PM signals.
D. By varying the tuning of a microphone preamplifier to produce AM signals.

Phase modulation is developed by ***varying the tuning of the amplifier tank circuit to produce phase modulation***, not in the microphone preamp circuit. **ANSWER C.**

E7E05 What circuit is added to an FM transmitter to boost the higher audio frequencies?

A. A de-emphasis network.
B. A heterodyne suppressor.
C. An audio prescaler.
D. A pre-emphasis network.

Pre-emphasis networks in modern FM transmitters help improve the signal-to-noise ratio by proportionally attenuating the less-important lower audio frequencies for voice and ***boosting*** voice intelligence in the ***higher audio frequencies***. **ANSWER D.**

E7E04 What is one way a single-sideband phone signal can be generated?

A. By using a balanced modulator followed by a filter.
B. By using a reactance modulator followed by a mixer.
C. By using a loop modulator followed by a mixer.
D. By driving a product detector with a DSB signal.

In a ***single-sideband transceiver***, a ***lower or upper sideband filter*** removes the unwanted sideband. **ANSWER A.**

E7F17 What do the letters I and Q in I/Q Modulation represent?

A. Inactive and Quiescent.
B. Instantaneous and Quasi-stable.
C. Instantaneous and Quenched.
D. In-phase and Quadrature.

While most of us at one time or another would like to have our IQ modulated (preferably in an upward direction) that is not what we're talking about here. ***I and Q stand for "In Phase" and "Quadrature."*** While I/Q modulation is at the heart of most modern digital modulation schemes, it's been around for a long, long time. The very first single sideband transmitters used the phasing method of SSB generation which uses two balanced modulators in an I/Q configuration. **ANSWER D.**

E7F04 What is a common method of generating an SSB signal using digital signal processing?

A. Mixing products are converted to voltages and subtracted by adder circuits.
B. A frequency synthesizer removes the unwanted sidebands.
C. Emulation of quartz crystal filter characteristics.
D. Combine signals with a quadrature phase relationship.

Digital signal processing has revolutionized ham radio transmitters and receivers and a host of after-market audio products by cleaning up signal reception. For SSB signal transmissions, ***DSP relies on phase-shifting*** the carrier and audio components by 90 degrees and sending this phased split to the SSB balanced modulator. The output is a DSP upper or lower sideband, with the carrier and unwanted sideband suppressed more than 40 dB. No heterodyning is required. All of this ***quadrature occurs in the DSP chip***. **ANSWER D.**

E7F03 What type of digital signal processing filter is used to generate an SSB signal?

A. An adaptive filter.
B. A notch filter.
C. A Hilbert-transform filter.
D. An elliptical filter.

A ***Hilbert transform filter*** is a ***DSP*** filter that performs an operation on an analytic signal, which is a signal containing both I and Q (In-phase and Quadrature) components. It is a "high tech" version of the old phasing method of ***single sideband generation***, popular before the crystal filter was readily available. All things old become new again in amateur radio! **ANSWER C.**

E7E07 What is meant by the term baseband in radio communications?

A. The lowest frequency band that the transmitter or receiver covers.
B. The frequency components present in the modulating signal.
C. The unmodulated bandwidth of the transmitted signal.
D. The basic oscillator frequency in an FM transmitter that is multiplied to increase the deviation and carrier frequency.

The term ***"baseband" refers to the range of frequencies occupied by the modulating signal*** in the transmitter. For digital modes and CW, it could be as narrow as 30 Hz. For sideband voice and FM, baseband frequency response could range from 2.5 kHz to 25 kHz. **ANSWER B.**

E1B01 Which of the following constitutes a spurious emission?

A. An amateur station transmission made at random without the proper call sign identification.
B. A signal transmitted to prevent its detection by any station other than the intended recipient.
C. Any transmitted signal that unintentionally interferes with another licensed radio station.
D. An emission outside its necessary bandwidth that can be reduced or eliminated without affecting the information transmitted.

May your spurious emissions be minimum! The FCC says so. A ***spurious emission*** could include random splatter to adjacent frequencies, second harmonics, and any other microvolt or millivolt emission that serves no purpose ***outside of your necessary bandwidth***. [97.3] **ANSWER D.**

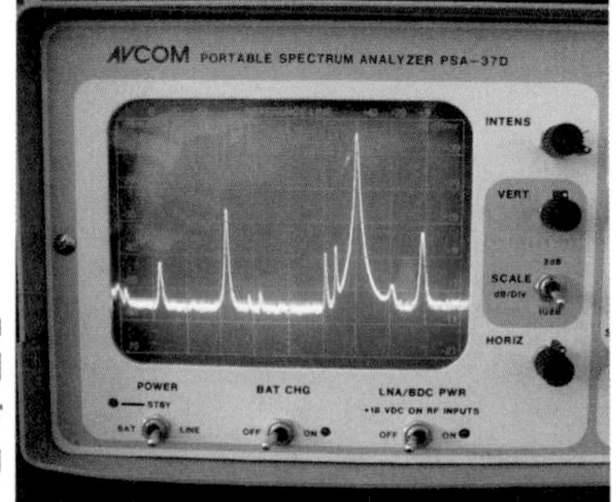

A close look at multiple spurious signals from a land mobile FM hand held not specifically tuned to the 2 meter ham band. All traces disappear when the hand held is un-keyed, proving these spurs are part of the main carrier.

E1B08 What limitations may the FCC place on an amateur station if its signal causes interference to domestic broadcast reception, assuming that the receivers involved are of good engineering design?

A. The amateur station must cease operation.

B. The amateur station must cease operation on all frequencies below 30 MHz.

C. The amateur station must cease operation on all frequencies above 30 MHz.

D. The amateur station must avoid transmitting during certain hours on frequencies that cause the interference.

If the FCC gets involved in an interference issue with your neighbor and it finds that your neighbor has a nice clean receiver of modern design, and your 1.5 kilowatt output is coming in over that receiver, ***the FCC could impose quiet hours on your transmissions on the specific band*** where you are causing the interference. Suggest you don't run the KW during ball games! [97.121] **ANSWER D.**

E7B12 What type of amplifier circuit is shown in Figure E7-1?

A. Common base.

B. Common collector.

C. Common emitter.

D. Emitter follower.

A common emitter amplifier is shown in Figure E7-1. It is the most common voltage amplifier circuit. While the emitter is not always connected directly to ground, it is also known as a grounded emitter amplifier. In a common emitter amplifier, the input signal is applied between the base and emitter while the output signal is taken between the collector and the emitter. The collector and the associated load resistor form a voltage divider, with the transistor acting as a variable resistor in series with the load resistance. At rest, the voltage at the collector of a properly functioning ***common emitter amplifier*** will be about half the supply voltage; this is a good quick test to perform when troubleshooting. **ANSWER C.**

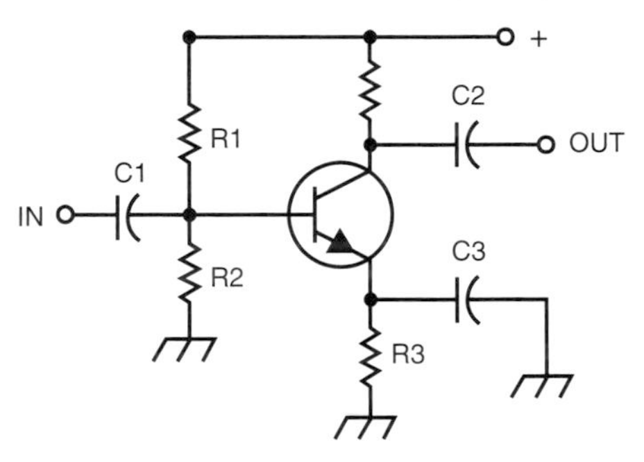

Figure E7-1

E7B10 In Figure E7-1, what is the purpose of R1 and R2?

A. Load resistors.

B. Fixed bias.

C. Self bias.

D. Feedback.

This reminds you a little bit of Thevenin's Theorem, doesn't it? ***R1 and R2*** form a voltage divider to set the voltage and the current at the base of the transistor for ***fixed bias***. **ANSWER B.**

E7B11 In Figure E7-1, what is the purpose of R3?

A. Fixed bias

B. Emitter bypass

C. Output load resistor.

D. Self bias.

Current through ***R3 produces*** a voltage drop that affects the base-emitter bias voltage. Since the transistor's collector and base current (the emitter current) determine the voltage at the emitter, it is called ***self bias***. **ANSWER D.**

E7B13 In Figure E7-2, what is the purpose of R?

A. Emitter load
B. Fixed bias
C. Collector load.
D. Voltage regulation.

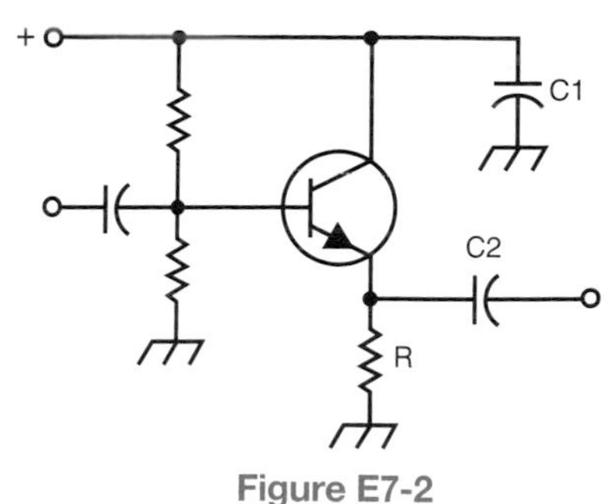

Figure E7-2

One end of the resistor R is connected to the emitter, the other end to ground. It is classified as an *emitter load resistor* in this circuit because the output is taken from the emitter. **ANSWER A.**

E7C01 How are the capacitors and inductors of a low-pass filter Pi-network arranged between the network's input and output?

A. Two inductors are in series between the input and output, and a capacitor is connected between the two inductors and ground.
B. Two capacitors are in series between the input and output, and an inductor is connected between the two capacitors and ground.
C. An inductor is connected between the input and ground, another inductor is connected between the output and ground, and a capacitor is connected between the input and output.
D. A capacitor is connected between the input and ground, another capacitor is connected between the output and ground, and an inductor is connected between input and output.

The *pi-network features a capacitor in parallel with the input*, a second *capacitor in parallel with the output*, and an *inductor in series between the two*. It looks like the Greek letter π. **ANSWER D.**

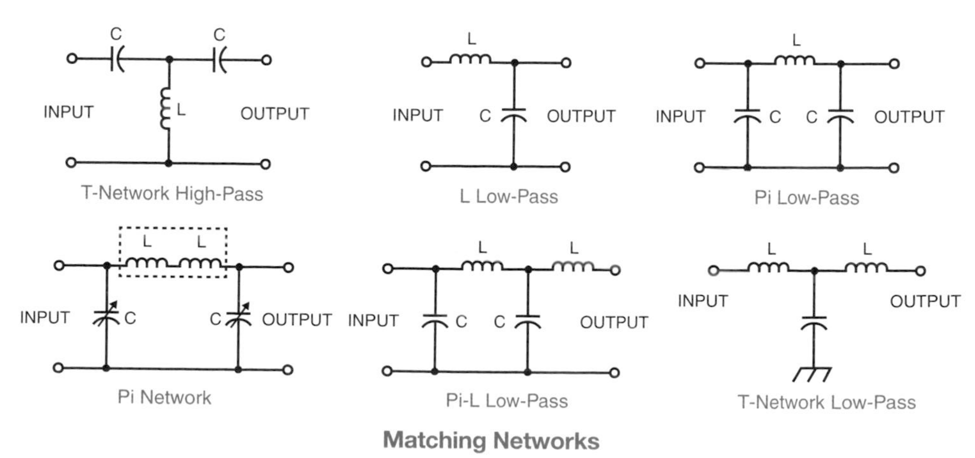

Matching Networks

E7C02 Which of the following is a property of a T-network with series capacitors and a parallel shunt inductor?

A. It is a low-pass filter.
B. It is a band-pass filter.
C. It is a high-pass filter.
D. It is a notch filter.

While a ***T-network*** is commonly used as an antenna tuner or transmatch, it can inadvertently become a ***high-pass filter*** under certain conditions, which is something you generally want to avoid in this application. Remember that capacitors decrease reactance with increasing frequency; therefore if these are in series with the signal path, they will pass higher frequencies with less attenuation. Likewise, an inductor has more reactance with increasing frequency, which means if they are in shunt parallel with the signal path, they will bypass less of the signal at higher frequencies. If you use a high-pass T-network as an antenna tuner, be sure your transmitter is clean, because the network will emphasize any harmonics you generate! **ANSWER C.**

E7C03 What advantage does a Pi-L-network have over a regular Pi-network for impedance matching between the final amplifier of a vacuum-tube transmitter and an antenna?

A. Greater harmonic suppression.
B. Higher efficiency.
C. Lower losses.
D. Greater transformation range.

Worldwide ham sets transmitting into a multi-band antenna system can generate harmonics. The ***Pi-L-network*** always ***gives us greater harmonic suppression*** inside the transceiver. **ANSWER A.**

E7C11 Which of the following is the common name for a filter network which is equivalent to two L-networks connected back-to-back with the two inductors in series and the capacitors in shunt at the input and output?

A. Pi-L.
B. Cascode.
C. Omega.
D. Pi.

The ***Pi Network*** is so named because it physically resembles the Greek letter π. Electrically, it looks like two L-networks connected back-to-back with the two inductors being "absorbed" into one inductor. This circuit, in addition to performing as a low-pass filter, is very commonly used as a transmitter tank circuit and impedance matching network. If the two capacitors are variable, as shown in the schematic on page 111, both the resonant frequency and the Q of the circuit can be adjusted at will, creating a versatile means of matching the transmitter output to a variety of loads. **ANSWER D.**

E7C12 Which describes a Pi-L-network used for matching a vacuum tube final amplifier to a 50 ohm unbalanced output?

A. A Phase Inverter Load network.
B. A Pi-network with an additional series inductor on the output.
C. A network with only three discrete parts.
D. A matching network in which all components are isolated from ground.

The Pi-L network is characterized by two series-adjustable inductors and two variable shunt-to-ground capacitors. You can spot the Pi-L network over a simple Pi network as the ***Pi-L network has an additional series inductor at the output***. **ANSWER B.**

E7C13 What is one advantage of a Pi-matching network over an L-matching network consisting of a single inductor and a single capacitor?

A. The Q of Pi-networks can be varied depending on the component values chosen.
B. L-networks cannot perform impedance transformation.
C. Pi-networks have fewer components.
D. Pi-networks are designed for balanced input and output.

The big advantage of a ***Pi matching network*** over the L network is the shunt-to-ground variable capacitors that ***can vary the Q***, depending on the value of the variable capacitors and adjustable tap coil chosen. **ANSWER A.**

E7B09 Which of the following describes how the loading and tuning capacitors are to be adjusted when tuning a vacuum tube RF power amplifier that employs a Pi-network output circuit?

A. The loading capacitor is set to maximum capacitance and the tuning capacitor is adjusted for minimum allowable plate current.
B. The tuning capacitor is set to maximum capacitance and the loading capacitor is adjusted for minimum plate permissible current.
C. The loading capacitor is adjusted to minimum plate current while alternately adjusting the tuning capacitor for maximum allowable plate current.
D. The tuning capacitor is adjusted for minimum plate current, and the loading capacitor is adjusted for maximum permissible plate current.

If you're going to be running a big power amplifier on the HF bands, chances are it will still use a vacuum tube for kilowatts of output power. Solid-state power amplifiers require no ***tuning***, but most solid-state power amps only put out about 800 watts maximum power. With a ***tube powered amplifier***, you alternately tune and ***increase the plate current with a loading capacitor***, and then ***dip the plate current with a tuning capacitor***. Do this for short periods of time, and do it on a frequency that no one else is transmitting on. First tune up into a dummy load to get the settings close, and then switch over to the antenna, select a frequency that no one is using, announce, "I'm going to quickly tune up," and then quickly tune up. **ANSWER D.**

E7C04 How does an impedance-matching circuit transform a complex impedance to a resistive impedance?

A. It introduces negative resistance to cancel the resistive part of impedance.
B. It introduces transconductance to cancel the reactive part of impedance.
C. It cancels the reactive part of the impedance and changes the resistive part to a desired value.
D. Network resistances are substituted for load resistances and reactances are matched to the resistances.

In older radios, you could vary the matching network to ***cancel the reactive (X) part of an impedance and change the value of the resistive part (R) of an impedance***. Newer transistorized radios have a fixed output, and there is no manual control of the fixed matching devices. However, manufacturers have come to the rescue in antenna matching by providing new radios with built-in automatic antenna tuners. **ANSWER C.**

E4D03 How can intermodulation interference between two repeaters occur?

A. When the repeaters are in close proximity and the signals cause feedback in the final amplifier of one or both transmitters.
B. When the repeaters are in close proximity and the signals mix in the final amplifier of one or both transmitters.
C. When the signals from the transmitters are reflected out of phase from airplanes passing overhead.
D. When the signals from the transmitters are reflected in phase from airplanes passing overhead.

"***Intermod***," in its true sense, ***occurs within the repeater's final amplifier***. When signals mix in one or both repeater amplifiers, sum and difference signals are heard on many different frequencies and as significant noise on the frequencies of the repeaters. **ANSWER B.**

Repeater sites like this one require hams to use added cavities, circulators, filters , and high-end duplexers to minimize intermodulation interference.

E4D08 What causes intermodulation in an electronic circuit?

A. Too little gain.
B. Lack of neutralization.
C. Nonlinear circuits or devices.
D. Positive feedback.

Intermodulation interference is minimized by a careful selection of linear circuits and components. If that newly-designed transceiver for 2 meters or 440 MHz or that repaired repeater is experiencing high levels of intermodulation, the problem could be caused by ***non-linear circuits, devices, and components*** inside the receiver. **ANSWER C.**

E4D06 What is the term for unwanted signals generated by the mixing of two or more signals?

A. Amplifier desensitization.
B. Neutralization.
C. Adjacent channel interference.
D. Intermodulation interference.

Repeaters are fun to operate through, but a real headache to keep on the air with a clean signal. The problem is ***intermodulation*** caused by other close-proximity transmitters. What you may hear are ***two signals coming in at once***. The resulting noise is commonly called "intermod." **ANSWER D.**

E7C10 Which of the following filters would be the best choice for use in a 2 meter repeater duplexer?

A. A crystal filter. C. A DSP filter.
B. A cavity filter. D. An L-C filter.

For a 2 meter repeater duplexer system, we need big filters. We call them ***cavity filters*** because they are hollowed out to form extremely narrow-band tuned resonant circuits. **ANSWER B.**

The professional cavity filter system on a 2 meter repeater offers minimal attenuation to received signals in the pass band, and major attenuation to the repeater's transmitter frequency. It is a job to filter out your transmit signals only 600 kHz away! Make sure these filter cavities are in a dry, warm cabinet to prevent condensation on the gleaming copper insides!

E4D04 Which of the following may reduce or eliminate intermodulation interference in a repeater caused by another transmitter operating in close proximity?

A. A band-pass filter in the feed line between the transmitter and receiver.
B. A properly terminated circulator at the output of the transmitter.
C. A Class C final amplifier.
D. A Class D final amplifier.

Your ham friends who regularly trek to mountaintops to maintain ham radio VHF/UHF repeaters deserve your major support. Mountaintop repeaters are located in the worst RF environment possible – in the midst of hundreds of nearby signals that may desensitize the receiver or create "gar-bage" on the input. Everything was fine when your team left the hill, but a new land-mobile radio system on the same tower is now causing ***intermodulation*** interference. One ***solution is to install a terminated circulator*** or ferrite isolator between your own transmitter and duplexer, which helps zero in on your own receive frequency and cancel out any other off-channel frequencies that seem to creep in only when your equipment is on transmit. **ANSWER B.**

E4E08 What type of signal is picked up by electrical wiring near a radio antenna?

A. A common-mode signal at the frequency of the radio transmitter.
B. An electrical-sparking signal.
C. A differential-mode signal at the AC power line frequency.
D. Harmonics of the AC power line frequency.

Electrical wiring within your shack will many times pick up ***common mode radio signals*** coming off your nearby transmitting antenna. Make sure your radio is well grounded and always suspect interference coming directly off the antenna when you are transmitting. **ANSWER A.**

HERE ARE SOME WEB ADDRESSES OF USEFUL SITES RELATED TO MODULATE YOUR TRANSMITTERS:

www.dx-code.org/
www.qrz.com/
www.hamcall.net/call
www.dxzone.com/catalog/DX_Resources/Callsigns/

CD 3 TRACK 2

Amps & Power Supplies

E1F03 Under what circumstances may a dealer sell an external RF power amplifier capable of operation below 144 MHz if it has not been granted FCC certification?

A. It was purchased in used condition from an amateur operator and is sold to another amateur operator for use at that operator's station,
B. The equipment dealer assembled it from a kit.
C. It was imported from a manufacturer in a country that does not require certification of RF power amplifiers.
D. It was imported from a manufacturer in another country and was certificated by that country's government.

Amplifiers for frequencies below 144 MHz could be ***purchased in used condition*** by an equipment dealer and ***then sold by that dealer to another amateur operator***. This rule prevents the sale of amplifiers really intended for CB use that are constructed without the FCC's blessing. [97.315] **ANSWER A.**

E1F11 Which of the following best describes one of the standards that must be met by an external RF power amplifier if it is to qualify for a grant of FCC certification?

A. It must produce full legal output when driven by not more than 5 watts of mean RF input power.
B. It must be capable of external RF switching between its input and output networks.
C. It must exhibit a gain of 0 dB or less over its full output range.
D. It must satisfy the FCC's spurious emission standards when operated at the lesser of 1500 watts or its full output power.

Recently, the FCC revised its policy on RF amplifiers. In order for an RF power amplifier to be granted FCC certification, ***the amp must meet tight spurious emission standards*** when it is operating at full power. Tube amplifiers with large band switching filter networks meet the specs more easily than smaller all-solid-state amps. This may be why the solid state amplifiers are still relatively expensive compared to tube amps. [97.317]
ANSWER D.

This little 2-meter linear power amplifier meets stringent FCC spurious emission standards.

E1B11 What is the permitted mean power of any spurious emission relative to the mean power of the fundamental emission from a station transmitter or external RF amplifier installed after January 1, 2003 and transmitting on a frequency below 30 MHZ?

A. At least 43 dB below.
B. At least 53 dB below.
C. At least 63 dB below.
D. At least 73 dB below.

Down on high frequency (below 30 MHz), the FCC requires any spurious emission to be at least 43 dB below the mean power output of that amplifier. Amplifiers contain both harmonic and band pass filters to reduce ***spurious emissions down at least 43 dB***. [97.307] **ANSWER A.**

E7B04 Where on the load line of a Class A common emitter amplifier would bias normally be set?

A. Approximately half-way between saturation and cutoff.
B. Where the load line intersects the voltage axis.
C. At a point where the bias resistor equals the load resistor.
D. At a point where the load line intersects the zero bias current curve.

The ***Class A amplifier*** offers the greatest linearity but provides only marginal efficiency, usually less than 25%. We normally find the Class A amplifier as a "small signal" amplifier. It is sometimes used in modulation circuits where perfect linearity is required. The input signal continuously varies the input current over the complete 360 degrees of the sine-wave signal, with ***bias normally set approximately half-way between saturation and cutoff***. **ANSWER A.**

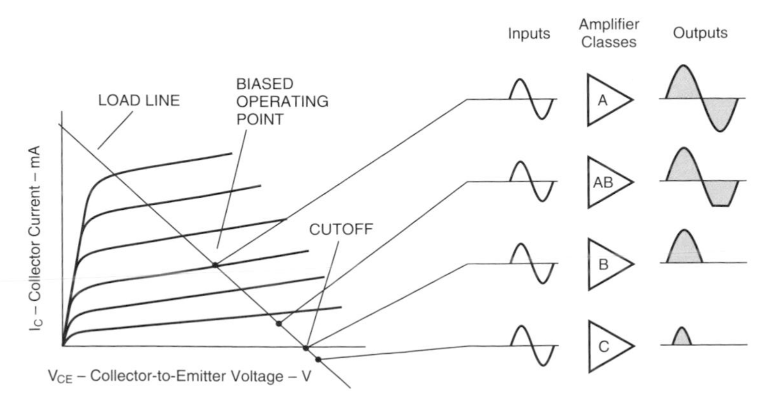

Various Classes of Transistorized Amplifiers

E7B01 For what portion of a signal cycle does a Class AB amplifier operate?

A. More than 180 degrees but less than 360 degrees.
B. Exactly 180 degrees.
C. The entire cycle.
D. Less than 180 degrees.

The ***Class AB amplifier*** has some of the properties of both the Class A and the Class B amplifier. It has output for ***more than 180 degrees*** of the cycle, but ***less than 360 degrees*** of the cycle. **ANSWER A.**

E7B06 Which of the following amplifier types reduces or eliminates even order harmonics?

A. Push-push.
B. Push-pull.
C. Class C.
D. Class AB.

The Class-B amplifier may operate a pair of matched tubes in "push-pull" configuration, and operates only over one half of an input cycle. In transistorized Class B amplifiers, the circuit may be called "complimentary-symmetry" where NPN and PNP transistors each conduct over half the cycle. The advantage of the ***push-pull circuit is the cancellation of even-order harmonics*** – second, fourth, etc. In tube-type Class B amplifiers, it is important that you always change both output tubes together, never replace one tube and leave the older tube. **ANSWER B.**

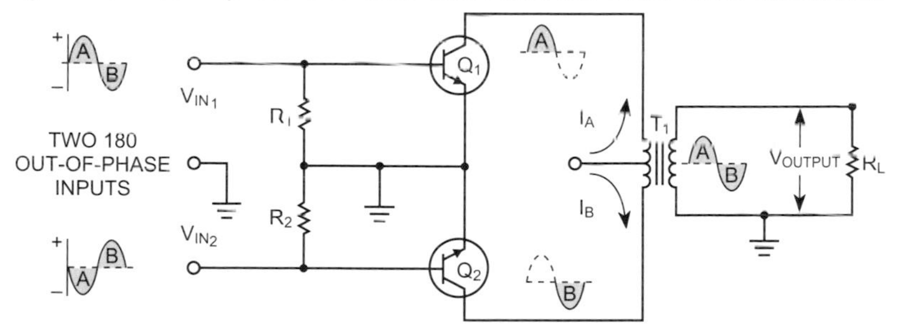

Transistorized Push-Pull Amplifier

E7B18 What is a characteristic of a grounded-grid amplifier?

A. High power gain.
B. High filament voltage.
C. Low input impedance.
D. Low bandwidth.

The ***grounded-grid amplifier*** remains one of the most popular methods of achieving legal limit HF power output. Yes, vacuum tubes like the 8877 are still available, and the power supply may only require B1 plate voltage and a tap off the transformer for AC low filament voltage. Another benefit is a 50 ohm ***low input impedance***, closely matching the output circuit of your HF transceiver. **ANSWER C.**

E7B08 How can an RF power amplifier be neutralized?

A. By increasing the driving power.
B. By reducing the driving power.
C. By feeding a 180-degree out-of-phase portion of the output back to the input.
D. By feeding an in-phase component of the output back to the input.

We can keep our power-amplified signal clean if the power ***amplifier is neutralized*** following the instructions in your amplifier's manual. Chances are you'll be ***feeding a 180-degree out-of-phase portion of the output back to the input*** with neutralizing capacitors chosen to obtain the proper amount of feedback. **ANSWER C.**

E7B17 Why are odd-order rather than even-order intermodulation distortion products of concern in linear power amplifiers?

A. Because they are relatively close in frequency to the desired signal.
B. Because they are relatively far in frequency from the desired signal.
C. Because they invert the sidebands causing distortion.
D. Because they maintain the sidebands, thus causing multiple duplicate signals.

Third-order intermodulation distortion can be a big problem with power amplifiers creating a small amount of in-band ***spurious emissions close to your operating frequency***. Third-order IMD will likely elicit ham operators claiming your signal is wide or splattering. **ANSWER A.**

E7B05 What can be done to prevent unwanted oscillations in an RF power amplifier?

A. Tune the stage for maximum SWR.
B. Tune both the input and output for maximum power.
C. Install parasitic suppressors and/or neutralize the stage.
D. Use a phase inverter in the output filter.

As a new Extra soon, you will likely be operating with a tower and beam tied into a power amplifier to boost your transceiver's 100 watt output signal. But cautiously work yourself into high-power amplifier operation. Big amps may cause interference to your home electronics, and big amps need to be tuned precisely to minimize parasitic oscillations. When viewed on a spectrum analyzer, parasitic oscillations look like a forest of trees surrounding your transmitted signal. ***In tube amplifiers, RF chokes in the tube plate circuit and small-value bypass and suppressor capacitors in the grid circuit may help reduce unwanted oscillations***. With tube amplifiers, step-by-step neutralization procedures are detailed in the instruction book. If the amplifier is new, you likely won't need to go through the neutralization process until you change out the final amplifier tube. **ANSWER C.**

E7B07 Which of the following is a likely result when a Class C amplifier is used to amplify a single-sideband phone signal?

A. Reduced intermodulation products.
B. Increased overall intelligibility.
C. Signal inversion.
D. Signal distortion and excessive bandwidth.

The Class AB amplifier will take an input signal and cause a field effect transistor or bipolar transistor to operate near cutoff, but not into cutoff. This will lead to a nearly linear reproduction of the input signal and using Class AB is common with careful attention to setting the bias just above the point where the signal base current goes into cutoff at zero. Once the transistor is biased beyond cutoff, efficiency continues to climb but at the cost of distortion and excessive bandwidth. This is why satellite operators using SSB caution against the use of non-linear FM-only solid state power amplifiers. The Class C amp is fine for data and FM, but for single sideband voice emissions, the ***Class C amplifier*** is operating outside of the linear portion of its operation. This ***causes the SSB voice signal to sound harsh, distorted, and occupying excessive bandwidth***. **ANSWER D.**

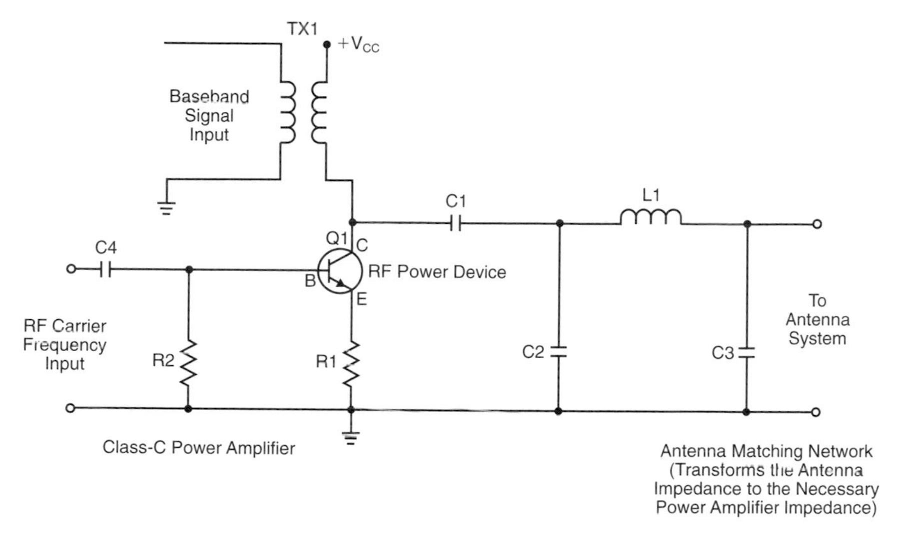

Class-C power amplifier-modulator being modulated with baseband signal.

Source: *Basic Communications Electronics*, Hudson & Luecke, © 1999 Master Publishing, Inc., Niles, IL

E7B02 What is a Class D amplifier?

A. A type of amplifier that uses switching technology to achieve high efficiency.
B. A low power amplifier that uses a differential amplifier for improved linearity.
C. An amplifier that uses drift-mode FETs for high efficiency.
D. A frequency doubling amplifier.

A less commonly known amplifier is called Class D and is made up of two matched transistors driven with enough input power to produce a square wave output. The ***Class D amplifier employs switching technology to offer high efficiency*** from its balanced twin devices. **ANSWER A.**

E7B03 Which of the following components form the output of a class D amplifier circuit?

A. A low-pass filter to remove switching signal components.
B. A high-pass filter to compensate for low gain at low frequencies.
C. A matched load resistor to prevent damage by switching transients.
D. A temperature compensating load resistor to improve linearity.

The ***Class D amplifier*** output may be rich in unwanted switching signal components. A ***low-pass filter*** allows the square wave to pass on to the next stage while attenuating the unwanted higher frequency components. **ANSWER A.**

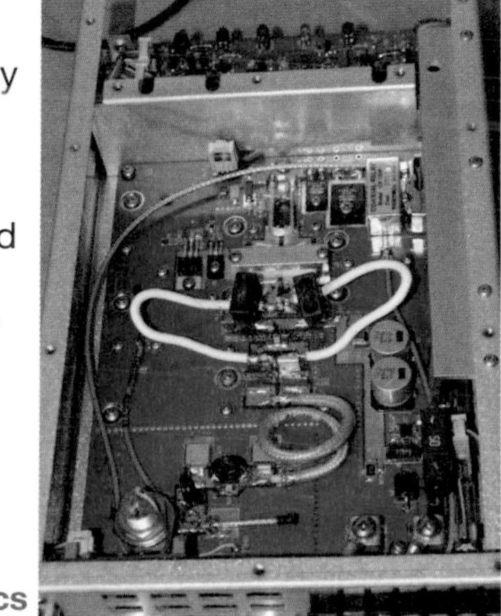

This small power amplifier from M² Electronics puts out more than a kilowatt!

E7B14 Why are switching amplifiers more efficient than linear amplifiers?

A. Switching amplifiers operate at higher voltages.
B. The power transistor is at saturation or cut off most of the time, resulting in low power dissipation.
C. Linear amplifiers have high gain resulting in higher harmonic content.
D. Switching amplifiers use push-pull circuits.

The speed at which an active device switches from the fully-on to fully-off state determines its efficiency. ***In either state the device dissipates zero power***; the only time heat is generated is during the transition. The faster the transition can be made, the less the heat. If the transition can be made instantaneously, the circuit can be 100% efficient. In reality, switching amplifiers can approach 95% efficiency at audio frequencies. Such performance has not yet entered the high-power RF realm, but it's getting there. **ANSWER B.**

E7B15 What is one way to prevent thermal runaway in a bipolar transistor amplifier?

A. Neutralization.
B. Select transistors with high beta.
C. Use a resistor in series with the emitter.
D. All of these choices are correct.

UHF and microwave operators who homebrew their own amplifiers are cautious to protect against transistor thermal runaway. Adding a fixed emitter resistor may cut down slightly on the transistor voltage gain to control transistor current without sacrificing linearity. In tube equipment, it is called cathode degeneration. In your transistorized amplifier, ***thermal protection*** is provided using degenerative emitter feedback, through the ***use of a resistor (R_3 in the diagram) in series with the emitter***, to ground. **ANSWER C.**

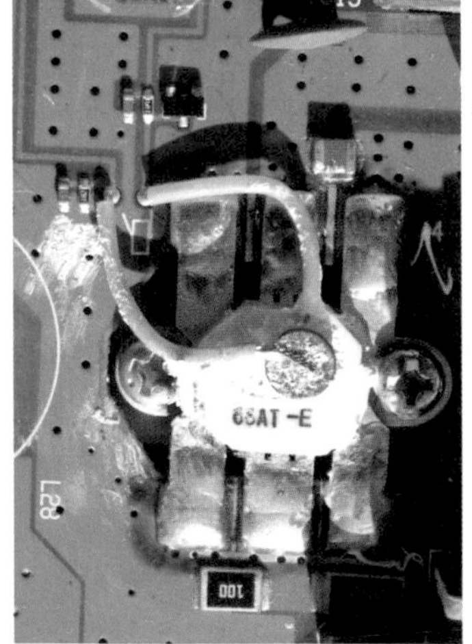

Transistors do not like to get hot! Small ones may use an emitter resistor. Large power transistors like this one use a thermistor bonded to it for thermal protection.

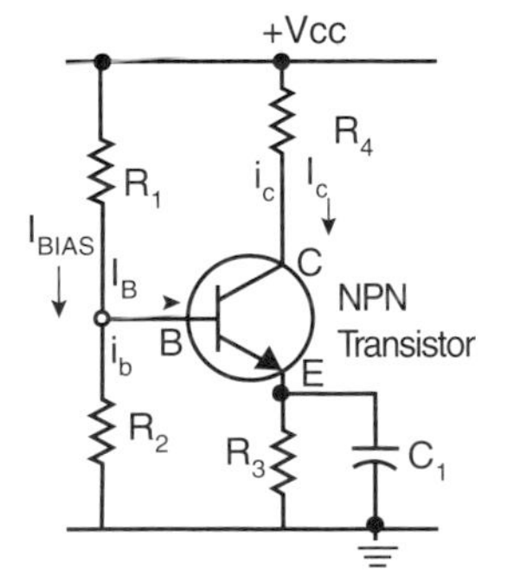

Active Device — Provides Gain
NPN Transistor
Current Gain = $h_{fe} = \frac{\Delta\, ic}{\Delta\, ib}$

Passive Device — No Gain
Resistors R_1, R_2, R_3, R_4 and C_1

Δ means "change in"

Small-signal amplifier with an NPN transistor as the active device and resistors and a capacitor as passive devices that set the "no-signal" (DC) operating point.
Source: *Basic Communications Electronics*, Hudson & Luecke, © 1999 Master Publishing, Inc., Niles, IL

E6D09 What devices are commonly used as VHF and UHF parasitic suppressors at the input and output terminals of a transistor HF amplifier?

A. Electrolytic capacitors.
B. Butterworth filters.
C. Ferrite beads.
D. Steel-core toroids.

If you look into the modern ***VHF/UHF*** single-band or dual-band ***amplifier***, you'll see many leads dressed with small ***ferrite beads*** to minimize parasitics coming down voltage or control lines. **ANSWER C.**

E7B16 What is the effect of intermodulation products in a linear power amplifier?

A. Transmission of spurious signals.
B. Creation of parasitic oscillations.
C. Low efficiency.
D. All of these choices are correct.

The effects of intermodulation, abbreviated IM, ***are spurious signal products*** usually traced to power amplifier stages or linear amplifiers with incorrect bias settings. Old time hams called any station with strong IM products "monkey chatter" that may travel up to 200 kHz up and down the band from the main transmitted signal. As a new Extra, with that new power amplifier turned on, start out on extremely low power input to the amp, and suspect IM if another station drops down to your frequency and tells you that your signal is wide, splattering, and you sound like "monkey chatter." With careful tuning and monitoring of amp input drive power from your transceiver, intermodulation can usually be solved without having to completely field strip your station for repair. **ANSWER A.**

E7D12 What is the drop-out voltage of an analog voltage regulator?

A. Minimum input voltage for rated power dissipation.
B. Maximum amount that the output voltage drops when the input voltage is varied over its specified range.
C. Minimum input-to-output voltage required to maintain regulation.
D. Maximum amount that the output voltage may decrease at rated load.

Unlike switching regulators, analog regulators can only reduce voltage from the unregulated value. This means that you need to have ***a bit more voltage going into the regulator than you want at the output.*** Typically, you want the input voltage 30% to 50% greater than the output voltage for reliable regulation. **ANSWER C.**

E7D01 What is one characteristic of a linear electronic voltage regulator?

A. It has a ramp voltage as its output.
B. It eliminates the need for a pass transistor.
C. The control element duty cycle is proportional to the line or load conditions.
D. The conduction of a control element is varied to maintain a constant output voltage.

A ***linear electronic voltage regulator varies the conduction of a circuit*** in direct proportion to variations in the line voltage to, or the load current from, the device. You will find a sophisticated voltage regulation circuit inside your base station power supply. **ANSWER D.**

A pair of three terminal regulators for a dual voltage power supply.

E7D03 What device is typically used as a stable reference voltage in a linear voltage regulator?

A. A Zener diode.
B. A tunnel diode.
C. An SCR.
D. A varactor diode.

A ***Zener diode*** is primarily used as a ***voltage reference***. You can recognize the symbol of a Zener by the zig-zag bar used in its schematic symbol instead of the straight line designating the cathode of a standard diode. Remember: zig-zag = Zener! A Zener diode is normally operated *backwards*, in the normally non-conducting direction. At a very precise voltage, the Zener will conduct in the *reverse direction*, becoming a very low resistance. A Zener can serve as a voltage regulator by itself, or as a voltage reference for a more elaborate voltage regulator. Before Zener diodes were around, the voltage reference of choice was a gas filled tube. **ANSWER A.**

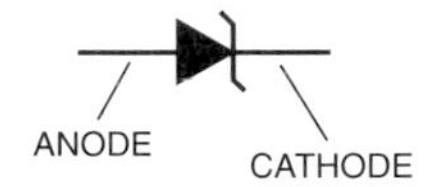

Here is the schematic symbol of a Zener diode. Since a diode only passes energy in one direction, look for that one-way arrow, plus a "Z" indicating it is a Zener diode. Doesn't that vertical line look like a tiny "Z".

Zener Diode

E7D02 What is one characteristic of a switching electronic voltage regulator?

A. The resistance of a control element is varied in direct proportion to the line voltage or load current.
B. It is generally less efficient than a linear regulator.
C. The controlled device's duty cycle is changed to produce a constant average output voltage.
D. It gives a ramp voltage at its output.

The ***switching voltage regulator*** actually ***switches the control device*** completely ***on or off***. **ANSWER C.**

E7D04 Which of the following types of linear voltage regulator usually make the most efficient use of the primary power source?

A. A series current source.
B. A series regulator.
C. A shunt regulator.
D. A shunt current source.

The very common ***series voltage regulator*** is used in many high-power ***linear power supplies***. With no load, the series regulator dissipates essentially zero power, while under full load it can dissipate up to 50% of the total power. By contrast, for lower power applications the *shunt regulator* is generally simpler. However, the shunt regulator dissipates the same amount of power regardless of the load, while the series regulator's power loss depends on the current draw. With no load, the shunt regulator is 0% efficient! **ANSWER B.**

E7D13 What is the equation for calculating power dissipation by a series connected linear voltage regulator?

A. Input voltage multiplied by input current.
B. Input voltage divided by output current.
C. Voltage difference from input to output multiplied by output current.
D. Output voltage multiplied by output current.

Linear or *analog* voltage regulators are very clean, but they can be quite inefficient. A linear voltage regulator must always have a supply voltage significantly greater than the desired output voltage. In many high power regulated supplies, as much as half the supply voltage is dropped in the regulation circuitry, which typically includes large series pass transistors mounted on massive heat sinks. As in any ***series circuit the power dissipation of any component is equal to the voltage drop across that component times the current through the component***. **ANSWER C.**

E7D11 What circuit element is controlled by a series analog voltage regulator to maintain a constant output voltage?

A. Reference voltage.
B. Switching inductance.
C. Error amplifier.
D. Pass transistor.

A pass transistor is in series with the load and must be able to handle the entire current that your powered device draws. The ***pass transistor is controlled by a Zener diode*** or other electronic regulator and is usually capable of dissipating a lot of heat. Pass transistors are perhaps the most vulnerable components in a regulated power supply. **ANSWER D.**

E7D05 Which of the following types of linear voltage regulator places a constant load on the unregulated voltage source?

A. A constant current source.
B. A series regulator.
C. A shunt current source.
D. A shunt regulator.

In circuits where the ***load on the unregulated input source must be kept constant***, we use the slightly less efficient ***shunt regulator***. These can sometimes get quite warm and are usually mounted to the chassis of the equipment which acts as a heat sink. **ANSWER D.**

E7D06 What is the purpose of Q1 in the circuit shown in Figure E7-3?

A. It provides negative feedback to improve regulation.
B. It provides a constant load for the voltage source.
C. It increases the current-handling capability of the regulator.
D. It provides D1 with current.

This ***transistor can handle greater current*** than a Zener diode. They sometimes call these transistors "series-pass transistors." **ANSWER C.**

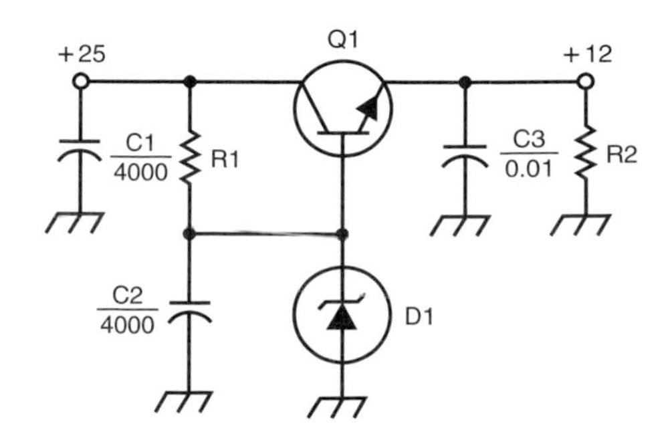

Figure E7-3

E7D07 What is the purpose of C2 in the circuit shown in Figure E7-3?

A. It bypasses hum around D1.
B. It is a brute force filter for the output.
C. To self-resonate at the hum frequency.
D. To provide fixed DC bias for Q1.

C2 provides additional filtering for the ripple frequency. It *bypasses voltage* changes *around* the Zener diode, *D1*, to keep the reference voltage stable. **ANSWER A.**

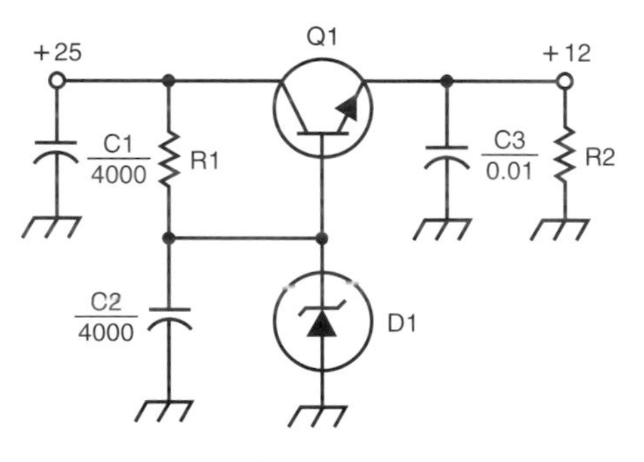

Figure E7-3

E7D08 What type of circuit is shown in Figure E7-3?

A. Switching voltage regulator.
B. Grounded emitter amplifier.
C. Linear voltage regulator.
D. Emitter follower.

Components D1 and Q1 give away the answer – *a linear voltage regulator*. This should give you a good, solid, regulated voltage output, even though the load current may be changing. **ANSWER C.**

E7D14 What is one purpose of a "bleeder" resistor in a conventional unregulated power supply?

A. To cut down on waste heat generated by the power supply.
B. To balance the low-voltage filament windings.
C. To improve output voltage regulation.
D. To boost the amount of output current.

The *bleeder resistor* provides a stable load level in the power supply and bleeds off current from the charged capacitors when the power supply is shut off. Good *output voltage regulation* is maintained by the bleeder resistor shunt which is grounded from the charged capacitors to chassis ground. **ANSWER C.**

A large wire wound power resistor.

E7D15 What is the purpose of a "step-start" circuit in a high voltage power supply?

A. To provide a dual-voltage output for reduced power applications.
B. To compensate for variations of the incoming line voltage.
C. To allow for remote control of the power supply.
D. To allow the filter capacitors to charge gradually.

The modern, high-voltage power supply likely contains a series resistor and capacitor in parallel in the high-voltage plate circuit of the tube in order to buffer the initial in-rush of startup current. This ***allows the filter capacitors to slowly build their charges***, minimizing that dreaded arc sound when you turn on a high-voltage supply that has been turned off for several years! Some ham operators bring up high-voltage on older equipment with a variable AC transformer. **ANSWER D.**

E7D16 When several electrolytic filter capacitors are connected in series to increase the operating voltage of a power supply filter circuit, why should resistors be connected across each capacitor?

A. To equalize, as much as possible, the voltage drop across each capacitor.
B. To provide a safety bleeder to discharge the capacitors when the supply is off.
C. To provide a minimum load current to reduce voltage excursions at light loads.
D. All of these choices are correct.

Those ***resistors*** in parallel across power supply electrolytic filter capacitors help ***equalize voltage drops*** across each of the capacitors, ***act as bleeder resistors*** when the power supply is turned off, and ***add a constant load*** if there is minimal current required out of the power supply circuits. **ANSWER D.**

Electrolytic capacitors work properly for 15 to 20 years, but after that they begin to dry out and need replacement.

E6D15 What is current in the primary winding of a transformer called if no load is attached to the secondary?

A. Magnetizing current.
B. Direct current.
C. Excitation current.
D. Stabilizing current.

With no load on the secondary of a transformer, the device works essentially as a pure inductor or choke. No power is dissipated except for the resistance of the primary winding and some core (or hysteresis) losses, if a core is used. All the ***current is converted into a magnetic field*** instead of usable work. **ANSWER A.**

E6D16 What is the common name for a capacitor connected across a transformer secondary that is used to absorb transient voltage spikes?

A. Clipper capacitor.
B. Trimmer capacitor.
C. Feedback capacitor.
D. Snubber capacitor.

When an AC voltage is rectified, harmonics are always generated as a result of the process. These harmonics can travel through power supply buses and cause interference. In addition, the rectification process can cause "ringing" in a power transformer, allowing excessive peak voltages to build up. A ***snubber network*** gets rid of these undesirable harmonics and affords a degree of protection to the entire device. **ANSWER D.**

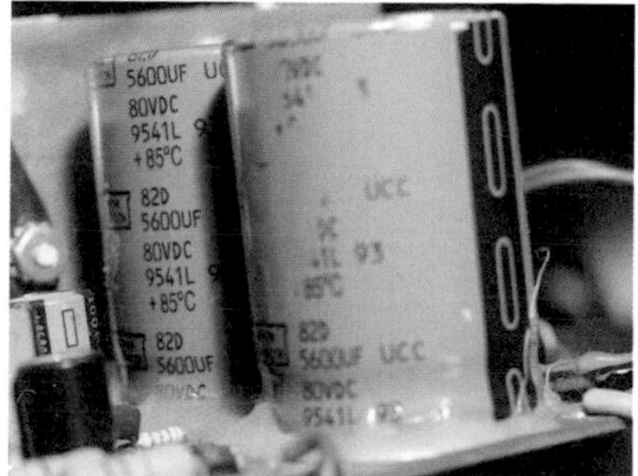

Snubber capacitors help minimize peak voltage spikes and harmonics from the rectifier section of the power supply.

E7D10 What is the primary reason that a high-frequency switching type high voltage power supply can be both less expensive and lighter in weight than a conventional power supply?

A. The inverter design does not require any output filtering.
B. It uses a diode bridge rectifier for increased output.
C. The high frequency inverter design uses much smaller transformers and filter components for an equivalent power output.
D. It uses a large power factor compensation capacitor to create free power from the unused portion of the AC cycle.

Conventional, heavy-transformer power supplies that convert home AC power down to 12 volts DC weigh in at about 2 lbs. for every 12 volt DC amp output. Gordo has plenty of 20 amp power supplies that weigh about 40 lbs. each. These power supplies are characterized by a huge step down transformer, diodes and big filter capacitors. The best efficiency you can hope for is about 50%. And, ohhhh, do they get warm under load! The new power supply technology is called "switcher." These supplies are lightweight, small in size, give off little heat, and offer almost 90% efficiency. The output of a switching power supply (whether converting house power down to DC or switching house power up to various high-voltage levels) is clean and pure thanks to high-speed field-effect transistors taking the place of that big heavy transformer. New switching power supplies also have minimal EMI (RF interference), and some may even take any detectable emitted spur and let you adjust it out of your favorite listening spot. The ***switching power supply*** is about the same cost as the old, big-transformer supplies, and the ***very small transformer and filter components*** easily fit in a relatively small package, either inside or outside the radio equipment. **ANSWER C.**

High-frequency switching power supplies like this one weigh just a couple of pounds! Transformer type power supplies may weigh 10 times as much! But remember to keep inexpensive "no-name" switchers away from coax lines and far away from your antenna system as these produce noise over HF frequencies. Modern name-brand switching power supplies are quiet on HF and VHF/UHF frequencies. The Jetstream power supply pictured here is nice and quiet on HF.

Receivers with Great Filters

E7E12 What is a frequency discriminator stage in an FM receiver?

A. An FM generator circuit.
B. A circuit for filtering two closely adjacent signals.
C. An automatic band-switching circuit.
D. A circuit for detecting FM signals.

Anytime you see the words ***frequency discriminator*** you know that you are dealing with ***signal detection or demodulation*** in an ***FM*** transceiver. **ANSWER D.**

E4C03 What is the term for the blocking of one FM phone signal by another, stronger FM phone signal?

A. Desensitization.
B. Cross-modulation interference.
C. Capture effect.
D. Frequency discrimination.

A unique property of your FM receiver is the ability to hear only the strongest incoming FM signal. If a distant station is transmitting, and a strong local station comes on the air, you will only hear the local station. ***The stronger signal is "capturing" your receiver***. (The reason AM is still used on aircraft VHF radio is specifically to avoid the capture effect. Everybody needs to hear everybody else! This is the last major use of AM in the VHF world.) **ANSWER C.**

E4D12 What is the term for the reduction in receiver sensitivity caused by a strong signal near the received frequency?

A. Desensitization.
B. Quieting.
C. Cross-modulation interference.
D. Squelch gain rollback.

If signals mysteriously come in strong then abruptly get weak and then come back strong again, chances are there is someone nearby transmitting on an adjacent frequency. Their signal is ***desensitizing your receiver***. You can check for adjacent signals with a frequency counter or a spectrum analyzer. **ANSWER A.**

E4D13 Which of the following can cause receiver desensitization?

A. Audio gain adjusted too low.
B. Strong adjacent channel signals.
C. Audio bias adjusted too high.
D. Squelch gain misadjusted.

Adjacent frequency signals may ***overload*** a receiver's front end causing the receiver to momentarily lose sensitivity. **ANSWER B.**

E4D14 Which of the following is a way to reduce the likelihood of receiver desensitization?

A. Decrease the RF bandwidth of the receiver.
B. Raise the receiver IF frequency.
C. Increase the receiver front end gain.
D. Switch from fast AGC to slow AGC.

If you have a dual-band handheld transceiver, you'll notice that the insides are all enclosed in tiny metal cans. This shielding is necessary to keep the receiver from desensitizing on one band when you are transmitting on another band. Shielding is a good way to minimize internal desensitization. Desensitization occurs when nearby signals get into your receiver and cause the gain to be reduced. By ***decreasing the RF bandwidth of the receiver***, you reduce the likelihood of nearby signals affecting your receiver. Only signals that are within the passband of the RF are likely to have an effect. **ANSWER A.**

E7E06 Why is de-emphasis commonly used in FM communications receivers?

A. For compatibility with transmitters using phase modulation.
B. To reduce impulse noise reception.
C. For higher efficiency.
D. To remove third-order distortion products.

De-emphasis in an FM receiver rolls off and attenuates the higher modulating frequency response. This compensates for the ***FM transmitter using phase modulation***, where the transmitter rolls off the lower frequencies. This provides ***compatibility*** in the FM communications systems. **ANSWER A.**

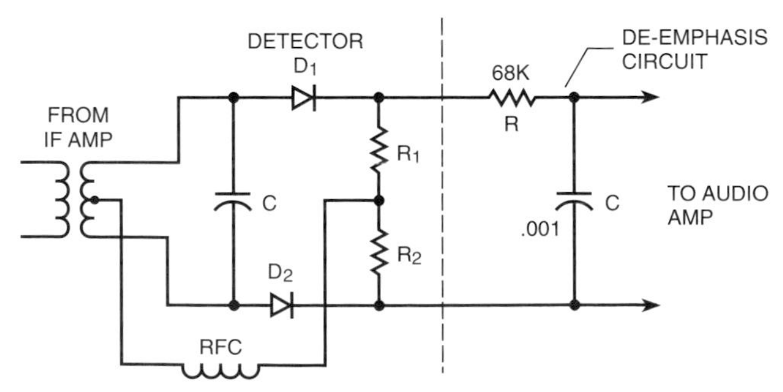

FM De-emphasis Circuit

E4D05 What transmitter frequencies would cause an intermodulation-product signal in a receiver tuned to 146.70 MHz when a nearby station transmits on 146.52 MHz?

A. 146.34 MHz and 146.61 MHz.
B. 146.88 MHz and 146.34 MHz.
C. 146.10 MHz and 147.30 MHz.
D. 173.35 MHz and 139.40 MHz.

To calculate the possible sources of intermodulation interference, we check for the signals that lead to this type of problem that are common at base and repeater stations. There are four calculations to find the intermodulation frequencies that could be causing the problem – two that add frequencies and two that subtract the frequencies. The subtracted calculations are the only ones that produce frequencies close enough to cause significant interference. As the question states, the intermodulation product f_{IMD} is received on 146.70 MHz when a signal at 146.52 MHz is transmitted. The calculations when the frequencies are subtracted are:

$$f_{IMD(2)} = 2f_1 - f_2$$

Rearranging,

$$f_2 = 2f_1 - f_{IMD(2)}$$

$$f_2 = 2(146.52) - 146.70$$

$$f_2 = 146.34 \text{ MHz}$$

$$f_{IMD(4)} = 2f_2 - f_1$$

Rearranging,

$$2f_2 = f_{IMD(4)} + f_1$$

$$f_2 = \frac{f_{IMD(4)} + f_1}{2}$$

$$f_2 = \frac{146.70 + 146.52}{2}$$

$$f_2 = 146.61 \text{ MHz}$$

ANSWER A.

E4D07 Which describes the most significant effect of an off-frequency signal when it is causing cross-modulation interference to a desired signal?

A. A large increase in background noise.
B. A reduction in apparent signal strength.
C. The desired signal can no longer be heard.
D. The off-frequency unwanted signal is heard in addition to the desired signal.

You are ***listening*** to a pal ***on the repeater*** and his ***signal is joined by a simultaneous*** dispatch of Peter's Perfect Plumbing Service. This is a classic example of ***cross-modulation interference*** to your friend's signal. **ANSWER D.**

E4D02 Which of the following describes two problems caused by poor dynamic range in a communications receiver?

A. Cross-modulation of the desired signal and desensitization from strong adjacent signals.
B. Oscillator instability requiring frequent retuning and loss of ability to recover the opposite sideband.
C. Cross-modulation of the desired signal and insufficient audio power to operate the speaker.
D. Oscillator instability and severe audio distortion of all but the strongest received signals.

In a receiver that does not have good dynamic range, ***cross-modulation*** of the desired signal can be encountered, and the receiver can be ***desensitized*** by strong adjacent frequency signals. **ANSWER A.**

E4C05 What does a value of -174 dBm/Hz represent with regard to the noise floor of a receiver?

A. The minimum detectable signal as a function of receive frequency.
B. The theoretical noise at the input of a perfect receiver at room temperature.
C. The noise figure of a 1 Hz bandwidth receiver.
D. The galactic noise contribution to minimum detectable signal.

Noise occurs any time objects are warmed above absolute zero. If you had a perfect receiver with no internal noise, it would still detect noise when connected to an antenna. ***At room temperature, that noise level is about -174 dBm*** for every 1 Hz of bandwidth. In addition to the natural noise, man-made noise also occurs and is much more prevalent at some frequencies than others. On HF bands, man-made noise is so common that a receiver with a very good noise floor is not nearly as important as it would be for a receiver used at 10 GHz trying to detect an EME signal. **ANSWER B.**

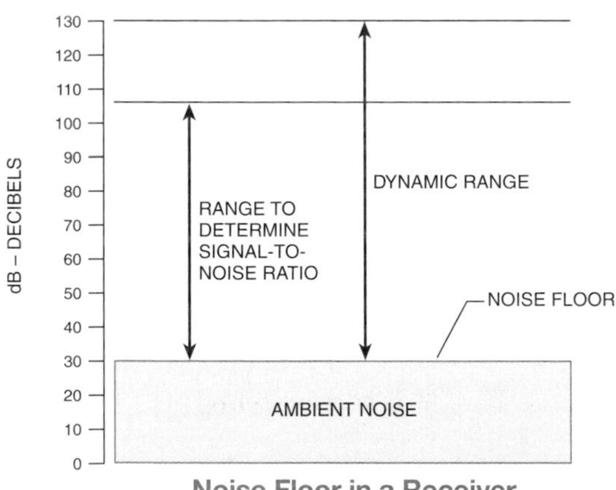

Noise Floor in a Receiver
Source: *Installing and Maintaining Sound Systems*, G. McComb, ©1996, Master Publishing, Inc.

Modern HF receivers are so sensitive that their internal noise floor can't be appreciated due to the many external noises picked up from your own home electronics.

E4C06 A CW receiver with the AGC off has an equivalent input noise power density of -174 dBm/Hz. What would be the level of an unmodulated carrier input to this receiver that would yield an audio output SNR of 0 dB in a 400 Hz noise bandwidth?

A. -174 dBm.
B. -164 dBm.
C. -155 dBm.
D. -148 dBm.

The weakest signal a receiver can receive is one that is just above the total of all noise in the receiver. The background noise the receiver will see at room temperature is -174 dBm/Hz. Add the noise caused by the receiver's bandwidth to the noise floor of the receiver and you have the total noise. The receiver described in this question matches the -174 dBm/Hz noise of the background, so it is a theoretically-perfect receiver, one with no noise floor of its own. Thus, the weakest signal it can receive is -174 + 10 log (bandwidth) = noise floor = -174 + 10 log (400) = ***-174 + 26 dBm = -148 dBm.*** **ANSWER D.**

E4C07 What does the MDS of a receiver represent?

A. The meter display sensitivity.
B. The minimum discernible signal.
C. The multiplex distortion stability.
D. The maximum detectable spectrum.

The *Minimum Discernible Signal* is tested in the SSB or CW mode with internal RF gain full ON. Modern HF equipment has achieved extraordinarily low noise floors through product designs that minimize phase noise. **ANSWER B.**

E6E05 Which of the following noise figure values is typical of a low-noise UHF preamplifier?

A. 2 dB.
B. -10 dB.
C. 44 dBm.
D. -20 dBm.

At UHF frequencies and above, the sensitivity of a receiver is generally limited by the internal noise generated by the front end amplifying device or devices. This is because any noise generated in the front end along with the desired signal is amplified by all the subsequent stages. The formula for Noise Figure is:

$$NF = 10\log_{10}(F) = 10\log_{10}\left(\frac{SNR_{in}}{SNR_{out}}\right) = SNR_{in,dB} = SNR_{out,dB}$$

where SNR is *signal to noise* ratio. When all the number crunching is done, we'll find that a pretty decent front end will have a noise figure of about ***2 dB***. **ANSWER A.**

E7C15 What is a crystal lattice filter?

A. A power supply filter made with interlaced quartz crystals.
B. An audio filter made with four quartz crystals that resonate at 1 kHz intervals.
C. A filter with wide bandwidth and shallow skirts made using quartz crystals.
D. A filter with narrow bandwidth and steep skirts made using quartz crystals.

The new worldwide transceiver you purchase will probably have some good SSB and CW filters built in. The filters are made of ***quartz crystals*** and have a nice ***narrow bandwidth and steep skirts***. You can buy accessory filters for a tighter response, but many times these will make your received signals sound pinched. Unless you do a lot of CW work, stay with the filters that come with the set. While crystal *lattice* and *ladder* networks are similar in purpose, they have different topologies. Lattice networks use pairs of crystals where the series resonant mode of one crystal is matched to the parallel resonant mode of the second one, with the combined output forming the desired bandpass shape. **ANSWER D.**

E7C08 Which of the following factors has the greatest effect in helping determine the bandwidth and response shape of a crystal ladder filter?

A. The relative frequencies of the individual crystals.
B. The DC voltage applied to the quartz crystal.
C. The gain of the RF stage preceding the filter.
D. The amplitude of the signals passing through the filter.

The ***bandwidth and response shape of a crystal ladder filter*** will be determined by the ***frequency of the individual crystals*** within that filter unit. A quartz crystal has an *equivalent* circuit to a resonant coil and capacitor circuit. While a quartz crystal has both a parallel and a series resonant mode, it is generally the series resonant mode that is used in a ladder filter. Several crystals can be strung together in a ladder configuration to form the equivalent of a filter circuit using a great number of coils and capacitors, but with much higher Q factors. **ANSWER A.**

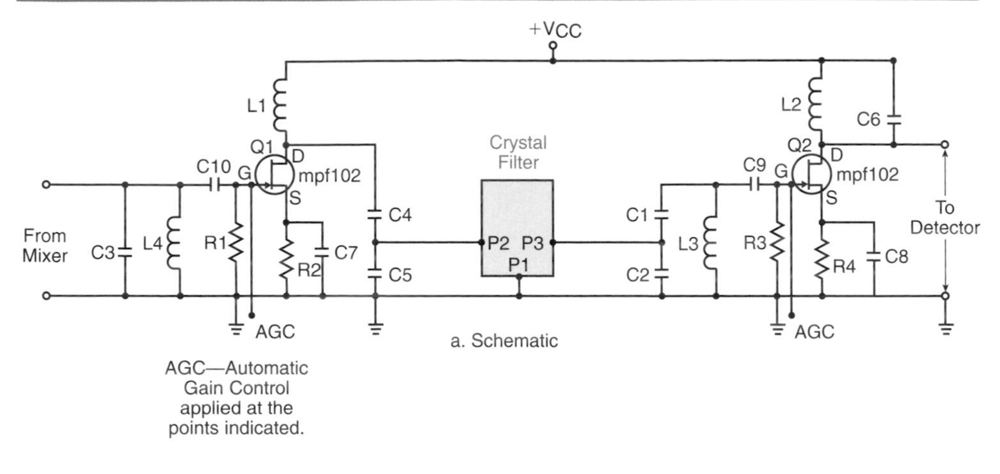

a. Schematic

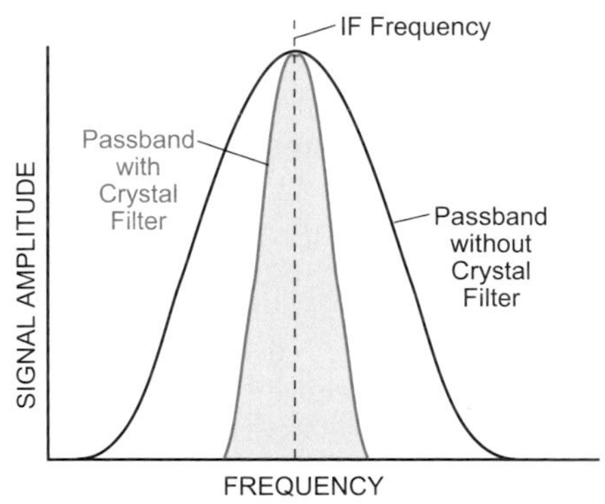

b. Effect of Crystal Filter on Passband

Cascaded IF amplifiers. The active devices are JFETs. The value of most components depends upon the frequency of operation.

Source: *Basic Communications Electronics*, Hudson & Luecke, © 1999 Master Publishing, Inc., Niles, IL

E7C05 Which filter type is described as having ripple in the passband and a sharp cutoff?

A. A Butterworth filter.
B. An active LC filter.
C. A passive op-amp filter.
D. A Chebyshev filter.

Look at the word "***Chebyshev***." It almost looks like it has a ***ripple***, doesn't it? **ANSWER D.**

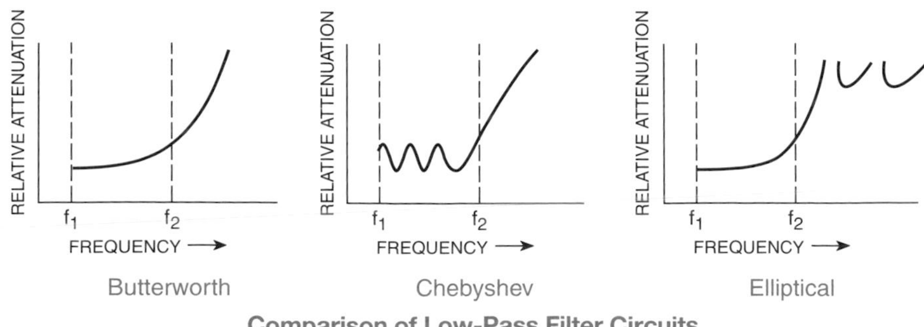

Comparison of Low-Pass Filter Circuits

E7C06 What are the distinguishing features of an elliptical filter?

A. Gradual passband rolloff with minimal stop band ripple.
B. Extremely flat response over its pass band with gradually rounded stop band corners.
C. Extremely sharp cutoff with one or more notches in the stop band.
D. Gradual passband rolloff with extreme stop band ripple.

Different types of filters fulfill different priorities with regard to the *cutoff* or *bandpass* response. In audio applications, for instance, it's important to have very smooth transitions between bandpass and bandstop frequencies, with no abrupt phase changes. A Butterworth filter is commonly used for audio equalization. For radio frequency work, it's generally important to have abrupt cutoff response, consistent with not too much ripple within the desired bandpass. The ***elliptical filter*** has the best performance with regard to ***sharpness of cutoff***, but can have significant ripple in the bandpass. A less extreme response can be achieved with a Chebychev type filter. Often, different types of filters are used in *cascade* to achieve the best overall response. **ANSWER C.**

E4C12 What is an undesirable effect of using too wide a filter bandwidth in the IF section of a receiver?

A. Output-offset overshoot.
B. Filter ringing.
C. Thermal-noise distortion.
D. Undesired signals may be heard.

Using an AM 10 kHz IF filter for pulling in 2.3 kHz SSB signals would result in ***receiving undesired signals*** that lie beyond the normal bandwidth of the desired signal. **ANSWER D.**

E4C10 Which of the following Is a desirable amount of selectivity for an amateur RTTY HF receiver?

A. 100 Hz.
B. 300 Hz.
C. 6000 Hz.
D. 2400 Hz.

For ***RTTY, 300 Hz is good selectivity***. **ANSWER B.**

E4C11 Which of the following is a desirable amount of selectivity for an amateur SSB phone receiver?

A. 1 kHz.
B. 2.4 kHz.
C. 4.2 kHz.
D. 4.8 kHz.

For ***SSB, a 2.4-kHz*** filter bandwidth offers good fidelity audio. **ANSWER B.**

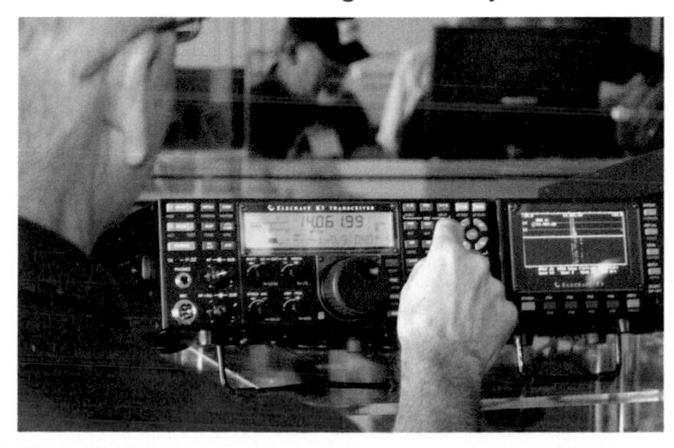

Set your SSB selectivity to 2.4 kHz for a natural sounding incoming voice signal. When set at 1.8 kHz, incoming signals lack a natural bass response.

E4C13 How does a narrow-band roofing filter affect receiver performance?

A. It improves sensitivity by reducing front end noise.
B. It improves intelligibility by using low Q circuitry to reduce ringing.
C. It improves dynamic range by attenuating strong signals near the receive frequency.
D. All of these choices are correct.

Roofing filters in the IF minimize contesting receiver "swamping" when you have another ham running a kilowatt to chat with his buddy, well up the band. IF stage roofing filters, like automatically-selected 15 kHz, 6 kHz, and 3 kHz filters, may feature 4-pole fundamental-mode monolithic crystal filter design to produce an ideal shape factor. The filters usually are positioned after the first mixer and significantly improve the 3rd-order intercept point performance for all stages that follow. Gordo appreciates the performance of roofing filters on 40 meters early in the morning when a nearby shortwave station numbs his receiver. ***Roofing filters*** eliminate this problem thanks to an ***improvement in the rig's dynamic range***. **ANSWER C.**

E4C02 Which of the following portions of a receiver can be effective in eliminating image signal interference?

A. A front-end filter or pre-selector.
B. A narrow IF filter.
C. A notch filter.
D. A properly adjusted product detector.

An effective means of ***eliminating image signal interference*** is with a ***front end filter or pre-selector filter network***. These filters are much wider than typical IF filters and roofing filters. A higher IF also separates the image further from the incoming signal, thus easing the front end filter requirements. The front end filters are electrically close to the antenna input jack, as opposed to the IF and roofing filters, which are in the middle of the receiver section. The front end filters must be band switched when there is a band change. **ANSWER A.**

E4C09 Which of the following choices is a good reason for selecting a high frequency for the design of the IF in a conventional HF or VHF communications receiver?

A. Fewer components in the receiver.
B. Reduced drift.
C. Easier for front-end circuitry to eliminate image responses.
D. Improved receiver noise figure.

The intermediate frequency (IF) stage in a conventional receiver is where much of the amplification occurs. Selecting a high frequency for the IF allows for improved selectivity with minimum noise. This allows ***easier to build front end circuitry to eliminate image responses*** from nearby strong signals. **ANSWER C.**

E4C14 What transmit frequency might generate an image response signal in a receiver tuned to 14.300 MHz and which uses a 455 kHz IF frequency?

A. 13.845 MHz.
B. 14.755 MHz.
C. 14.445 MHz.
D. 15.210 MHz.

An image frequency of an incoming desired signal may be plus or minus two times the IF. In this example, the incoming signal is at 14.300 MHz, and the local oscillator is at 14.755 MHz. The incoming 14.300 MHz mixes with 14.755 MHz, producing a different intermediate frequency of 455 kHz and so does an incoming frequency of 15.210 MHz. (15.210 - 14.300 = 455 kHz, twice.) 15.210 MHz would be the image frequency. Various technologies, including good RF input tuned circuits, can minimize the image frequency thus minimizing the problem. Again, *14.300 MHz + 0.455 + 0.455 = 15.210 MHz*. You double the IF to come up with the correct answer. **ANSWER D.**

E7C09 What is a Jones filter as used as part of an HF receiver IF stage?

A. An automatic notch filter.
B. A variable bandwidth crystal lattice filter.
C. A special filter that emphasizes image responses.
D. A filter that removes impulse noise.

The Jones filter is a variable-bandwidth crystal filter network, where bandwidth may be changed yet the center frequency remains constant. This filter network allows narrow bandwidth on one side of center and wide bandwidth on the other side. Shunt varactors operate much like a voltage-controlled oscillator allowing for variable bandwidth selection. This filter system is found in TenTec equipment and carries a U.S. patent! *Variable bandwidth* is how we remember the *Jones filter.* **ANSWER B.**

E4C15 What is usually the primary source of noise that is heard from an HF receiver with an antenna connected?

A. Detector noise.
B. Induction motor noise.
C. Receiver front-end noise.
D. Atmospheric noise.

Unlike the case of UHF and above, the limit of useful receiver sensitivity at HF frequencies is determined by factors external to the receiver itself. Two of these factors are thermal agitation noise caused by the random motion of molecules in the antenna wire itself and *atmospheric noise, commonly known as static*. On lower HF bands, atmospheric noise is the main factor, while on quiet days on the upper HF bands, thermal agitation noise may be the limiting factor. **ANSWER D.**

E7F02 What kind of digital signal processing audio filter is used to remove unwanted noise from a received SSB signal?

A. An adaptive filter.
B. A crystal-lattice filter.
C. A Hilbert-transform filter.
D. A phase-inverting filter.

The *adaptive filter* found in digital signal processing (DSP) circuits *removes unwanted noise* from incoming SSB signals. **ANSWER A.**

Modern High Frequency rigs have built-in adaptive DSP filter options. If you have an old HF transceiver, you can add an external amplified DSP speaker and dial in your own level of noise elimination. Listen to this on Gordo's audio CD course.

E4E12 What is one disadvantage of using some types of automatic DSP notch-filters when attempting to copy CW signals?

A. A DSP filter can remove the desired signal at the same time as it removes interfering signals.
B. Any nearby signal passing through the DSP system will overwhelm the desired signal.
C. Received CW signals will appear to be modulated at the DSP clock frequency.
D. Ringing in the DSP filter will completely remove the spaces between the CW characters.

Digital signal processing (DSP) can relieve your eardrums of constant static roar while waiting for a station to come up on the air. DSP subtracts recurring "hash," allowing only voice, data, and CW tones to pass through to the receiver. The voice may sound slightly pinched with DSP turned on, but you won't hear the static roar between each syllable. A good DSP filter will work magic on CW, as long as the operator is sending at medium to moderate speeds – above 5 wpm. Any ***CW signal*** sent slower ***may be interpreted as an annoying heterodyne by the DSP circuit***, and the circuit will take any continuous steady tone and cancel it – not what you want when copying CW. Just tell the operator to speed up their code sending. The modern DSP notch-filter will recognize the spaces between the dots and dashes and bring the signal through clearly. **ANSWER A.**

E7C07 What kind of filter would you use to attenuate an interfering carrier signal while receiving an SSB transmission?

A. A band-pass filter.
B. A notch filter.
C. A Pi-network filter.
D. An all-pass filter.

On a band such as 40 meters, where radio amateurs share the spectrum with many powerful shortwave broadcasters, multiple carriers or heterodynes can make operation challenging, to say the least. The narrow-band, tunable ***notch filter can be used to null out an offending carrier***. Generally this is done in the IF stages of a receiver, but can also be performed with an outboard audio filter unit. Modern software defined radios (SDRs) are capable of performing the notch function automatically, even on several carriers at a time. This is one outstanding feature of digital signal processing that many of us old timers really appreciate! **ANSWER B.**

E4C04 How is the noise figure of a receiver defined?

A. The ratio of atmospheric noise to phase noise.
B. The ratio of the noise bandwidth in Hertz to the theoretical bandwidth of a resistive network.
C. The ratio of thermal noise to atmospheric noise.
D. The ratio in dB of the noise generated by the receiver to the theoretical minimum noise.

For VHF and UHF weak signal work on SSB and CW, you want to operate a receiver with an extremely low noise floor. An extremely low noise floor allows weak signals to be detected above the receiver's internal noise. A ***receiver with a good noise floor*** will yield nearly ***identical noise*** power ***when switched between a dummy load and an antenna***, measured 3 dB S (the desired signal power) + N/N, single tone, at 500 Hz and 2400 Hz. This works out to be a ratio, in dB, of the internal noise generated by the receiver itself compared to the theoretical minimum noise that the best receiver in the world would generate! Up on UHF, where we do a lot of weak signal operation like Moon bounce, this is an important consideration. But down on HF, atmospheric noise would drown out any slight differences in a receiver's own receiver-generated noise, so the noise floor it is a less important ratio to consider in that new HF transceiver! **ANSWER D.**

E4D01 What is meant by the blocking dynamic range of a receiver?

A. The difference in dB between the noise floor and the level of an incoming signal which will cause 1 dB of gain compression.

B. The minimum difference in dB between the levels of two FM signals which will cause one signal to block the other.

C. The difference in dB between the noise floor and the third order intercept point.

D. The minimum difference in dB between two signals which produce third order intermodulation products greater than the noise floor.

When you are working a contest on HF and tune in to a rare station just above the noise floor, you may hear that station request responses to transmit 5 kHz "up." This is contest "split operation" and it leaves the station in the clear without people calling right on top of his ultra weak signal. Powerful stations just a few kilohertz away will sometimes cause the weak station's signal to disappear in the noise, because your receiver is saturated with powerful signals nearby on the band. A receiver with good dynamic range will minimize desensitization when nearby signals blast your front end circuitry. ***Blocking dynamic range*** is measured in decibels at your ***receiver noise floor***, with AGC turned off, and a ***nearby signal*** that leads to ***1 dB of gain compression in the receiver***, causing that weak signal to dip slightly into the noise. Blocking dynamic range tests are more fairly calculated at 100 kHz off from the desired signal. **ANSWER A.**

E4D09 What is the purpose of the preselector in a communications receiver?

A. To store often-used frequencies.

B. To provide a range of AGC time constants.

C. To increase rejection of unwanted signals.

D. To allow selection of the optimum RF amplifier device.

You ask why those big contest HF transceivers are so large? On the inside, they may contain high-Q, motor-driven preselector coils that can automatically track, or may be manually adjusted, to your desired receive frequency. These preselector coils may achieve a Q of over 300 and provide steep passband filtering not available with smaller fixed components or older external bandpass boxes. The peak of these optional built-in ***preselector*** modules may even be adjusted away from your frequency to provide added ***isolation from a specific adjacent frequency*** transmitting station. The Q is so high that each preselector on each band needs its own micro-motor drive assembly. But for big contesters, the big radio can really cut through the QRM. **ANSWER C.**

E4D10 What does a third-order intercept level of 40 dBm mean with respect to receiver performance?

A. Signals less than 40 dBm will not generate audible third-order intermodulation products.
B. The receiver can tolerate signals up to 40 dB above the noise floor without producing third-order intermodulation products.
C. A pair of 40 dBm signals will theoretically generate a third-order intermodulation product with the same level as the input signals.
D. A pair of 1 mW input signals will produce a third-order intermodulation product which is 40 dB stronger than the input signal.

If you input two tones that are f Hz apart to the receiver input, the mixer will output not only those two tones shifted down to the intermediate frequency but also third-order intermodulation tones at f Hz above and below the original tones. At normal input signal levels, these third-order tones are much weaker than the original two tones. But as you increase the amplitude of the original two tones, the ***third-order tones*** increase in amplitude three times faster than the original tones. If you continue to increase the input tone signal levels, at some point the ***signal strength of the input tones and the third-order tones will be equal. This is called the third order intercept point***. A third-order intercept point of 40 dBm means that, if the input tones were 40 dBm, then the third order tones would also be 40 dBm. The 3rd order intercept point is not achievable in actual practice, as the receiver will saturate long before it reaches that point. **ANSWER C.**

E4D11 Why are third-order intermodulation products created within a receiver of particular interest compared to other products?

A. The third-order product of two signals which are in the band of interest is also likely to be within the band.
B. The third-order intercept is much higher than other orders.
C. Third-order products are an indication of poor image rejection.
D. Third-order intermodulation produces three products for every input signal within the band of interest.

Your new high frequency transceiver boasts extraordinary receiver sensitivity, ultra-tight selectivity, and a 130 dBm third order intercept point. Third order intermodulation could occur in the receiver IF or RF stages when a pair of contest stations near you are hammering your "front end" just a few kHz away. Improved equipment designs minimize the ***third order "phantom signals"*** that are non-desirable weak signal products from these other two stations, and are ***strong enough to interfere with the station you are trying to tune in***. Not only are you hammered by two strong signals of interference, but also artifacts of their clean transmitted signals. Physically larger inductors in the band pass filter stage may help, along with low-distortion diodes in the band pass frequency switching circuitry. Monolithic roofing filters, more expensive than overtone mode filters, will offer a better shape factor and are less susceptible to intermodulation distortion within your top quality transceiver's RF and IF stages. If you are considering a contest-capable transceiver to feed into your 100-foot-long boom HF antenna, the multi-thousand-dollar high end HF transceivers offer the ultimate in the rejection of third order phantom signals with a distortion-free, high-dynamic-range receiver system. **ANSWER A.**

E7E10 How does a diode detector function?

A. By rectification and filtering of RF signals.
B. By breakdown of the Zener voltage.
C. By mixing signals with noise in the transition region of the diode.
D. By sensing the change of reactance in the diode with respect to frequency.

Since a ***diode*** only conducts for half of the AC signal, it ***may be used as a rectifier***. By ***filtering out the radio frequency energy*** after rectification, detection is accomplished – the modulated signal is what remains. **ANSWER A.**

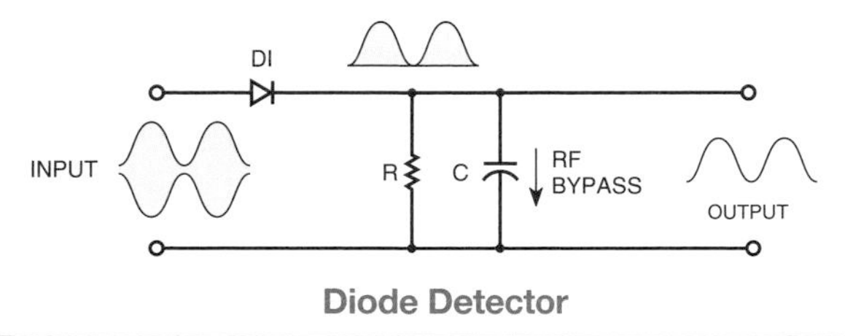

Diode Detector

E7E11 Which type of detector is used for demodulating SSB signals?

A. Discriminator.
B. Phase detector.
C. Product detector.
D. Phase comparator.

A ***product detector*** is found in ***SSB*** receivers. It mixes the incoming signal with a beat frequency oscillator signal that is a locally-generated carrier, which is mixed with the incoming signal. **ANSWER C.**

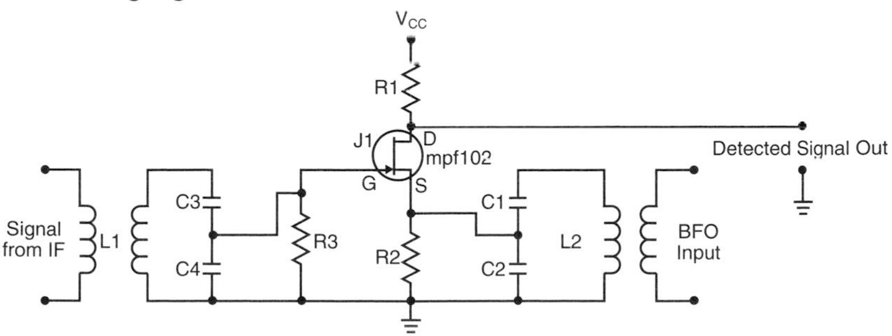

Product detector circuit used for SSB.

E7C14 Which mode is most affected by non-linear phase response in a receiver IF filter?

A. Meteor scatter.
B. Single-Sideband voice.
C. Digital.
D. Video.

Non-linear phase response within a receiver's IF filter is seldom noticed in CW, SSB, or video modes. Unfortunately, complex ***digital modes are most affected by non-linear phase response*** where individual data pulses incur unwanted shifts of time, disrupting the smooth flow of data within the IF filter. While most filters may produce a small amount of phase distortion, all but data can work through this problem. **ANSWER C.**

E7E08 What are the principal frequencies that appear at the output of a mixer circuit?

A. Two and four times the original frequency.
B. The sum, difference and square root of the input frequencies.
C. The two input frequencies along with their sum and difference frequencies.
D. 1.414 and 0.707 times the input frequency.

The mixer produces your ***original two frequencies and the sum and difference frequencies***. The more elaborate the transceiver, the more mixing stages in the set. Mixing stages help filter out unwanted or phantom signals that could cause interference. **ANSWER C.**

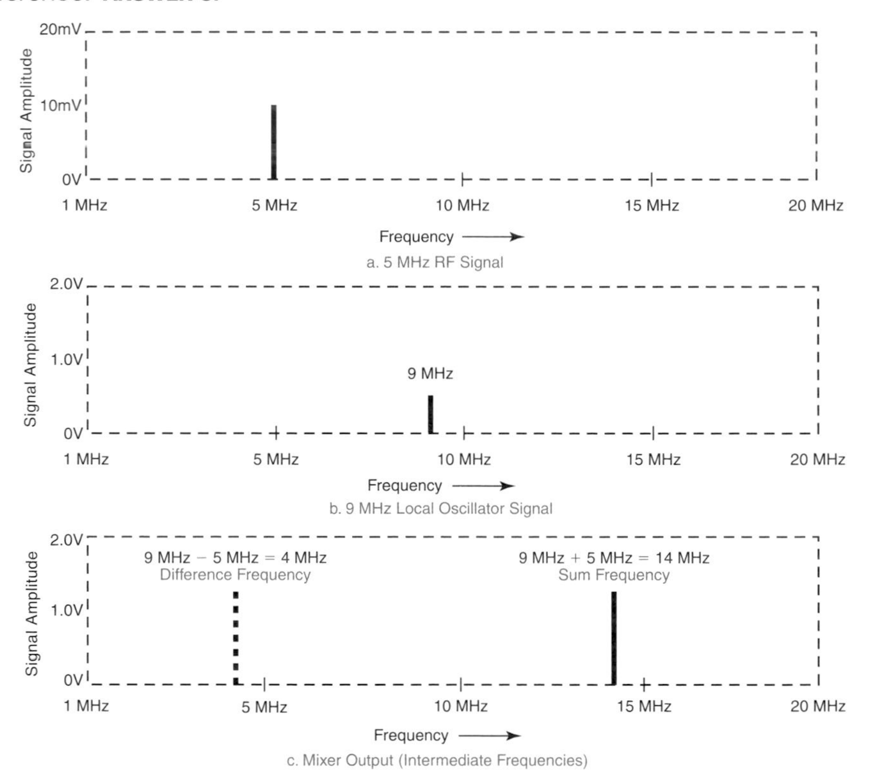

Frequency spectrum plot of mixer signals.
Source: *Basic Communications Electronics*, Hudson & Luecke, © 1999 Master Publishing, Inc., Niles, IL

E7E09 What occurs when an excessive amount of signal energy reaches a mixer circuit?

A. Spurious mixer products are generated.
B. Mixer blanking occurs.
C. Automatic limiting occurs.
D. A beat frequency is generated.

Some VHF power amplifiers may incorporate a hefty pre-amplifier circuit for use on receiving that boosts incoming signal levels. When the pre-amp is turned on, it could generate ***spurious mixer products as a result of the high signal level***. The spurious products are not really signals that you are trying to receive. If you use a handheld with a power amp, turn the pre-amp off to avoid these unwanted signals. **ANSWER A.**

E4E01 Which of the following types of receiver noise can often be reduced by use of a receiver noise blanker?

A. Ignition noise.
B. Broadband white noise.
C. Heterodyne interference.
D. All of these choices are correct.

Going mobile with your HF transceiver? A good *noise blanker will help minimize ignition noise* from recurring spark plug "pops." A manual repetition rate blanker or, better yet, an automatic noise blanker will synch with the plug "pops" and cancel them almost completely – along with noise from other sources. However, the noise blanker may result in any signal over S-9 sounding distorted. **ANSWER A.**

E4E03 Which of the following signals might a receiver noise blanker be able to remove from desired signals?

A. Signals which are constant at all IF levels.
B. Signals which appear across a wide bandwidth.
C. Signals which appear at one IF but not another.
D. Signals which have a sharply peaked frequency distribution.

The *noise blanker* works best on repetitive, correlated, *wide-bandwidth noise* like sparkplug noise, motor noise, and pesky home light dimmers. **ANSWER B.**

E4E02 Which of the following types of receiver noise can often be reduced with a DSP noise filter?

A. Broadband white noise.
B. Ignition noise.
C. Power line noise.
D. All of these choices are correct.

Most high frequency transceivers now include a digital signal processing filter. This filter will magically subtract broadband white noise, help minimize power line noise, may cancel out a steady frequency heterodyne signal, and help to reduce ignition noise. *All of these may be addressed with a DSP noise filter*. But, you know, We always recommend solving the noise problem at the source, rather than trying to mask it, or subtract it, with DSP. **ANSWER D.**

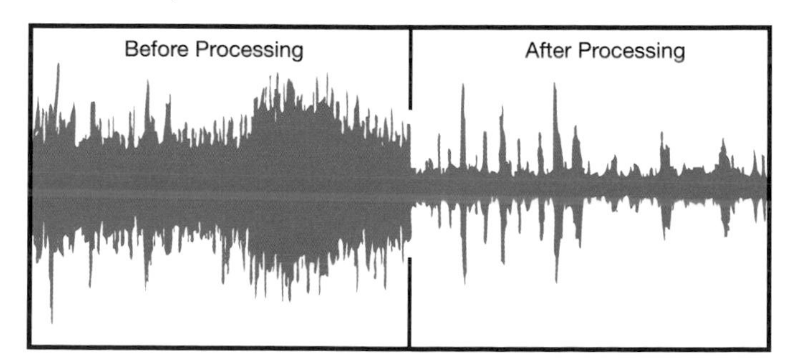

Car noise removed from hands-free communication.

Look at the improvement in reception when a DSP circuit is turned on, and the background white noise all but disappears, to reveal signals on the band! DSP noise elimination works well on atmospheric white noise.

E4E07 How can you determine if line noise interference is being generated within your home?

A. By checking the power line voltage with a time domain reflectometer.
B. By observing the AC power line waveform with an oscilloscope.
C. By turning off the AC power line main circuit breaker and listening on a battery operated radio.
D. By observing the AC power line voltage with a spectrum analyzer.

Home office equipment like fax machines or telephones with intercom, fish tank heaters, and ultrasonic bug repellers all can create whistles and rhythmical pulsing sounds on both HF and VHF/UHF station outputs. To find out whether the noise is being generated by your own home electronics, ***shut off your AC power main and continue listening as you run your rig off of a 12-volt battery source***. If the noise instantly disappears, use your 2-meter handi-talkie and start "sniffing" for the noise source when the power is turned on again. Of course, many home appliances will blink 12:00 after you turn the power back on! **ANSWER C.**

You can check your own house for noise by putting your HF rig on a battery and then shutting off circuit breakers to see where in the house your noise polluter is located!

E4E13 What might be the cause of a loud roaring or buzzing AC line interference that comes and goes at intervals?

A. Arcing contacts in a thermostatically controlled device.
B. A defective doorbell or doorbell transformer inside a nearby residence.
C. A malfunctioning illuminated advertising display.
D. All of these choices are correct.

With your new, big, directional HF antenna taking advantage of your new Extra Class privileges, you may pick up noise in one specific direction, likely when you're aimed at the local shopping center. ***All of these*** answers ***can generate a variety of interference*** that may come and go at intervals, just like that flashing neon sign down the street. There is no simple filter to cancel out many of these nuisance noise sources, other than aiming your beam to a point where the noise sources are minimized. **ANSWER D.**

E4E14 What is one type of electrical interference that might be caused by the operation of a nearby personal computer?

A. A loud AC hum in the audio output of your station receiver.
B. A clicking noise at intervals of a few seconds.
C. The appearance of unstable modulated or unmodulated signals at specific frequencies.
D. A whining type noise that continually pulses off and on.

Personal computers and other digital equipment use square-wave signals called ***"clocks"*** that run at very high frequencies. These clocks can cause problems for radio receivers because they are composed of all the odd harmonics of the fundamental clock frequency. As you tune your receiver up and down the band, you can expect to find these harmonics. They may be ***unstable*** because clocks are turned on and off as different operations take place in the computer, and they may appear modulated when they are affected by other signals or operations of the computer. If you suspect interference from a digital device, find a frequency where the interference is happening and then simply turn off the suspected device. If the interference goes away, you've identified the culprit. To reduce the noise from such devices, try moving them farther away from the receiver, or try adding shielding around the device to reduce the level of interference. **ANSWER C.**

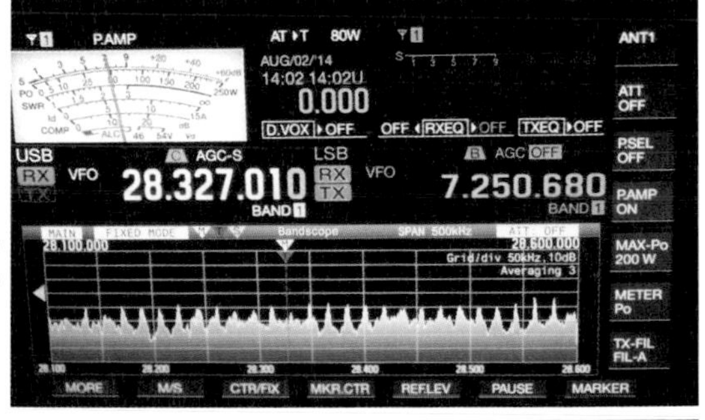

See the series of pulses on the spectrum scope, way up on 10 meters and 6 meters? You are seeing and hearing "birdies" generated by your home electronics.

E4E10 What is a common characteristic of interference caused by a touch controlled electrical device?

A. The interfering signal sounds like AC hum on an AM receiver or a carrier modulated by 60 Hz hum on a SSB or CW receiver.
B. The interfering signal may drift slowly across the HF spectrum.
C. The interfering signal can be several kHz in width and usually repeats at regular intervals across an HF band.
D. All of these choices are correct.

If you or your neighbor has a *touch lamp or an electronic light dimmer* control, be prepared to hear *pulse noise* created by these circuits. Sometimes the touch lamp interference will *drift*, sometimes it will have a *hum* on it, and it is *usually wide and repeats itself* up the entire HF band. The touch circuit usually radiates the energy directly through the air, so there is no magic filter to quiet the QRN. Buy your neighbor a new type of lamp control, and ban touch lamps from your home QTH! Oh yes, one more thing – don't be surprised when you operate on 40 meters and these turned off lamp circuits switch on with your transmissions! **ANSWER D.**

E4E09 What undesirable effect can occur when using an IF noise blanker?

A. Received audio in the speech range might have an echo effect.
B. The audio frequency bandwidth of the received signal might be compressed.
C. Nearby signals may appear to be excessively wide even if they meet emission standards.
D. FM signals can no longer be demodulated.

For mobile applications, a high frequency transceiver will incorporate an IF noise blanker as well as digital noise reduction (DNR), which is part of the transceiver's digital signal processing. Some transceivers offer blanker on or blanker off with no adjustable amount of noise blanking. Others offer variable noise blanker settings. This is better. It is possible for the *IF noise blanker to cause a very strong signal to sound distorted* when the noise blanker is set to maximum. It may also cause strong signals up to 30 kHz away to begin "chattering" your audio circuit on a frequency that is absolutely open. You may be tempted to tell the offending station up frequency that they are "splattering," but the actual problem is not their transmitter (well within specs), but the fact that you left your noise blanker turned on. **ANSWER C.**

E4E11 Which is the most likely cause if you are hearing combinations of local AM broadcast signals within one or more of the MF or HF ham bands?

A. The broadcast station is transmitting an over-modulated signal.
B. Nearby corroded metal joints are mixing and re-radiating the broadcast signals.
C. You are receiving sky wave signals from a distant station.
D. Your station receiver IF amplifier stage is defective.

Your modern, high frequency transceiver employs filter networks, many specifically tuned to minimize powerful double sideband local AM broadcast signals coming in on your lower ham bands. But don't blame the AM broadcaster just yet – ***corroded contacts*** on your tower or aluminum rain gutters might ***re-radiate the broadcast station signals***. This is why many hams with big towers use multiple insulators in their guy lines to minimize rectification of broadcast band signals due to corroded metal joints. **ANSWER B.**

E7F01 What is meant by direct digital conversion as applied to software defined radios?

A. Software is converted from source code to object code during operation of the receiver.
B. Incoming RF is converted to a control voltage for a voltage controlled oscillator.
C. Incoming RF is digitized by an analog-to-digital converter without being mixed with a local oscillator signal.
D. A switching mixer is used to generate I and Q signals directly from the RF input.

Direct Digital Conversion is a method that has only been technologically feasible in very recent years, as ***the analog-to-digital sampling process is done right at the antenna input*** at radio frequencies. This method allows for extremely high performance and versatility, but is subject to all the limitations of an A-to-D sampling scheme, such as aliasing. **ANSWER C.**

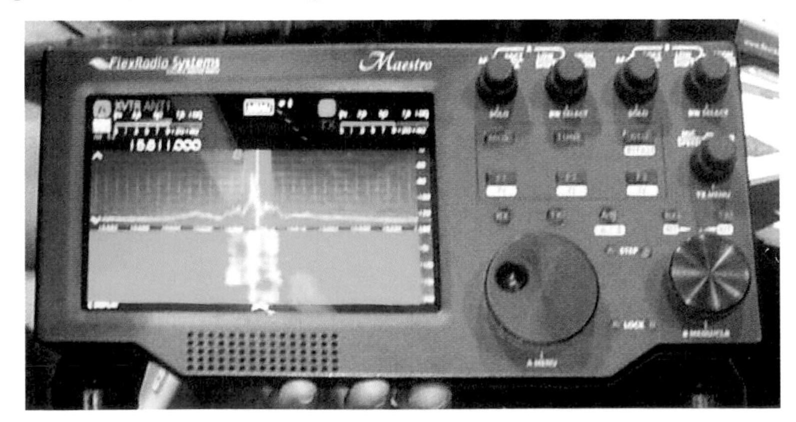

This HF radio head works Wi-Fi to the SDR black box in your shack. Great for lap top operation!

E8A08 Why would a direct or flash conversion analog-to-digital converter be useful for a software defined radio?

A. Very low power consumption decreases frequency drift.
B. Immunity to out of sequence coding reduces spurious responses.
C. Very high speed allows digitizing high frequencies.
D. All of these choices are correct.

Besides being the simplest kind of A-to-D converter to understand, the flash A-to-D converter is fast, hence its name. In the ***flash A-to-D converter, every bit is digitized*** at the same time. **ANSWER C.**

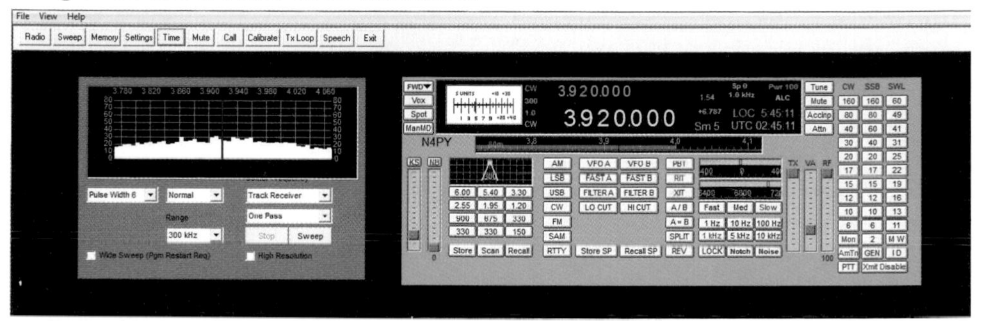

Software Defined Radio display.

E4C08 An SDR receiver is overloaded when input signals exceed what level?

A. One-half the maximum sample rate.
B. One-half the maximum sampling buffer size.
C. The maximum count value of the analog-to-digital converter.
D. The reference voltage of the analog-to-digital converter.

Unlike your conventional analog receiver, which tends to ***overload*** gradually in the presence of strong signals, your SDR receiver will hit a "brick wall" in the presence of overloading. The results of this are obvious and ugly. ***The A-to-D converter*** in your receiver's front end has nowhere to go when the "bit bucket" gets full, and ***simply ceases to function***. Happily, this state of being is only temporary, and as soon as you attenuate the signal to a proper level, your SDR receiver will be its old wonderful self again. **ANSWER C.**

E4C16 Which of the following is caused by missing codes in an SDR receiver's analog-to-digital converter?

A. Distortion.
B. Overload.
C. Loss of sensitivity.
D. Excess output level.

The binary system, just like our more familiar decimal system, relies on *place value*. Some bits are more *valuable* than others and will be much more sorely missed if something goes awry. A missing least significant bit (LSB) may not be noticeable; while a missing most significant bit (MSB) would most likely result in utterly unreadable data. In an SDR, ***a missing MSB would cause severe distortion*** at the very least. **ANSWER A.**

E8A04 What is "dither" with respect to analog to digital converters?

A. An abnormal condition where the converter cannot settle on a value to represent the signal.
B. A small amount of noise added to the input signal to allow more precise representation of a signal over time.
C. An error caused by irregular quantization step size.
D. A method of decimation by randomly skipping samples.

In one of life's strange paradoxes, we can reduce noise in a digital sampler by introducing noise! In reality, this simply converts one type of noise into a much less obnoxious form. ***Adding a tiny bit of random noise*** overcomes the natural hysteresis in any sampling system. **ANSWER B.**

E4C17 Which of the following has the largest effect on an SDR receiver's linearity?

A. CPU register width in bits.
B. Anti-aliasing input filter bandwidth.
C. RAM speed used for data storage.
D. Analog-to-digital converter sample width in bits.

As in any digitizing process, ***the more bits***, "decimal places," one has to work with, ***the more faithful the reproduction of the original signal***. Fewer bits means more jagged steps in the waveform. Of course, accuracy has its price in the name of processing speed and power. The number of bits you use also determines the dynamic range, which is the difference between the smallest detectable signal and the overload point. **ANSWER D.**

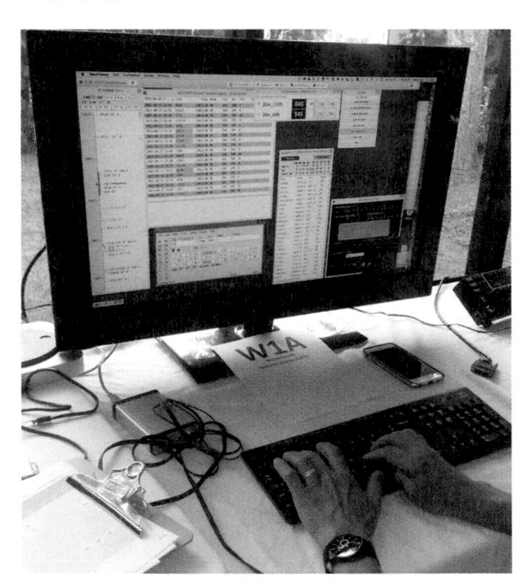

One color big screen does it all, from logging contacts, to checking propagation charts, checking for duplicate contacts, to working the radio's controls – all from one small SDR box hidden below the console.

Oscillate & Synthesize This!

E7H01 What are three oscillator circuits used in Amateur Radio equipment?

A. Taft, Pierce and negative feedback.
B. Pierce, Fenner and Beane.
C. Taft, Hartley and Pierce.
D. Colpitts, Hartley and Pierce.

Colpitts oscillators have a capacitor, just like C in the name. ***Hartley*** is tapped, and a ***Pierce*** oscillator uses a crystal. **ANSWER D.**

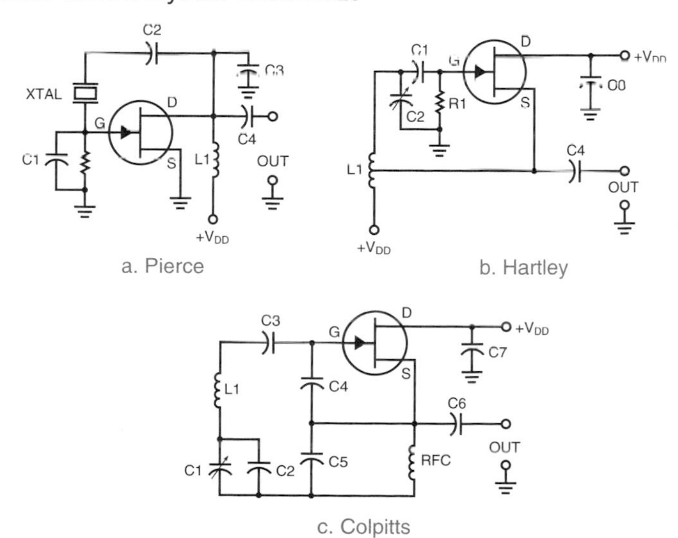

Three major types of Oscillators

E7H03 How is positive feedback supplied in a Hartley oscillator?

A. Through a tapped coil.
B. Through a capacitive divider.
C. Through link coupling.
D. Through a neutralizing capacitor.

Hartley is always tapped. The ***tapped coil*** provides inductive coupling for positive feedback. Look at the schematic above. **ANSWER A.**

E7H04 How is positive feedback supplied in a Colpitts oscillator?

A. Through a tapped coil.
B. Through link coupling.
C. Through a capacitive divider.
D. Through a neutralizing capacitor.

In the ***Colpitts*** oscillator, we use a ***capacitive divider*** to provide feedback. Look at the schematic on page 149. **ANSWER C.**

E7H05 How is positive feedback supplied in a Pierce oscillator?

A. Through a tapped coil.
B. Through link coupling.
C. Through a neutralizing capacitor.
D. Through a quartz crystal.

The ***Pierce*** oscillator uses a ***quartz crystal*** to obtain positive feedback. Look at the schematic on page 149. **ANSWER D.**

Each of these oscillators has its particular place in amateur radio circuitry. Hartley oscillators are often used in the local oscillator circuits of radio receivers, in BFO circuits, and some audio oscillators. They are capable of fairly wide tuning ranges.

The Colpitts oscillator is used in higher frequency circuits such as VFOs where stability becomes more important and a bit harder to achieve than with a Hartley. A Colpitts VFO, however, requires a ganged variable capacitor and can have a tendency to "quit" at certain frequencies if the ratio of capacitance between the two variable capacitor sections is not maintained uniformly across the tuning range. Therefore, most successful Colpitts VFO designs restrict the tuning range.

The Pierce oscillator is the simplest of all crystal oscillators and is used when the ultimate in frequency stability is required. However, the Pierce oscillator is not tunable.

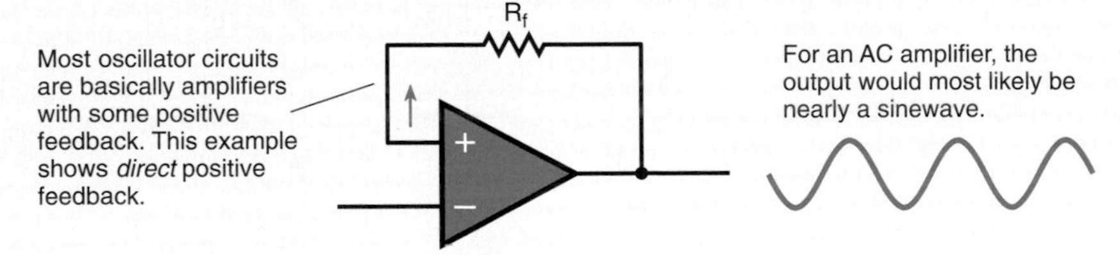

a. Amplifier with Positive Feedback – Oscillator

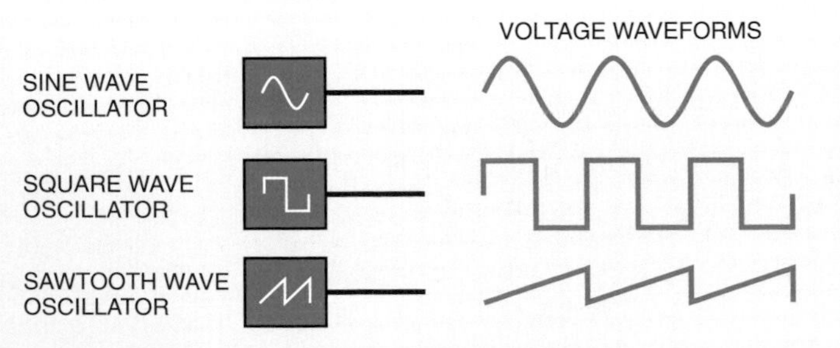

b. Oscillator Waveforms

Oscillators – An oscillator is basically an amplifier with positive feedback from output to input.

Source: *Basic Electronics* © 1994, Master Publishing, Inc., Niles, Illinois

E7H06 Which of the following oscillator circuits are commonly used in VFOs?

A. Pierce and Zener.
B. Colpitts and Hartley.
C. Armstrong and deforest.
D. Negative feedback and balanced feedback.

Since the ***Colpitts*** oscillator uses a big capacitor in a ***variable frequency oscillator*** (VFO), it's the most common oscillator circuit for older VFO radios. Newer radios, even though they say they have a VFO, are really digitally-controlled via an optical reader. They just behave as though they have a large capacitor behind that big tuning dial! **ANSWER B.**

E7H12 Which of the following must be done to insure that a crystal oscillator provides the frequency specified by the crystal manufacturer?

A. Provide the crystal with a specified parallel inductance.
B. Provide the crystal with a specified parallel capacitance.
C. Bias the crystal at a specified voltage.
D. Bias the crystal at a specified current.

There is plenty to influence the ***precise frequency*** of a crystal in the oscillator circuit – crystal impedance, crystal aging, the method of holding the crystal in place inside the can, temperature and, finally, the ***crystal's natural parallel*** or ***holder capacitance*** (over which you have no control), usually in the order of 6 pF. Your circuit's parallel capacitance, which you can control usually by means of a trimmer, is another influence to consider. **ANSWER B.**

Quartz crystals are still found in modern, synthesized ham radios. Some are enclosed in heated metal chambers to keep them temperature compensated for spot-on digital signal work. These two quartz crystals feed reference signals to various oscillator stages in the transceiver.

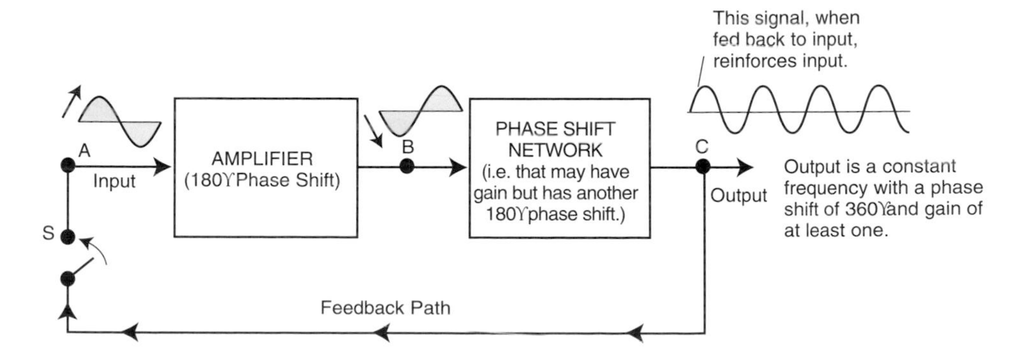

An oscillator outputs a signal of constant frequency.
Source: Basic *Communications Electronics*, Hudson & Luecke, © 1999 Master Publishing, Inc., Niles, IL

E7H08 Which of the following components can be used to reduce thermal drift in crystal oscillators?

A. NP0 capacitors.
B. Toroidal inductors.
C. Wirewound resistors.
D. Non-inductive resistors.

Well, this wobbles some old nostalgia glands. Many of us homebrewers spent many hours attempting to stabilize oscillators by means of ***temperature compensating capacitors***. This is almost a lost art. An ***NP0 capacitor*** is one that has a *neutral* thermal coefficient; it drifted neither higher nor lower in capacitance as the temperature changed. NP0 actually stands for Negative/Positive 0. Other capacitors were designed with either a positive or negative temperature coefficient to compensate for other components with the opposite temperature coefficient. Are we having fun yet? **ANSWER A.**

E6D02 What is the equivalent circuit of a quartz crystal?

A. Motional capacitance, motional inductance, and loss resistance in series, all in parallel with a shunt capacitor representing electrode and stray capacitance.
B. Motional capacitance, motional inductance, loss resistance, and a capacitor representing electrode and stray capacitance all in parallel.
C. Motional capacitance, motional inductance, loss resistance, and a capacitor representing electrode and stray capacitance all in series.
D. Motional inductance and loss resistance in series, paralleled with motional capacitance and a capacitor representing electrode and stray capacitance.

Motional capacitance and motional inductance of a quartz crystal refers to the natural series resonance. The motional inductance is typically huge, while the motional capacitance and series resistance are tiny, resulting in a very large Q. This crystal forms a series or parallel resonant circuit with a trimmer capacitor or inductor to give it spot-on frequency. To find the correct answer out of these four choices, commit to memory "***motional capacitance... motional inductance AND loss resistance in series***" and "shunt capacitance representing electrode and stray capacitance" will fall into place. The job of a quartz crystal is as important as a beating heart. Your question pool committee wants you to fully understand the complexities that go into the equivalent circuit of a quartz crystal. You can read more about this "heart" that is beating within your radio in the ***ARRL Handbook for Radio Communications.*** **ANSWER A.**

E6D03 Which of the following is an aspect of the piezoelectric effect?

A. Mechanical deformation of material by the application of a voltage.
B. Mechanical deformation of material by the application of a magnetic field.
C. Generation of electrical energy in the presence of light.
D. Increased conductivity in the presence of light.

The piezoelectric effect is crucial to nearly all we do in radio. Every quartz crystal in your computer, cell-phone or ham rig relies on this property. In light of this, it's amazing that for a long time in early amateur radio we somehow managed to get by without even knowing about this phenomenon! When you apply a mechanical stress to a piezoelectric material, such as quartz, it generates a small DC voltage. And if you ***apply an electric field to the crystal, it mechanically deforms***. Quartz crystals are at the heart of nearly all precision oscillators and a great number of electrical filters. They can even be used as transducers in microphones or loudspeakers. **ANSWER A.**

E7H13 Which of the following is a technique for providing highly accurate and stable oscillators needed for microwave transmission and reception?

A. Use a GPS signal reference.
B. Use a rubidium stabilized reference oscillator.
C. Use a temperature-controlled high Q dielectric resonator.
D. All of these choices are correct.

Up at 10 GHz and 24 GHz oscillator stability is of paramount importance. When Gordo is mountain topping with his 10 GHz setup, the sun warming his transmitter can throw him many kHz off frequency! Microwave operators who run beacons will many times use GPS satellite signals for a good stable reference, or may develop a Rubidium standard to keep the oscillator set on the nose. Also, a temperature-controlled high Q dielectric resonator can keep the beacon within a few cycles of being on frequency. ***All of these signal references are ways of providing good stability to your microwave system***. **ANSWER D.**

E7H02 Which describes a microphonic?

A. An IC used for amplifying microphone signals.
B. Distortion caused by RF pickup on the microphone cable.
C. Changes in oscillator frequency due to mechanical vibration.
D. Excess loading of the microphone by an oscillator.

Some of us old timers had a transmitter or two that you could modulate just by yelling at it... without a microphone! This was not a desirable condition, but it could be a lot of fun. Nevertheless, unless great care is taken, nearly any oscillator can become microphonic – this is where microscopic ***movements caused by vibrations*** of any frequency determining components can change the value of that component and thus frequency modulate the thing. Oscillators aren't the only possibly microphonic component in a radio, but they're the most common. Well-designed oscillators are mechanically rigid and well isolated from their outside environment. **ANSWER C.**

E7H07 How can an oscillator's microphonic responses be reduced?

A. Use of NP0 capacitors.
B. Eliminating noise on the oscillator's power supply.
C. Using the oscillator only for CW and digital signals.
D. Mechanically isolating the oscillator circuitry from its enclosure.

Reducing microphonics requires ***removing any source of mechanical vibration*** by making everything more rigid, and physically isolating it. **ANSWER D.**

E7H14 What is a phase-locked loop circuit?

A. An electronic servo loop consisting of a ratio detector, reactance modulator, and voltage-controlled oscillator.
B. An electronic circuit also known as a monostable multivibrator.
C. An electronic servo loop consisting of a phase detector, a low-pass filter, a voltage-controlled oscillator, and a stable reference oscillator.
D. An electronic circuit consisting of a precision push-pull amplifier with a differential input.

Few other circuits have changed the course of Amateur Radio equipment as much as the phase-locked loop (PLL) circuit. There is only one correct answer with the word "phase" in it, so ***look for the word phase when you read phase-locked loop*** in the question. **ANSWER C.**

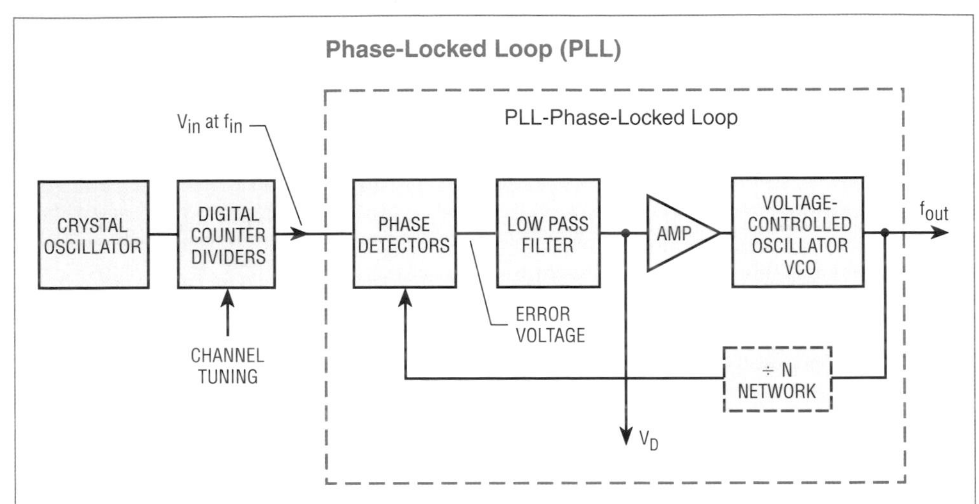

In a phase-locked loop the output frequency of the VCO is locked by phase and frequency to the frequency of the input signal V_{in}.

By using a very accurate crystal controlled source and dividing it down (or multiplying it up), the output frequency, f_{out}, of the PLL will be controlled very accurately and with rigid stability to the input frequency, f_{in}. The phase detector develops an error voltage that is determined by the frequencies and phase of the two inputs. The feedback system wants to reduce the error voltage to zero.

V_D is a voltage proportional to the frequency changes occuring in V_{in}; therefore, the PLL can be used as an FM demodulator.

With divide by N network, f_{out} will be f_{in} multiplied by N.

Channel tuning occurs in the Digital Counter Dividers.

E7H15 Which of these functions can be performed by a phase-locked loop?

A. Wide-band AF and RF power amplification.
B. Comparison of two digital input signals, digital pulse counter.
C. Photovoltaic conversion, optical coupling.
D. Frequency synthesis, FM demodulation.

When Gordo built his first external ***frequency synthesizer***, he was able to "synthesize" hundreds of channels with just a couple of crystals. How many of you remember being "rock bound" with an old rig? **ANSWER D.**

E7H09 What type of frequency synthesizer circuit uses a phase accumulator, lookup table, digital to analog converter, and a low-pass anti-alias filter?

A. A direct digital synthesizer.
B. A hybrid synthesizer.
C. A phase locked loop synthesizer.
D. A diode-switching matrix synthesizer.

Direct digital synthesis allows a single chip to handle multi-loop synthesizer circuits in your radio's local oscillator. Although the DDS does not output a pure sine wave, it offers fractions of a Hertz resolution with an extremely low noise floor. Within the chip is a ***phase accumulator, lookup table, digital-to-analog converter, and a low-pass anti-alias filter*** with a spectrally clean output. **ANSWER A.**

E7H10 What information is contained in the lookup table of a direct digital frequency synthesizer?

A. The phase relationship between a reference oscillator and the output waveform.
B. The amplitude values that represent a sine-wave output.
C. The phase relationship between a voltage-controlled oscillator and the output waveform.
D. The synthesizer frequency limits and frequency values stored in the radio memories.

Within the DDS, a lookup table will sample nearest available voltage levels in the digital-to-analog converter (DAC) ***that represent a sine-wave output***. **ANSWER B.** Here is what a block diagram of direct digital-frequency synthesizer (DDS) looks like:

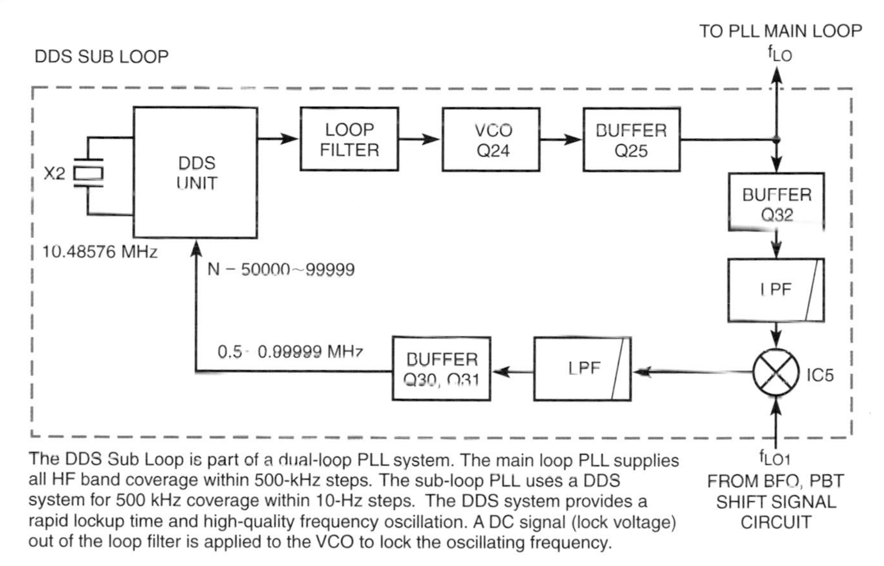

DDS Sub Loop of a Transmitter Dual-Loop PLL
Courtesy of ICOM America, Inc.

E7H11 What are the major spectral impurity components of direct digital synthesizers?

A. Broadband noise.
B. Digital conversion noise.
C. Spurious signals at discrete frequencies.
D. Nyquist limit noise.

Unwanted components of a DDS output are spurs at discrete frequencies – unwanted emissions. These are easier to filter out than broadband noise. However, filtering these numerous spurious signals generated by the direct digital-frequency-synthesizer is tricky because they are spread over a wide frequency range. Some signals are called "alias signals," which are usually in multiples on each side of the desired frequency and may extend several MHz above and below the desired frequency. Many times a low pass filter may help mitigate the problem with some of these ***spurs at discreet frequencies***. **ANSWER C.**

E4C01 What is an effect of excessive phase noise in the local oscillator section of a receiver?

A. It limits the receiver's ability to receive strong signals.
B. It reduces receiver sensitivity.
C. It decreases receiver third-order intermodulation distortion dynamic range.
D. It can cause strong signals on nearby frequencies to interfere with reception of weak signals.

Like every innovation in radio, each new concept can have undesirable side effects. The nearly universal adoption of Phase Locked Loop (PLL) technology greatly increased the versatility and stability of amateur radio receivers and transmitters, but created a few new problems in the process. All PLL circuits have a certain degree of ***phase noise***, which is a byproduct of the PLL's frequency correction process. If a PLL is used as a local oscillator in a receiver, any phase noise will also frequency modulate the incoming signals. This ***can interfere with reception of weak signals***, greatly reducing the usable sensitivity of the receiver. Phase noise is an important factor to consider when looking at the specifications of an amateur receiver or transceiver. Some transceivers are much better in this aspect than others. **ANSWER D.**

E7F05 How frequently must an analog signal be sampled by an analog-to-digital converter so that the signal can be accurately reproduced?

A. At half the rate of the highest frequency component of the signal.
B. At twice the rate of the highest frequency component of the signal.
C. At the same rate as the highest frequency component of the signal.
D. At four times the rate of the highest frequency component of the signal.

The Nyquist sampling theorem tells us that you need to ***sample the analog signal at twice the rate of the highest frequency signal***. Remember if you're sampling a *square wave,* you have odd harmonics well above the base frequency, and you need to include those too! **ANSWER B.**

Software Defined Radios provide rock-solid frequency stability with 1 Hz accuracy when operating in digital modes.

E7F06 What is the minimum number of bits required for an analog-to-digital converter to sample a signal with a range of 1 volt at a resolution of 1 millivolt?

A. 4 bits.
B. 6 bits.
C. 8 bits.
D. 10 bits.

The dynamic range of a digital converter is equal to the power ratio between the maximum and minimum signals it can handle. Using the formula $dB = 20 \log (V_2/V_1)$ where V_2 is the largest signal and V_1 is the smallest, we have a dynamic range of 60 dB. (Remember, when you are determining dB using VOLTAGE ratios, you need the "voltage dB" formula. This formula assumes the input and output impedances of the device are the same, which is a fair assumption in digital circuitry.) Now, we still need to go one step further to answer the question. The formula we use to convert dynamic range to the number of bits is:

Dynamic Range (dB) $= 20 \times \log (2^n)$

where: $n =$ the number of bits in the ADC.

We know the dynamic range (60 dB) so we work this formula backwards.

$60/20 = \log (2^n)$

$3 = \log (2^n)$

$\text{Antilog } (3) = 1000 = 2^n$

$2^{10} = 1024$

So, if we plug in 10 for n, we know we can do the job with ***10 bits***, with just a tiny bit left over. 8 bits won't do it, because $2^8 = 256$. We need at least 1000. **ANSWER D.**

E7F07 What function can a Fast Fourier Transform perform?

A. Converting analog signals to digital form.
B. Converting digital signals to analog form.
C. Converting digital signals from the time domain to the frequency domain.
D. Converting 8 bit data to 16 bit data.

The main "engine" of Digital Signal Processing is the Fast Fourier Transform which ***converts*** complicated ***frequency domain signals to*** simple ***time domain signals***, where the number crunching is much simpler. After the processing is done, the results are usually converted back to the frequency domain, using the inverse FFT function. **ANSWER C.**

E7F08 What is the function of decimation with regard to digital filters?

A. Converting data to binary code decimal form.
B. Reducing the effective sample rate by removing samples.
C. Attenuating the signal.
D. Removing unnecessary significant digits.

Digital sampling can be a labor intensive process, at least for the computer doing the thinking. It's sometimes effective to ***remove redundant or unnecessary samples*** to speed up the process. **ANSWER B.**

E7F09 Why is an anti-aliasing digital filter required in a digital decimator?

A. It removes high-frequency signal components which would otherwise be reproduced as lower frequency components.
B. It peaks the response of the decimator, improving bandwidth.
C. It removes low frequency signal components to eliminate the need for DC restoration.
D. It notches out the sampling frequency to avoid sampling errors.

The process of digitizing a signal is essentially identical to mixing in the analog universe and has the same byproduct, namely an ***image frequency***. In the A-to-D converter, the image ***is known as the alias***, but the physical process is identical. Images are usually not desirable and ***must be filtered out*** to avoid interfering with the "real" signal. **ANSWER A.**

E7F10 What aspect of receiver analog-to-digital conversion determines the maximum receive bandwidth of a Direct Digital Conversion SDR?

A. Sample rate.
B. Sample width in bits.
C. Sample clock phase noise.
D. Processor latency.

In any digital processing scenario, whether in a digital oscilloscope or software defined radio, Nyquist sampling theorem reigns supreme. The ***bandwidth*** of the system ***is dictated by the sampling rate***, which must be twice the frequency of the highest frequency signal component. **ANSWER A.**

E7F11 What sets the minimum detectable signal level for an SDR in the absence of atmospheric or thermal noise?

A. Sample clock phase noise.
B. Reference voltage level and sample width in bits.
C. Data storage transfer rate.
D. Missing codes and jitter.

As discussed earlier, the dynamic range is determined by the ***number of bits in the sample*** width, (more correctly termed *sample depth)*. The minimum detectable signal is the smallest "step" available in your Analog-to-Digital converter. **ANSWER B.**

E7F12 What digital process is applied to I and Q signals in order to recover the baseband modulation information?

A. Fast Fourier Transform.
B. Decimation.
C. Signal conditioning.
D. Quadrature mixing.

The ***Fast Fourier Transform*** (FFT) serves multiple functions in the SDR radio and many other digital radio systems. The FFT can serve as a ***demodulator*** for many types of digital signals. **ANSWER A.**

E7F13 What is the function of taps in a digital signal processing filter?

A. To reduce excess signal pressure levels.
B. Provide access for debugging software.
C. Select the point at which baseband signals are generated.
D. Provide incremental signal delays for filter algorithms.

In a digital filter, ***taps*** are equivalent to poles in an analog filter and ***provide incremental signal delays*** for filter algorithms. You can synthesize the effect of a lot of physical components by using logical taps in a modern digital filter. **ANSWER D.**

E7F14 Which of the following would allow a digital signal processing filter to create a sharper filter response?

A. Higher data rate.
B. More taps.
C. Complex phasor representations.
D. Double-precision math routines.

An analog filter with a lot of sections or poles can more closely approach the ideal frequency response than can a simpler filter. A digital filter can simulate a very complex (and expensive) analog filter through the use of ***more logical taps***, or iterations. **ANSWER B.**

E7F15 Which of the following is an advantage of a Finite Impulse Response (FIR) filter vs an Infinite Impulse Response (IIR) digital filter?

A. FIR filters delay all frequency components of the signal by the same amount.
B. FIR filters are easier to implement for a given set of passband rolloff requirements.
C. FIR filters can respond faster to impulses.
D. All of these choices are correct.

The ***FIR filter has predictable behavior***, regardless of the number of taps, and is generally easier to design. **ANSWER A.**

E7F16 How might the sampling rate of an existing digital signal be adjusted by a factor of 3/4?

A. Change the gain by a factor of 3 / 4.
B. Multiply each sample value by a factor of 3 / 4.
C. Add 3 to each input value and subtract 4 from each output value.
D. Interpolate by a factor of three, then decimate by a factor of four.

Sampling rates can only be decimated by a power of two, such as 2:1, 4:1, etc. Adjusting by an odd factor is a little more complex, but "doable." We can ***increase the sample rate by*** an arbitrary factor, such as ***3 (interpolation) and then*** we can ***decimate the result by a 4:1 factor***, for a net factor of 3/4. **ANSWER D.**

Circuits for Resonance for All!

E5A01 What can cause the voltage across reactances in series to be larger than the voltage applied to them?

A. Resonance.
B. Capacitance.
C. Conductance.
D. Resistance.

Coils (having inductance) and capacitors (having capacitance) will be at resonance when the capacitive reactance equals the inductive reactance. ***At resonance, voltages greater than the applied voltage*** can be present across these components in the circuit. You can see this with a mobile whip antenna – if something were to touch the whip while you are transmitting, you might see quite a spark! **ANSWER A.**

Equation for Resonant Frequency

X_L = Inductive Reactance

$X_L = 2\pi fL$

X_C = Capacitive Reactance

$$X_C = \frac{1}{2\pi fC}$$

L = Inductance in henrys

C = Capacitance in farads

f = Frequency in hertz

At Resonance $X_L = X_C$

$$2\pi f_r L = \frac{1}{2\pi f_r C}$$

Solving for f_r

$$f_r^2 = \frac{1}{(2\pi)^2 LC}$$

The Resonant Frequency, f_r, is:

$$f_r = \frac{1}{2\pi\sqrt{LC}}$$

On-line calculator site: www.calculatoredge.com/index.htm

E5A02 What is resonance in an electrical circuit?

A. The highest frequency that will pass current.
B. The lowest frequency that will pass current.
C. The frequency at which the capacitive reactance equals the inductive reactance.
D. The frequency at which the reactive impedance equals the resistive impedance.

At resonance, XL (inductive reactance) equals XC (capacitive reactance). In a mobile series resonant whip antenna, there will be maximum current through the loading coil when the whip is tuned to resonance. On your base station vertical trap antenna or trap beam antenna, parallel resonant circuits offer high impedance, trapping out a specific length of the antenna for a certain frequency. At resonance, the resonant parallel circuit looks like a high impedance to any of your power getting through to the longer length of the antenna system. **ANSWER C.**

E5A03 What is the magnitude of the impedance of a series RLC circuit at resonance?

A. High, as compared to the circuit resistance.
B. Approximately equal to capacitive reactance.
C. Approximately equal to inductive reactance.
D. Approximately equal to circuit resistance.

Did you ever wonder why the serious mobile ham always uses a gigantic loading coil for his series resonant antenna? The bigger the loading coil the less resistance. *At resonance, the magnitude of the impedance is equal to the circuit resistance*. **ANSWER D.**

E5A04 What is the magnitude of the impedance of a circuit with a resistor, an inductor and a capacitor all in parallel, at resonance?

A. Approximately equal to circuit resistance.
B. Approximately equal to inductive reactance.
C. Low, as compared to the circuit resistance.
D. Approximately equal to capacitive reactance.

Whether the system is parallel or series RLC resonant, the *impedance will always be equal to the circuit resistance* when the circuit is at resonance. **ANSWER A.**

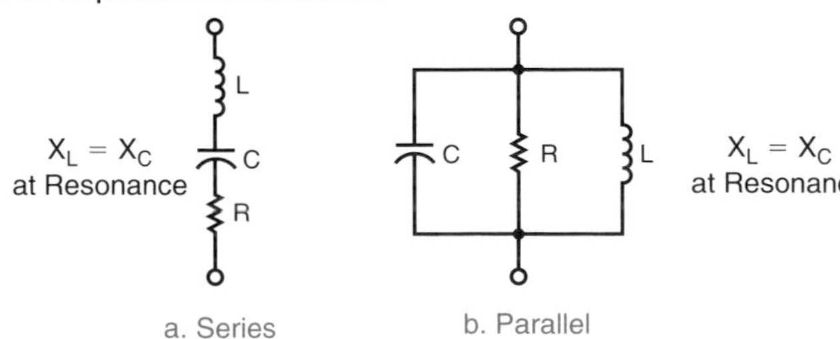

Series and Parallel Resonant Circuits

E5A05 What is the magnitude of the current at the input of a series RLC circuit as the frequency goes through resonance?

A. Minimum.
B. Maximum.
C. R/L.
D. L/R.

A mobile whip antenna will have *maximum current* in a *series* RLC circuit *at resonance*. One way you can tell whether or not a mobile whip is working properly is to feel the coil after the transmitter has been shut down. If it's warm, there has been current through the coil and up to the whip tip. **ANSWER B.**

E5A06 What is the magnitude of the circulating current within the components of a parallel LC circuit at resonance?

A. It is at a minimum.
B. It is at a maximum.
C. It equals 1 divided by the quantity 2 times Pi, multiplied by the square root of inductance L multiplied by capacitance C.
D. It equals 2 multiplied by Pi, multiplied by frequency, multiplied by inductance.

A ***parallel circuit at resonance*** is similar to a tuned trap in a multi-band trap antenna. At resonance, the trap keeps power from going any further. But be assured there is ***maximum circulating current*** within the parallel RLC circuit. Now don't get confused with this question – a parallel circuit has maximum current within it and minimum current going on to the next stage. That's why they call those parallel resonant circuits "traps." **ANSWER B.**

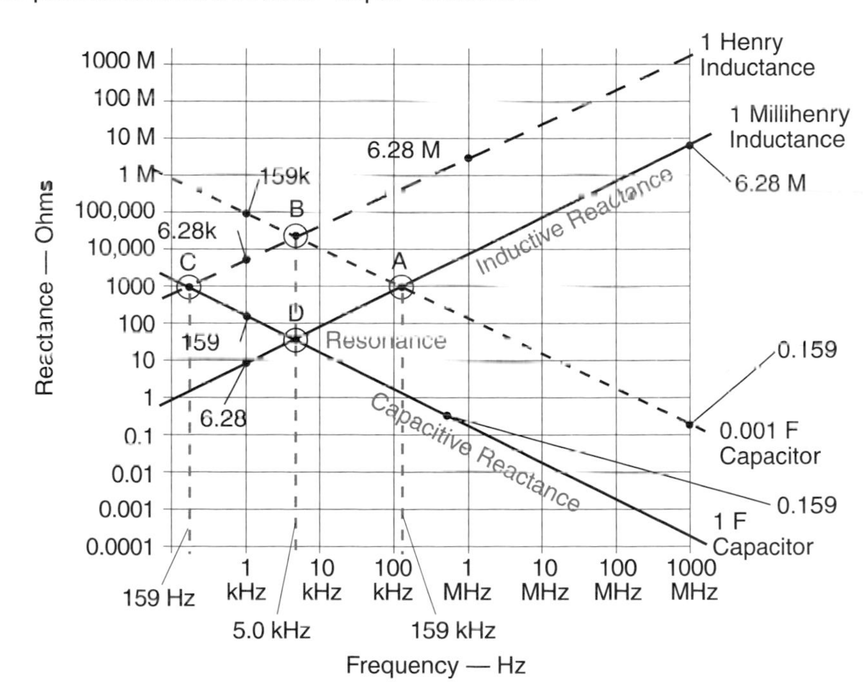

Variation of inductance and capacitive reactance with frequency (illustration not to exact log-log scale).

Source: Basic *Communications Electronics*, Hudson & Luecke, © 1999 Master Publishing, Inc., Niles, IL

E5A07 What is the magnitude of the current at the input of a parallel RLC circuit at resonance?

A. Minimum.
B. Maximum.
C. R/L.
D. L/R.

The total ***current into*** and out of a ***parallel RLC circuit at resonance is at a minimum***. On old radios and vacuum tube linear amplifiers, new or old, we would tune a plate control to see the current dip at resonance. **ANSWER A.**

E5A08 What is the phase relationship between the current through and the voltage across a series resonant circuit at resonance?

A. The voltage leads the current by 90 degrees.
B. The current leads the voltage by 90 degrees.
C. The voltage and current are in phase.
D. The voltage and current are 180 degrees out of phase.

At *Resonance*: $X_L = X_C$ when a circuit is ***at resonance, voltage and current are in phase***. **ANSWER C.**

Out of Resonance Examples

The two waveforms illustrate the relationship of voltage and current in an out-of-phase circuit.

ELI = voltage (E) leads current (I) in an inductive (L) circuit.
ICE = current (I) leads voltage (E) in a capacitive (C) circuit.

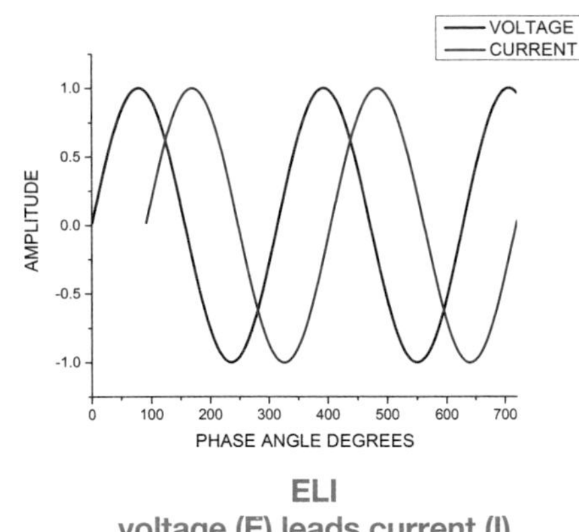

ELI
voltage (E) leads current (I)
(or current lags voltage)

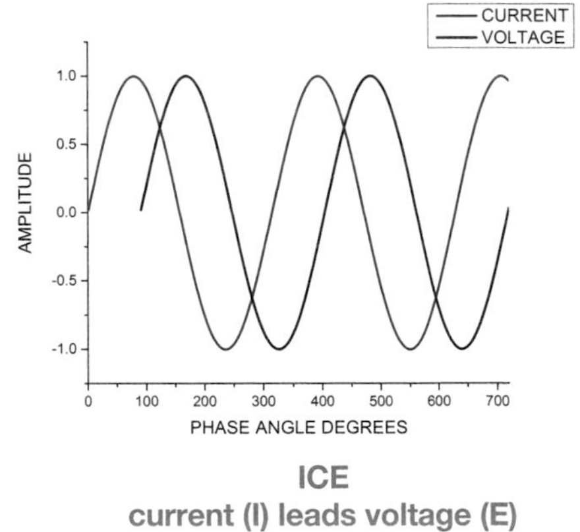

ICE
current (I) leads voltage (E)
(or voltage lags current)

E5B07 What is the phase angle between the voltage across and the current through a series RLC circuit if X_C is 500 ohms, R is 1 kilohm, and X_L is 250 ohms?

A. 68.2 degrees with the voltage leading the current.
B. 14.0 degrees with the voltage leading the current.
C. 14.0 degrees with the voltage lagging the current.
D. 68.2 degrees with the voltage lagging the current.

Here is a question on the phase angle between voltage and current in a series RLC circuit. When an applied voltage produces a current in a series RLC circuit, the voltage and current will be out of phase due to the reactance (X_L and X_C) in the circuit. Refer to the schematic below.

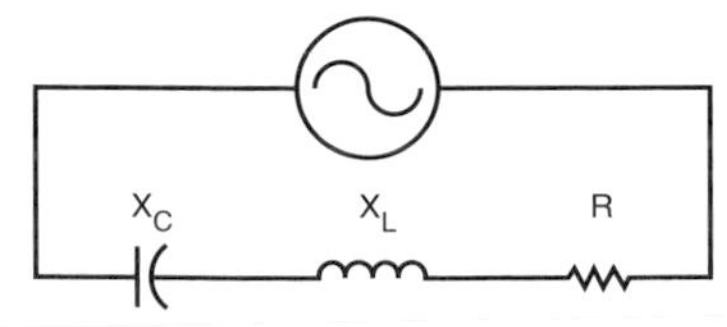

Series RLC Circuit Schematic for Phase Angle Calculations

Here is the formula, memorize it:

$\text{Tan } \phi = \frac{X}{R}$ where: ϕ is the phase angle in **degrees**
$X = (X_L - X_C)$ is total reactance in the circuit in **ohms**
R is circuit resistance in **ohms**

Since X_L and X_C are in opposition, we simply subtract one from the other. Now here's a fun way to remember the relationship of current and voltage in non-resonant circuits: ELI the ICE man! Say it out loud, "ELI the ICE man." In a more inductive (L) circuit, voltage (E) leads current (I) just like in the word "ELI" where E is before I. In a more capacitive (C) circuit, current (I) leads voltage (E), just like the word "ICE" where I comes before E. This is an ICE circuit, because X_C is 500 ohms and X_L is only 250 ohms. The total reactance (X) is 500 - 250 = 250. By dividing 1000 (the resistance) into 250, the tangent of ϕ comes out 0.25. A tangent 0.25 represents a phase angle of 14 degrees. This is an ICE circuit with voltage lagging current. **ANSWER C.**

E5B08 What is the phase angle between the voltage across and the current through a series RLC circuit if X_C is 100 ohms, R is 100 ohms, and X_L is 75 ohms?

A. 14 degrees with the voltage lagging the current.
B. 14 degrees with the voltage leading the current.
C. 76 degrees with the voltage leading the current.
D. 76 degrees with the voltage lagging the current.

This is more ICE than ELI because X_C is greater than X_L, and the tangent of ϕ of 0.25 gives us a ***14-degree phase angle***. Since the ***circuit is ICE, voltage (E) is lagging current (I)***. **ANSWER A.**

E5B11 What is the phase angle between the voltage across and the current through a series RLC circuit if X_C is 25 ohms, R is 100 ohms, and X_L is 50 ohms?

A. 14 degrees with the voltage lagging the current.
B. 14 degrees with the voltage leading the current.
C. 76 degrees with the voltage lagging the current.
D. 76 degrees with the voltage leading the current.

When an applied voltage produces a current in a series RLC circuit, the voltage and current will be out of phase due to the reactance (X_L and X_C) in the circuit. Right off the bat, you can zero in on 2 out of 4 answers by looking to see whether voltage is lagging the current, or voltage is leading the current. Voltage lagging is an ICE (more capacitive) circuit. Voltage leading the current is an ELI (more inductive) circuit. Now let's take a look at this problem where X_C is 25 ohms, X_L is 50 ohms, and R is 100 ohms. The reactance (X) is 50 - 25 = 25. Simple subtraction. Easy stuff. We find the tangent by dividing 100 (the resistance) into 25 = 0.25. From the trig table stored in our brain, the tangent of 14 degrees is 0.25. This gives us a ***phase angle of 14 degrees, and voltage is leading the current (ELI)***. After your tangent calculations you will find that each of the three ELI the ICE man questions has a phase angle of 14 degrees. In this question, the circuit is ELI with voltage leading current. **ANSWER B.**

A Neat Trick for Calculating Parallel Impedances

Graphical solutions to many electronics problems can often make things a lot simpler than working out complicated formulas, and they can also give you valuable insight into principles that you can't get just by "crunching numbers." Working out complex parallel impedances can be somewhat involved, and even seasoned electronics folks like us sometimes forget the formulas. By now, you should know how to work out series RLC networks using Cartesian coordinates. Working out parallel impedances is similar, but with an interesting twist. Let's figure out the total impedance (Z) and phase angle of a parallel 400 ohm resistor and 800 ohm inductor, using this clever graphical method. Take a look at the graph below.

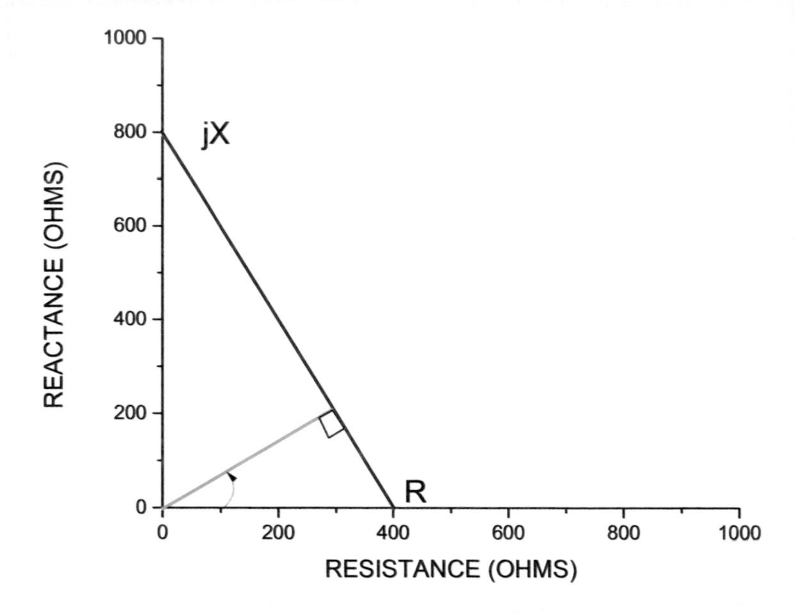

The first thing you'll need to do is plot the two component values on their representative axis. Make a small mark at 400 on the horizontal resistance axis, designated by R, and a small mark at 800 on the vertical reactance axis, designated jX. Remember the j prefix tells us the value is an imaginary number.

Now draw a straight line segment between R and jX, (shown in red). Next, draw a line segment from the origin (0,0) to the previous diagonal, at the point where the two segments are perpendicular, (shown in green). You'll need a square of some kind to do this...don't try to just eyeball it. The length of the green line segment is the impedance magnitude, and the angle the green segment makes with the horizontal axis is the phase angle. If you've drawn your graph carefully, the length of the green line should be 357. The phase angle should be 26.6 degrees. Nothing to it!

Now you can do the calculations the "hard" way and see if you get the same results. You should always know how to work a problem at least two different ways!

Vector Addition

Here is more detail. In an AC circuit, when calculating the impedance of the circuit, the reactance and resistance must be added vectorially rather than algebraically. This vector addition can be understood best by looking at the following diagram:

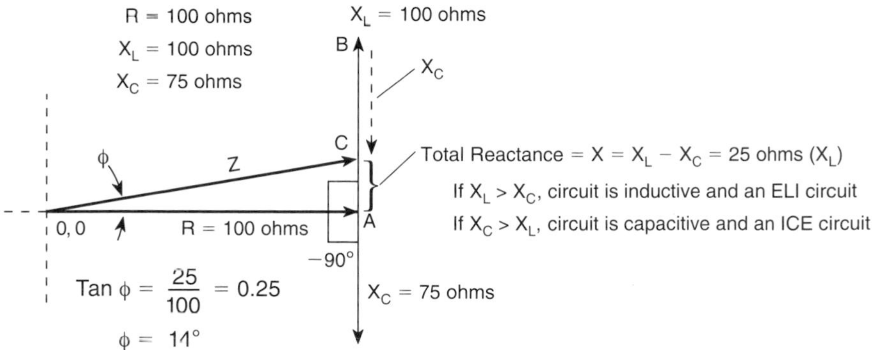

The resistance vector (of a size proportional to its value) is plotted horizontally (along x axis). Any reactance vector (using same scale as for resistance value) is plotted at right angles (90) to the resistance vector at the tip of the resistance vector (Point A). Inductive reactance (X_L) would be plotted at +90, while capacitive reactance (X_C) would be plotted at −90. The total reactance is determined by plotting the X_C vector from the tip (Point B) of the X_L vector. The result is Point C on the X_L vector. The total impedance vector, Z, is drawn from the origin (tail of R vector) to point C. The triangle formed is a right triangle; therefore, by trigonometry and the Pythagorean theorem:

$$Z = \sqrt{R^2 + (X_L - X_C)^2} = \sqrt{100^2 + 25^2}$$

$$= \sqrt{10625} = 103.08 \text{ ohms}$$

and:

$$\text{Tan } \phi = \frac{X}{R} = \frac{25}{100} = 0.25 \qquad \phi = 14°$$

The tangent of the angle φ is found by dividing the value of the side opposite (X) vector by the value of the side adjacent (R) vector. φ is the phase angle of the impedance.

E5B09 What is the relationship between the current through a capacitor and the voltage across a capacitor?

A. Voltage and current are in phase.
B. Voltage and current are 180 degrees out of phase.
C. Voltage leads current by 90 degrees.
D. Current leads voltage by 90 degrees.

A ***capacitor is an "ICE" circuit where current leads the voltage***, as you see in "ICE" where the letter "I" is leading the letter "E." **ANSWER D.**

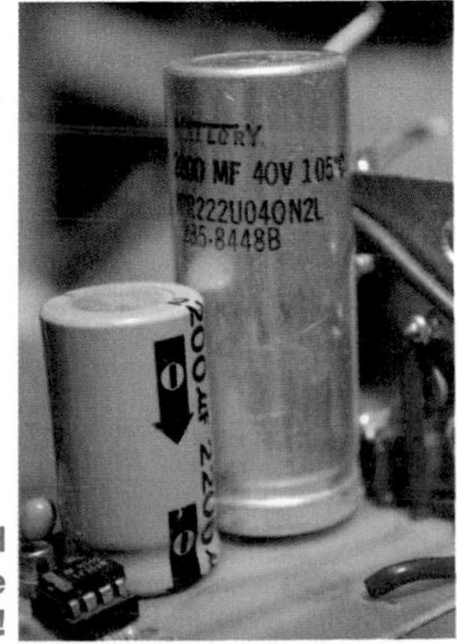

When re-capping an older tube transceiver, be sure and watch polarity on these types of capacitors. They make a one-time loud POOF BANG if you get them wrong!

E5B10 What is the relationship between the current through an inductor and the voltage across an inductor?

A. Voltage leads current by 90 degrees.
B. Current leads voltage by 90 degrees.
C. Voltage and current are 180 degrees out of phase.
D. Voltage and current are in phase.

An inductor is an ***"ELI" condition where voltage leads the current*** as we see in "ELI" where the letter "E" is ahead of the letter "I." **ANSWER A.**

Notice these coils are spaced apart, and the upper right ones oriented 90 degrees apart to prevent mutual coupling.

E5C13 What coordinate system is often used to display the resistive, inductive, and/or capacitive reactance components of impedance?

A. Maidenhead grid.
B. Faraday grid.
C. Elliptical coordinates.
D. Rectangular coordinates.

The ***rectangular coordinate system*** is used to display resistive, inductive, and/or capacitive reactance components of an impedance. It represents the reactance or resistance of each component as a vector whose length represents the magnitude of the reactance or resistance. **ANSWER D.**

E5C09 When using rectangular coordinates to graph the impedance of a circuit, what does the horizontal axis represent?

A. Resistive component.
B. Reactive component.
C. The sum of the reactive and resistive components.
D. The difference between the resistive and reactive components.

The horizontal axis represents the resistive component of the impedance. When the impedance is in a circuit and voltage is applied, there will be a current, and the voltage drop across the ***resistance is plotted on the horizontal axis***. **ANSWER A.**

E5C01 Which of the following represents a capacitive reactance in rectangular notation?

A. –jX.
B. +jX.
C. X.
D. Omega.

In rectangular notation, the horizontal axis represents resistance and the vertical axis represents reactance. Inductive reactance is in the positive direction, and capacitive reactance is in the negative direction. If you have a hard time remembering this, remember that inductive reactance INCREASES with frequency (it has a POSITIVE correlation), while ***capacitive reactance DECREASES with frequency*** (it has a NEGATIVE correlation). ***The reactance*** or vertical ***axis is designated by the "j operator,"*** which should immediately clue you to the fact that we're working with COMPLEX numbers. From your high school math class, you probably remember the imaginary component of a complex number prefixed with an "i", but since i is already used in electronics, we use the letter j. But it means the same thing, an imaginary value. So, the ***capacitive reactance is represented by –jX***. **ANSWER A.**

Representing an AC Quantity

Rectangular Coordinates

A quantity can be represented by points that are measured on an X-Y plane with an X and Y axis perpendicular to each other. The point where the axes cross is the origin, the points on the X and Y axis are called coordinates, and the magnitude is represented by the length of a vector from the origin to the unique point defined by the *rectangular coordinates.*

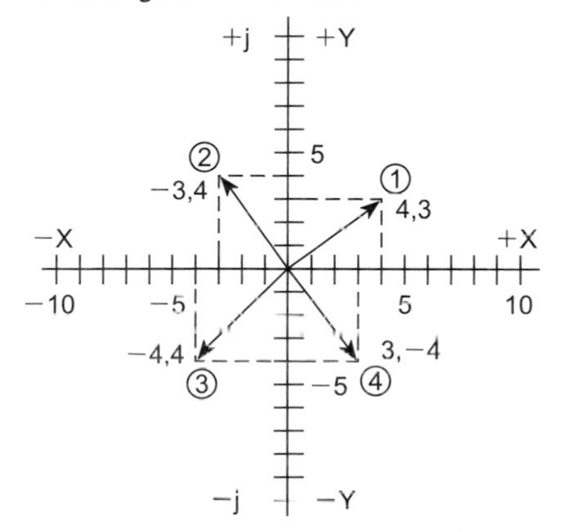

To be able to identify AC quantities and their phase angle, the X axis is called the real axis and the Y axis is called the imaginary axis. There are positive and negative real axis coordinates and positive and negative imaginary axis coordinates. The imaginary axis coordinates have a **j** operator to identify that they are imaginary coordinates.

In rectangular coordinates:

Vector 1 is 4 + j3

Vector 2 is −3 + j4

Vector 3 is −4 − j4

Vector 4 is 3 − j4

Polar Coordinates

A quantity can be represented by a vector starting at an origin, the length of which is the magnitude of the quantity, and rotated from a zero axis through 360°.

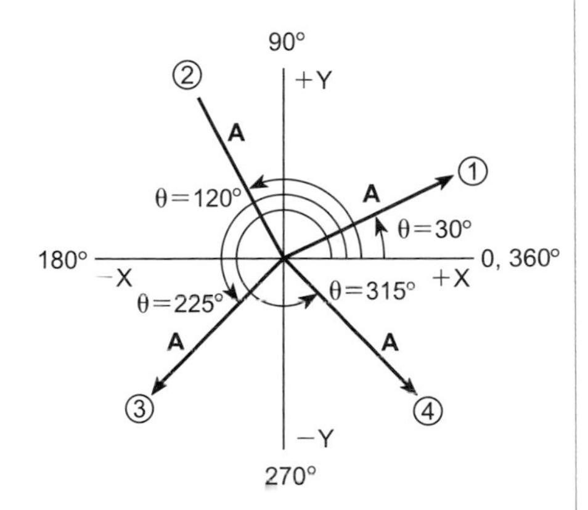

The position of the Vector A can be defined by the angle through which it is rotated. When we do so we are defining the AC quantity in *polar coordinates.*

In polar coordinates:

Vector 1 is $A\underline{/30°}$

Vector 2 is $A\underline{/120°}$

Vector 3 is $A\underline{/225°}$

Vector 4 is $A\underline{/315°}$

E5C05 What is the name of the diagram used to show the phase relationship between impedances at a given frequency?

A. Venn diagram.
B. Near field diagram.
C. Phasor diagram.
D. Far field diagram.

A ***phasor is simply an arrow or vector that shows the magnitude and phase of an electrical current in a circuit***. The length of the phasor is the magnitude, and the direction of the arrow is the phase angle between voltage and current. **ANSWER C.**

E5C07 What is a vector?

A. The value of a quantity that changes over time.
B. A quantity with both magnitude and an angular component.
C. The inverse of the tangent function.
D. The inverse of the sine function.

Distance and direction are the two things you need to know to get from here to there. Knowing just the distance often isn't enough. We all know people who could take a wrong turn in an elevator. A vector is an arrow that points the way and tells you how far you need to go. Vectors are used in all kinds of sciences, but they're especially useful in radio circuits where a ***vector*** is used to ***represent a quantity with both magnitude and an angular component***. **ANSWER B.**

E5C06 What does the impedance 50 -j25 represent?

A. 50 ohms resistance in series with 25 ohms inductive reactance.
B. 50 ohms resistance in series with 25 ohms capacitive reactance.
C. 25 ohms resistance in series with 50 ohms inductive reactance.
D. 25 ohms resistance in series with 50 ohms capacitive reactance.

In a complex impedance, the "j operator" always designates the reactive component. We know that + reactance is inductive and – reactance is capacitive. So ***50 -j25 ohms is 50 ohms of resistance and 25 ohms of capacitive reactance***. **ANSWER B.**

E5C10 When using rectangular coordinates to graph the impedance of a circuit, what does the vertical axis represent?

A. Resistive component.
B. Reactive component.
C. The sum of the reactive and resistive components.
D. The difference between the resistive and reactive components.

When graphing rectangular coordinates, ***the vertical axis represents the reactive component of the impedance***. When the impedance is in a circuit and voltage is applied, there will be a current, and the voltage across the reactance is plotted perpendicular to the horizontal axis and parallel to the vertical axis. Remember – horizontal is resistive, and vertical is reactive. **ANSWER B.**

E5C11 What do the two numbers that are used to define a point on a graph using rectangular coordinates represent?

A. The magnitude and phase of the point.
B. The sine and cosine values.
C. The coordinate values along the horizontal and vertical axes.
D. The tangent and cotangent values.

When we see the two numbers defining a point on a rectangular coordinate graph, these two numbers are ***the coordinate values along the horizontal and vertical axes***, respectively. The horizontal coordinate is given first. **ANSWER C.**

E5C12 If you plot the impedance of a circuit using the rectangular coordinate system and find the impedance point falls on the right side of the graph on the horizontal axis, what do you know about the circuit?

A. It has to be a direct current circuit.
B. It contains resistance and capacitive reactance.
C. It contains resistance and inductive reactance.
D. It is equivalent to a pure resistance.

If you are computing the impedance using rectangular coordinates, and the ***impedance point falls to the right side and directly on the horizontal line***, the impedance is equivalent to a ***pure resistance***. **ANSWER D.**

E5C14 Which point on Figure E5-2 best represents the impedance of a series circuit consisting of a 400 ohm resistor and a 38 picofarad capacitor at 14 MHz?

A. Point 2. C. Point 5.
B. Point 4. D. Point 6.

Look at Figure E5-2 and notice it is a rectangular-coordinate system with axes at right angles to each other. The indicated points on the plane have X and Y coordinate values. Watch out for Point 8 on +X axis – this is an 8, and not to be confused with an answer "B." In this question, the resistance of 400 ohms can be found along the +X axis; and since the circuit is capacitive, we go down to 300 on the -Y axis and project a line parallel to the X axis to intersect at Point 4 a projection from 400 on the +X axis. You can work the problem all the way out, but in this question you can visually spot it at a glance. **ANSWER B.**

To check yourself, solve the formula $X_C = 1/2\pi fC$ as follows:

$$X_C = \frac{1}{2\pi fC} = \frac{1}{6.28 \times 14 \times 10^{6} \times 38 \times 10^{-12}}$$

$$= \frac{1}{3.341 \times 10^{3} \times 10^{-6}} = 0.2993 \times 10^{3} = 299.3 \text{ ohms or } 300 \text{ ohms}$$

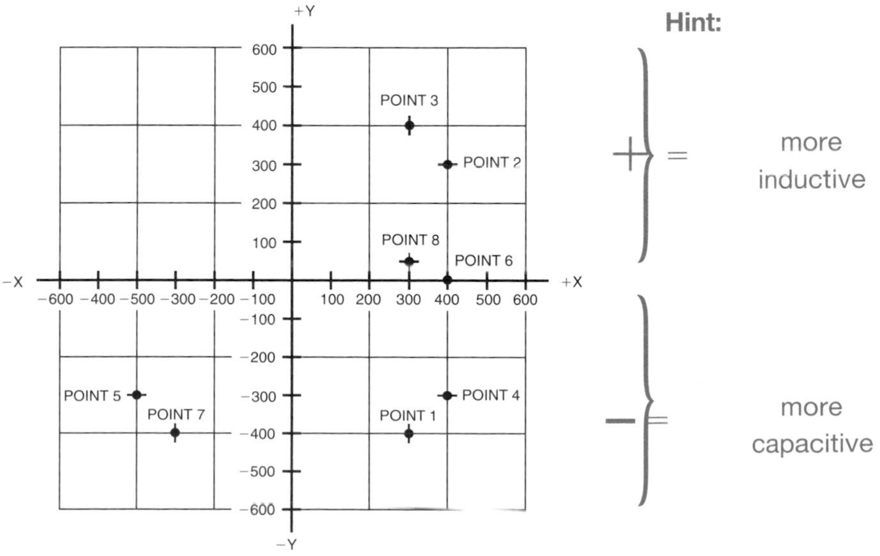

Hint:

+ = more inductive

− = more capacitive

Figure E5-2

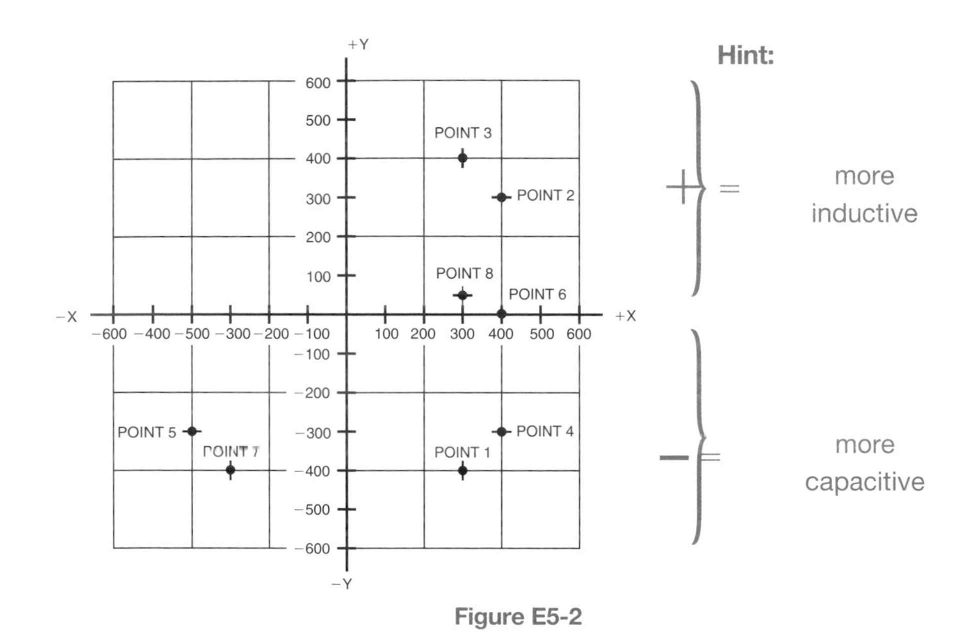

Figure E5-2

E5C15 Which point in Figure E5-2 best represents the impedance of a series circuit consisting of a 300 ohm resistor and an 18 microhenry inductor at 3.505 MHz?

A. Point 1. C. Point 7.
B. Point 3. D. Point 8.

The inductive reactance at 3.505 MHz is $X_L = 2\pi \text{ X } 3.505 \text{ X } 10^6 \text{ X } 18 \text{ X } 10^{-6} = 396.2$ or approximately 400 ohms of inductive reactance. ***Follow the +X axis out to 300*** for the resistance value. Move up on a line parallel to the ***+Y axis*** until you intersect the ***400 line***. You need to go in a positive Y direction because the circuit is inductive. ***Point 3*** is the answer. **ANSWER B.**

E5C16 Which point on Figure E5-2 best represents the impedance of a series circuit consisting of a 300 ohm resistor and a 19 picofarad capacitor at 21.200 MHz?

A. Point 1. C. Point 7.
B. Point 3. D. Point 8.

The capacitive reactance is $X_C = 1/2\pi fC = 1/6.28 \text{ X } 21.2 \text{ X } 10^6 \text{ X } 19 \text{ X } 10^{-12} = 1/2.53 \text{ X } 10^{-3} = 395$ ohms. We first move to the right along the ***+X axis*** to the value of the ***300 ohm resistor***, and move ***DOWN*** on a line perpendicular to the X axis (parallel to ***the -Y axis***) ***to Point 1*** at the intersection with the minus 400 line. 395 ohms is considered as 400 ohms. (If we went up, rather than down, we would be incorrect. Going up the +Y axis would mean the circuit is inductive. 19 picofarads makes the circuit capacitive, so we go down.) **ANSWER A.**

E5C17 Which point on Figure E5-2 best represents the impedance of a series circuit consisting of a 300 ohm resistor, a 0.64-microhenry inductor and an 85-picofarad capacitor at 24.900 MHz?

A. Point 1. C. Point 5.
B. Point 3. D. Point 8.

In this problem we must calculate both inductive and capacitive reactance.

$X_L = 2\pi \times 24.9 \times 10^6 \times 0.64 \times 10^{-6} = 100$ ohms.

$$X_C = \frac{1}{2\pi \times 24.9 \times 10^6 \times 85 \times 10^{-12}} = \frac{1}{1.33 \times 10^{-2}} = 75 \text{ ohms.}$$

X_L, inductive reactance of 100 ohms, minus X_C, capacitive reactance of 75 ohms, results in 25 ohms total reactance for this series circuit. To plot this on the rectangular coordinate system shown in figure E5-2 move to the right along the *+X axis to 300 ohms resistance*. Then, since the circuit is slightly inductive ($X_L > X_C$), move up the *+Y axis to 25 ohms*, where you find *Point 8*, correct answer. **ANSWER D.**

E5C08 What coordinate system is often used to display the phase angle of a circuit containing resistance, inductive and/or capacitive reactance?

A. Maidenhead grid.
B. Faraday grid.
C. Elliptical coordinates.
D. Polar coordinates.

To display the impedance of a circuit with its phase angle, you use a polar coordinate system. *Polar coordinates provide a visual representation of the value of the impedance and its phase angle*. **ANSWER D.**

E5C02 How are impedances described in polar coordinates?

A. By X and R values.
B. By real and imaginary parts.
C. By phase angle and amplitude.
D. By Y and G values.

Most electrical principles can be expressed in a number of ways. A complex impedance can be described either in rectangular coordinates, or polar notation. Just two ways of representing the same thing through different characteristics of it. A rectangular coordinate system uses resistance and total reactance to represent inductance. *In polar notation, impedance is expressed by magnitude and phase angle.* **ANSWER C.**

E5C03 Which of the following represents an inductive reactance in polar coordinates?

A. A positive real part.
B. A negative real part.
C. A positive phase angle.
D. A negative phase angle.

In polar notation, the "starting point" is the positive horizontal axis, which represents a pure resistance with no reactance. Two things can give you a pure resistance; a pure resistor (naturally) or an RLC circuit at its resonant frequency. In the latter case, as you increase the frequency above resonance, the inductive reactance increases. As a result, the *impedance will want to rotate counterclockwise toward the +j axis (vertical) giving us a positive phase angle*. **ANSWER C.**

E5C04 Which of the following represents a capacitive reactance in polar coordinates?

A. A positive real part.
B. A negative real part.
C. A positive phase angle.
D. A negative phase angle.

In a complex circuit (one containing resistance and reactance) the *capacitive reactance is represented by the negative -j axis*. The more capacitive reactance it has, the closer it will approach the -j axis. **ANSWER D.**

E5B12 What is admittance?

A. The inverse of impedance.
B. The term for the gain of a field effect transistor.
C. The turns ratio of a transformer.
D. The unit used for Q factor.

In circuits containing many parallel components, it can be much easier to do calculations based on the ***reciprocal values of*** resistance, reactance or ***impedance***, which are conductance, susceptance, and ***admittance,*** respectively. The unit of conductance, susceptance, or admittance is the siemens, designated by the upside down OMEGA symbol ℧. Before 1971, the MHO (ohm spelled backwards) was the official unit of conductance, susceptance, and admittance, and many of us old timers still believe mho is mo' better. It's easy to remember that with mo' MHO, you have mo' current flow! But siemens (spelled with a small s) is now the official term. **ANSWER A.**

E5B06 What is susceptance?

A. The magnetic impedance of a circuit.
B. The ratio of magnetic field to electric field.
C. The inverse of reactance.
D. A measure of the efficiency of a transformer.

A circuit or component has ***susceptance*** because it is ***susceptible to current flow***. This is ***the opposite of reactance***, which generally reacts negatively to current flow. So, all these fancy electrical terms actually have a simple grammatical meaning too! This fact can be a great help when your mind goes blank during an exam. **ANSWER C.**

E5B13 What letter is commonly used to represent susceptance?

A. G.
B. X.
C. Y.
D. B.

It's easy to confuse units with symbols. The ***symbol for susceptance is B***, while the unit of susceptance is the mho (or siemens). **ANSWER D.**

E5B03 What happens to the phase angle of a reactance when it is converted to a susceptance?

A. It is unchanged.
B. The sign is reversed.
C. It is shifted by 90 degrees.
D. The susceptance phase angle is the inverse of the reactance phase angle.

While not too commonly used in amateur radio, you should become familiar with the reciprocal functions of resistance (R), reactance (X), and impedance (Z), namely, *conductance (G), susceptance (B),* and *admittance (Y),* respectively. These reciprocal functions are immensely useful in parallel complex circuits. The UNIT associated with all three reciprocals is the mho, which is ohm spelled backwards. (Since 1971, the mho has been renamed the siemens, but you may prefer to continue using the mho, because, frankly, mho is mo' better.) In any case, ***to convert from any of the fundamental units to the reciprocal simply divide into 1, which reverses the sign*** so + becomes -. The reciprocal of Reactance is Susceptance ($B=1/X$). **ANSWER B.**

E5B05 What happens to the magnitude of a reactance when it is converted to a susceptance?

A. It is unchanged.
B. The sign is reversed.
C. It is shifted by 90 degrees.
D. The magnitude of the susceptance is the reciprocal of the magnitude of the reactance.

Susceptance (B) is the reciprocal of reactance (X). The value or ***magnitude*** of susceptance can be calculated as the ***reciprocal*** of reactance using B=1/X. However, reactance also has a *sign*. When you take the reciprocal of reactance, you also have to ***flip the sign over***, so + becomes -. However, answer B is incorrect, because sign is not a magnitude! Be sure to answer the actual question; don't presume! **ANSWER D.**

E4B15 Which of the following can be used as a relative measurement of the Q for a series-tuned circuit?

A. The inductance to capacitance ratio.
B. The frequency shift.
C. The bandwidth of the circuit's frequency response.
D. The resonant frequency of the circuit.

In an antenna just as in a series-tuned circuit, the higher the Q the narrower the bandwidth of frequencies it will pass. A high-frequency mobile whip antenna with a small helical-coil has a relatively low Q, offering good ***bandwidth***, but with less efficiency. **ANSWER C.**

E5A10 How is the Q of an RLC series resonant circuit calculated?

A. Reactance of either the inductance or capacitance divided by the resistance.
B. Reactance of either the inductance or capacitance times the resistance.
C. Resistance divided by the reactance of either the inductance or capacitance.
D. Reactance of the inductance times the reactance of the capacitance.

For a ***series RLC circuit, Q = X/R*** where X is total inductance and R is resistance. **ANSWER A.**

E5A09 How is the Q of an RLC parallel resonant circuit calculated?

A. Reactance of either the inductance or capacitance divided by the resistance.
B. Reactance of either the inductance or capacitance multiplied by the resistance.
C. Resistance divided by the reactance of either the inductance or capacitance.
D. Reactance of the inductance multiplied by the reactance of the capacitance.

By definition, in a resonant circuit, inductive reactance and capacitive reactance are equal, so either can be used for the calculation of Q. In a ***parallel circuit, Q = R/X*** where R is resistance and X is inductance. **ANSWER C.**

E5A15 Which of the following can increase Q for inductors and capacitors?

A. Lower losses.
B. Lower reactance.
C. Lower self-resonant frequency.
D. Higher self-resonant frequency.

Never argue with a definition! Q is defined as the opposite of circuit loss. ***Less loss = More Q***. **ANSWER A.**

E5A17 What is the result of increasing the Q of an impedance-matching circuit?

A. Matching bandwidth is decreased.
B. Matching bandwidth is increased.
C. Matching range is increased.
D. It has no effect on impedance matching.

One of the several definitions of Q is the "sharpness" of tuning. Since the purpose of a tuning circuit in the first place is to be *selective*, the better it is at being selective the higher the quality (Q) of the circuit. Quite logical, isn't it? Bandwidth is the reciprocal of selectivity, so ***the greater the Q, the lower the bandwidth***. **ANSWER A.**

E5A13 What is an effect of increasing Q in a resonant circuit?

A. Fewer components are needed for the same performance.
B. Parasitic effects are minimized.
C. Internal voltages and circulating currents increase.
D. Phase shift can become uncontrolled.

The Q of a resonant circuit is defined as the ratio of reactance to resistance (X/R). A resonant circuit with no internal resistance can have infinite circulating current, as well as infinite voltage across the inductor or capacitor. In the real world, you will never have this situation, but you can still have some surprisingly ***high circulating currents and voltages in high Q circuits*** such as antenna tuners or antenna traps. Q is an important factor to consider in any radio frequency circuit design. **ANSWER C.**

Half-Power Bandwidth

An amplifier's voltage gain will vary with frequency as shown. At the cutoff frequencies, the voltage gain has dropped to 0.707 of what it is in the mid-band. These frequencies, f_1 and f_2, are sometimes called the half-power frequencies.

If the output voltage is 10 volts across a 100-ohm load when the gain is A at the mid-band, then the power output, P_O, at mid-band is:

$$P_O = \frac{E^2}{R} = \frac{10^2}{100} = \frac{100}{100} = 1 \text{ Watt}$$

At the cutoff frequency, the output voltage will be 0.707 of what it is at the mid-band; therefore, 7.07 volts. The power output is:

$$P_{0.707} = \frac{(7.07)^2}{100} = \frac{50}{100} = 0.5 \text{ Watt}$$

The power output at the cutoff frequency points is one-half the mid-band power. The half-power bandwidth is between the frequencies f_1 and f_2.

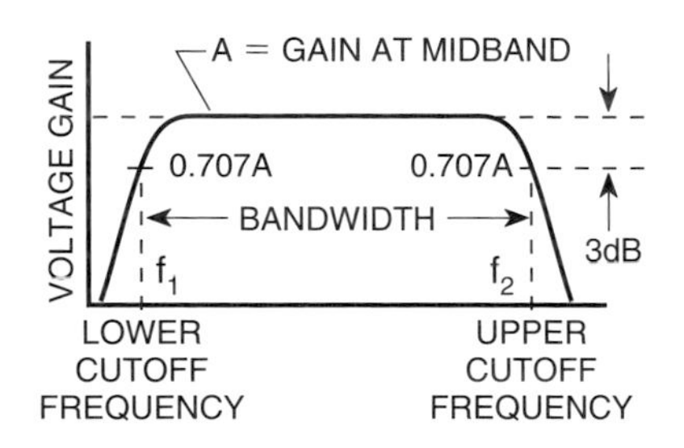

Power Ratio in dB

$$P_{dB} = 10 \log_{10} \frac{P_O}{P_{0.707}}$$
$$= 10 \log_{10} \frac{1}{0.5}$$
$$= 10 \log_{10} 2$$
$$= 10 \times 0.301$$
$$= 3 \text{ db}$$

The power output at the 0.707 frequencies is 3 db down from the mid-band power.

E5A11 What is the half-power bandwidth of a parallel resonant circuit that has a resonant frequency of 7.1 MHz and a Q of 150?

A. 157.8 Hz.
B. 315.6 Hz.
C. 47.3 kHz.
D. 23.67 kHz.

To find the half-power bandwidth, divide the resonant frequency (in kHz) by Q. 7100 kHz divided by 150 equals 47.33. Remember, to convert MHz to kHz, move the decimal three places to the right. On your calculator, press clear, *7100 ÷ 150 = 47.33*. **ANSWER C.**

Here is the formula:

$$\text{Half-power bandwidth} = \frac{f_r}{Q}$$

where: f_r is the resonant frequency in **kHz**
Q is a **quality factor** of the circuit

E5A12 What is the half-power bandwidth of a parallel resonant circuit that has a resonant frequency of 3.7 MHz and a Q of 118?

A. 436.6 kHz.
B. 218.3 kHz.
C. 31.4 kHz.
D. 15.7 kHz.

With this question, go from MHz to kHz, and then divide it by the Q. Clear, enter *3700 ÷ 118 = 31.36*. Round the answer to 31.4. **ANSWER C.**

E5A14 What is the resonant frequency of a series RLC circuit if R is 22 ohms, L is 50 microhenries and C is 40 picofarads?

A. 44.72 MHz.
B. 22.36 MHz.
C. 3.56 MHz.
D. 1.78 MHz.

The schematic shown in the diagram is a simple series resonant circuit. Don't panic. The following resonance formula will carry you through many questions on series and parallel circuits – one of which may be on your exam. First, you can throw away the resistance found in the parallel or series circuit. It is not needed for the resonance formula. Here is the resonance formula that you should memorize:

$$f_r = \frac{10^6}{2\pi\sqrt{LC}}$$

where: f_r is resonant frequency in **kHz**
$2\pi = 6.28$
L is inductance in **microhenries**
C is capacitance in **picofarads**

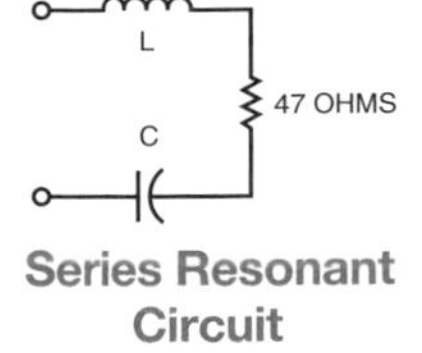

Series Resonant Circuit

First, multiply L X C. For this question, that is $50 \times 40 = 2000$. Now we need the square root of this, since in the resonance formula L and C are under the square root sign. With 2000 showing on your calculator, simply press the square root key. Presto, you have 44.72. Now multiply by 2π, which is $6.28 \times 44.72 = 280.84$. The final calculation for the formula is 1,000,000 ÷ 280.84. Press your clear button, and enter 1,000,000 ÷ 280.84 = 3560.7463 kHz. Your answer came out in kHz, but they are looking for MHz in the answer. No problem here – simply move the decimal point three places to the left, and you get answer C, *3.56 MHz*. That's all there is to it! **ANSWER C.**

E5A16 What is the resonant frequency of a parallel RLC circuit if R is 33 ohms, L is 50 microhenries and C is 10 picofarads?

A. 23.5 MHz.
B. 23.5 kHz.
C. 7.12 kHz.
D. 7.12 MHz.

Use the formula at E5A14. First multiply 50 microhenries and 10 picofarads. Take the square root of the result. Then multiply by 6.28. Divide that answer into 1,000,000. Finally, change kHz to MHz. Easy! Here are the keystrokes: Clear, 50 × 10 = √ × 6.28 = 140.4. Remember 140.4. Clear, *1,000,000 ÷ 140.4 = 7122 kHz which is 7.12 MHz*. **ANSWER D.**

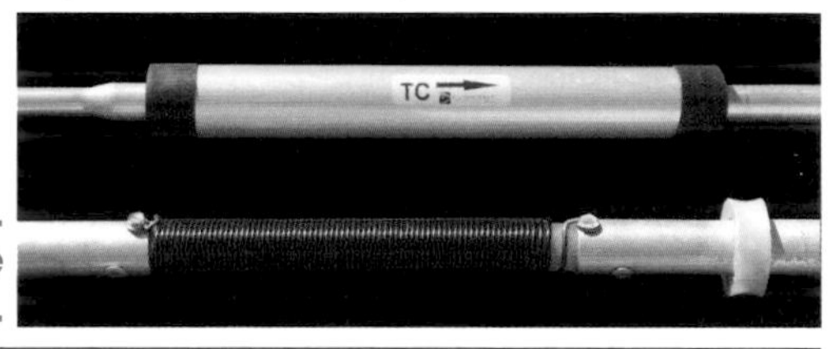

Antenna traps are parallel RLC circuits. The lower trap has the capacitive cover removed to see the coil.

E5B01 What is the term for the time required for the capacitor in an RC circuit to be charged to 63.2% of the applied voltage?

A. An exponential rate of one.
B. One time constant.
C. One exponential period.
D. A time factor of one.

In an RC circuit, assuming there is no initial charge on the capacitor, it takes *one time constant to charge a capacitor to 63.2 percent* of its final supply voltage value. **ANSWER B.**

E5B02 What is the term for the time it takes for a charged capacitor in an RC circuit to discharge to 36.8% of its initial voltage?

A. One discharge period.
B. An exponential discharge rate of one.
C. A discharge factor of one.
D. One time constant.

In one time constant, the capacitor charges to 63.2% or discharges to 36.8%. Notice the relationship: 63.2 + 36.8 = 100%. One time constant involves 63.2%; the resulting percentage just depends on whether it's charging from 0% or discharging from 100%. Consider the filter capacitors in a power supply where the capacitors discharge when the input power is removed. The time it takes a capacitor in an RC circuit to *discharge to 36.8 percent of its initial value* of stored charge *is one time constant*. **ANSWER D.**

E5B04 What is the time constant of a circuit having two 220 microfarad capacitors and two 1 megohm resistors, all in parallel?

A. 55 seconds.
B. 110 seconds.
C. 440 seconds.
D. 220 seconds.

Here is an RC circuit with all components in parallel. It will be easy if you remember that two equal capacitors in parallel have a total capacitance of twice the value of one of the capacitors, and that two equal value resistors in parallel have a total resistance equal to one-half the value of one of the resistors. More good news – "megs" (10^6) are being multiplied by "micros" (10^{-6}) so the powers of ten are cancelled in the calculation. Two 220 microfarad capacitors in parallel equal a larger 440 microfarad capacitor. Two 1-megohm resistors in parallel equal a resistance of 0.5 megohms. *Solve for the time constant using the simple formula T = R X C*. Multiply *0.5 × 440 and you end up with 220 seconds* as the correct answer. The calculator keystrokes are: Clear, 0.5 × 440 = 220. **ANSWER D.**

GOOD TO KNOW:

To calculate the percentage of the supply voltage after two time constants in an RC circuit, write down the percent after a single time constant as 63.2 percent. The remaining percent of charge that the capacitor must charge is 36.8 percent. It is found by deducting 63.2 percent from 100 percent. Since 36.8 percent is the final value to which the capacitor will charge after it has charged to 63.2 percent, in the next time constant the capacitor will charge to 63.2 percent of the 36.8 percent. In other words, another percent charge is added to the capacitor in the second time constant equal to:

$$36.8\% \times 63.2\% = 23.26\% \text{ (rounded to 23.3\%)}$$

This really needs to be calculated in decimal format as:

$$0.368 \times 0.632 = 0.2326$$

Percent values are converted to decimals by moving the decimal point two places to the left. Decimal values are converted to percent values by multiplying by 100 (moving decimal point to right two places.)

Since the final percent of charge in two time constants is desired, it is only necessary to add the two percent charges together:

	Percent	Decimal
Charge in first time constant =	63.2%	0.632
Charge in second time constant =	23.3%	0.233
Total charge after two time constants	86.5%	0.865

The process can be continued for the third time constant. Since the capacitor has charged to 86.5 percent in two time constants, if you start from that point the value of the third charge would be another 13.5 percent. Therefore, in the next time constant, the third, the capacitor would charge another:

$$13.5\% \times 63.2\% = 8.53\% \text{ (rounded to 8.5\%)}$$

After the third time constant the total charge is:

$$86.5\% + 8.5\% = 95\%$$

If the process is continued through five time constants, the capacitor charge is 99.3 percent. Therefore, in electronic calculations, the capacitor is considered to be fully charged (or discharged) after five time constants

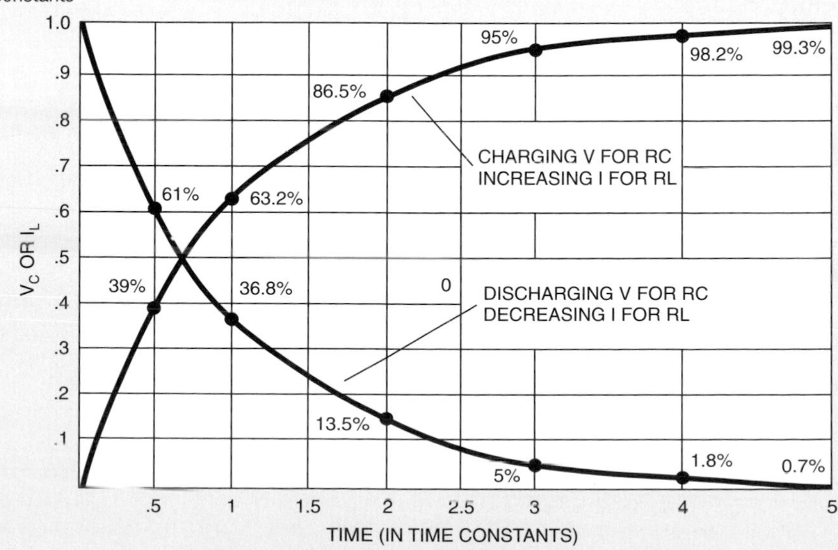

Time Constants Showing V and I Percentages

E5D01 What is the result of skin effect?

A. As frequency increases, RF current flows in a thinner layer of the conductor, closer to the surface.
B. As frequency decreases, RF current flows in a thinner layer of the conductor, closer to the surface.
C. Thermal effects on the surface of the conductor increase the impedance.
D. Thermal effects on the surface of the conductor decrease the impedance.

If you ever had a chance to inspect an old wireless station, you would have seen that the antenna "plumbing" leading out of the transmitter was usually constructed of hollow copper tubing. ***Radio frequency current always travels along the thinner outside layer of a conductor***, and as frequency increases, there is almost no current in the center of the conductor. The higher the frequency the greater the ***skin effect***. **ANSWER A.**

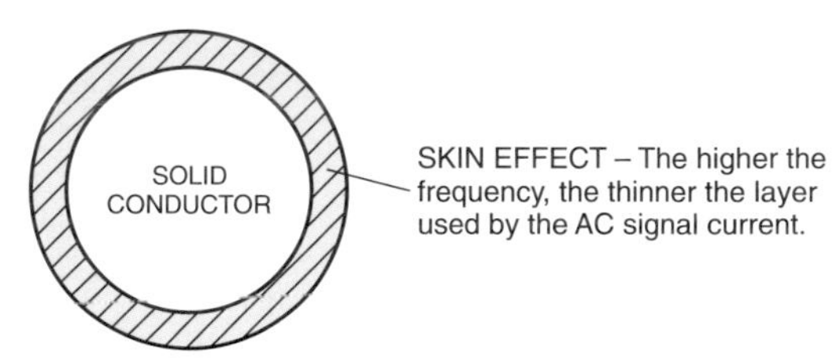

Cross-Section of Wire Conductor Illustrating Skin Effect

E5D02 Why is it important to keep lead lengths short for components used in circuits for VHF and above?

A. To increase the thermal time constant.
B. To avoid unwanted inductive reactance.
C. To maintain component lifetime.
D. All of these choices are correct.

VHF circuitry can sometimes look pretty ugly. This is because a straight line between two points isn't always the prettiest path. But for VHF circuitry, it is the path that works the best. Any time a piece of wire is a significant fraction of a wavelength long, it can become an antenna. Or, short of that, it can act as an unwelcome inductor. So, ***to reduce unwanted inductance keep all your leads in a VHF circuit as short as possible***. If you can do this and make it look pretty, you're doing very well! **ANSWER B.**

Notice no long lead lengths on this VHF amplifier wire-wound toroidal inductor.

E5D05 Which parasitic characteristic increases with conductor length?

A. Inductance.
B. Permeability.
C. Permittivity.
D. Malleability.

Any length of wire has some inductance in it. If you *want* this inductance, it's a good thing. If you *don't* want this inductance, it's generally considered "parasitic." In either case, the physical laws are the same: ***more wire = more inductance***. **ANSWER A.**

E5D03 What is microstrip?

A. Lightweight transmission line made of common zip cord.
B. Miniature coax used for low power applications.
C. Short lengths of coax mounted on printed circuit boards to minimize time delay between microwave circuits.
D. Precision printed circuit conductors above a ground plane that provide constant impedance interconnects at microwave frequencies.

In VHF radio circuits, it's not always possible to reduce the length of wires so that they don't act as inductors (See question E5D02). But sometimes you can do the next best thing: make them look like transmission lines. If you ***place a wire close to a ground plane***, such as a copper clad circuit board, it will act as a short length of transmission line or ***"microstrip,"*** which will cancel out the inductance. This construction practice is fairly standard in most VHF and UHF radio equipment nowadays. **ANSWER D.**

E5D04 Why are short connections necessary at microwave frequencies?

A. To increase neutralizing resistance.
B. To reduce phase shift along the connection.
C. Because of ground reflections.
D. To reduce noise figure.

A transmission line, whether in the form of a run of coaxial cable or a copper trace on a circuit board, will introduce some degree of phase shift, since it has a finite velocity factor. Whether the phase shift causes a problem or not depends on the application. For a given length of transmission line, the ***phase shift*** will be greater as the frequency increases. It's usually safest to keep transmission lines as short as possible unless there is a compelling reason to do otherwise. **ANSWER B.**

See the two surface mount components to the left of the toroidal inductor? These components have very short leads to minimize unwanted phase shifts along the conductors.

E5D07 What determines the strength of the magnetic field around a conductor?

A. The resistance divided by the current.
B. The ratio of the current to the resistance.
C. The diameter of the conductor.
D. The amount of current flowing through the conductor.

As ***increasing current*** passes through a conductor the ***strength of the magnetic field*** will continue to build. Have you ever been to a junk yard and watched the electromagnet coil magically lift a car off the ground and drop it into a nearby railroad car? When the coil is energized, it becomes a magnet. As quickly as the energy is cut off, magnetism stops and the car drops off. **ANSWER D.**

E5D08 What type of energy is stored in an electromagnetic or electrostatic field?

A. Electromechanical energy.
B. Potential energy.
C. Thermodynamic energy.
D. Kinetic energy.

Energy stored in an ***electromagnetic field*** in an inductor, or in an ***electrostatic field*** in a capacitor, is called ***potential energy***. The term *stored* in the question is your best clue. If you store something, it is pretty stationary; it's not moving. Kinetic means moving, so we can eliminate that answer. Answers A and C are entirely off base. So that leaves us with potential energy... something we can do actual work with once we take it out of storage. **ANSWER B.**

E5D06 In what direction is the magnetic field oriented about a conductor in relation to the direction of electron flow?

A. In the same direction as the current.
B. In a direction opposite to the current.
C. In all directions; omni-directional.
D. In a direction determined by the left-hand rule.

If you grab a big straight cable with your left hand, with your thumb pointing in the direction the electrons are flowing in the conductor (opposite from conventional current), your fingers will naturally point in the ***direction of the magnetic lines*** of force. This is known as the ***left-hand rule***. **ANSWER D.**

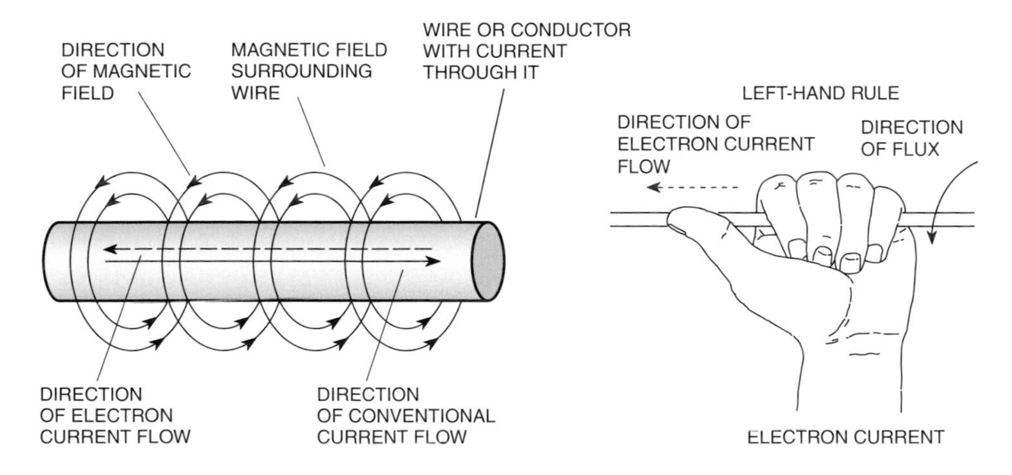

Left-Hand Rule

E6D04 Which materials are commonly used as a slug core in a variable inductor?

A. Polystyrene and polyethylene.
B. Ferrite and brass.
C. Teflon and Delrin.
D. Cobalt and aluminum.

Ferrite is a ceramic compound of iron that has particularly high permeability, that is, it concentrates magnetic fields around an inductor and so increases the inductance. ***Brass***, on the other hand, has negative permeability, and can decrease the inductance of a coil if you insert a slug of the alloy into a coil. Both properties allow convenient tuning of radio frequency inductors. **ANSWER B.**

E6D14 Which type of slug material decreases inductance when inserted into a coil?

A. Ceramic.
B. Brass.
C. Ferrite.
D. Powdered-iron.

Brass has a negative permeability and ***can be used to reduce the inductance*** of an inductor. While you are not too likely to see brass slugs in amateur radio gear, high powered broadcast transmitters often used moveable brass slugs in the tuning networks. **ANSWER B.**

Veteran land mobile radio guys know to carefully tune these inductors, as the core "slugs" are fragile! Only the proper plastic tuning tool will do!

E6D12 What is the definition of saturation in a ferrite core inductor?

A. The inductor windings are over coupled.
B. The inductor's voltage rating is exceeded causing a flashover.
C. The ability of the inductor's core to store magnetic energy has been exceeded.
D. Adjacent inductors become over-coupled.

An ideal inductor creates a magnetic field that is proportional to the current applied. However, some materials, such as ***ferrite, can saturate***, meaning that at some level of applied current the ***magnetic field no longer increases***. This means the inductor is no longer in a linear state and can create distortion and harmonics, which are not normally desirable in radio circuits. Saturable reactors are used occasionally in specialized circuits such as magnetic amplifiers or sweep generators. **ANSWER C.**

E6D17 Why should core saturation of a conventional impedance matching transformer be avoided?

A. Harmonics and distortion could result.
B. Magnetic flux would increase with frequency.
C. RF susceptance would increase.
D. Temporary changes of the core permeability could result.

Any time ***saturation*** of an inductor's core occurs, the relationship between voltage and current is no longer linear. Any non-linear device ***will create harmonics and distortion***. In addition, heating of the core can cause fracturing, especially in the case of a ferrite or powdered iron core. Core choice is an important consideration in many radio frequency design tasks. **ANSWER A.**

E6D13 What is the primary cause of inductor self-resonance?

A. Inter-turn capacitance.
B. The skin effect.
C. Inductive kickback.
D. Non-linear core hysteresis.

A perfect inductor should have no resonant frequency. However, most practical inductors, especially ***multi-turn inductors, have capacitance between the windings***. This causes the inductor to have a resonant frequency of its own, which can sometimes cause severe problems in a radio circuit. Certain types of inductor windings are designed to minimize parasitic capacitance, such as the honeycomb winding, that assures that each time a wire crosses another it does so at right angles, which greatly reduces capacitance. Many hams in the do-it-yourself era had elaborate coil winding apparatus just for doing this sort of thing. **ANSWER A.**

E5D09 What happens to reactive power in an AC circuit that has both ideal inductors and ideal capacitors?

A. It is dissipated as heat in the circuit.
B. It is repeatedly exchanged between the associated magnetic and electric fields, but is not dissipated.
C. It is dissipated as kinetic energy in the circuit.
D. It is dissipated in the formation of inductive and capacitive fields.

In a circuit that has both capacitors and inductors, the ***reactive power*** oscillates back and forth between the magnetic and electric fields and ***is NOT dissipated***. **ANSWER B.**

E5D10 How can the true power be determined in an AC circuit where the voltage and current are out of phase?

A. By multiplying the apparent power times the power factor.
B. By dividing the reactive power by the power factor.
C. By dividing the apparent power by the power factor.
D. By multiplying the reactive power times the power factor.

If we ***multiply apparent power times the power factor*** of an AC voltage and current out of phase, we can ***determine true power***. **ANSWER A.**

Apparent and True Power

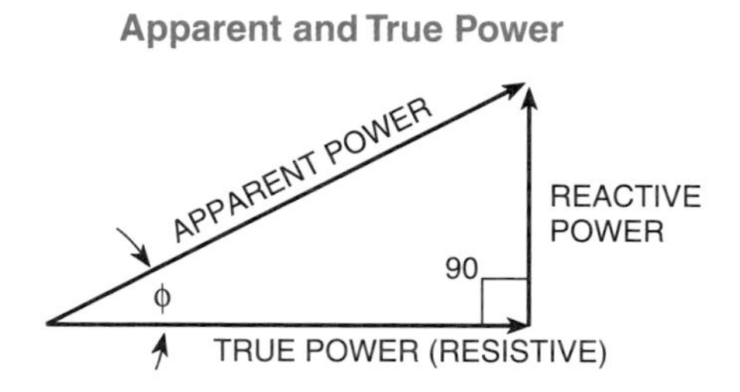

True power is power dissipated as heat, while reactive power of a circuit is stored in inductances or capacitances and then returned to the circuit.

Apparent power is voltage times current without taking into account the phase angle between them. According to right angle trigonometry,

$$\text{cosine } \phi = \frac{\text{side adjacent}}{\text{hypotenuse}}$$

therefore,

$$\text{cosine } \phi = \frac{\text{True Power}}{\text{Apparent Power}}$$

therefore,

True Power = Apparent Power × Cosine ϕ

Cosine ϕ is called the *power factor* (PF) of a circuit.

Since apparent power is E × I,

True Power = E × I × Cos ϕ

or

True Power = E × I × PF

E5D11 What is the power factor of an R-L circuit having a 60 degree phase angle between the voltage and the current?

A. 1.414.
B. 0.866.
C. 0.5.
D. 1.73.

The power factor of any circuit is always ***the COSINE of the phase angle between voltage and current***. So, if the given phase angle is 60 degrees, all you need to do is take the COSINE of 60 degrees. If you're using a scientific calculator, it's as simple as typing in 60 and hitting the COS button! ***The answer is 0.5***.
ANSWER C.

ϕ	$\cos \phi$
30°	0.866
45°	0.707
60°	0.5

E5D12 How many watts are consumed in a circuit having a power factor of 0.2 if the input is 100-VAC at 4 amperes?

A. 400 watts.
B. 80 watts.
C. 2000 watts.
D. 50 watts.

The solution to this question uses the formula:

True power = E (volts) × I (amps) × PF (power factor).

True power = ***100 volts × 4 amps × 0.2 = 80 watts***. Easy enough, right? **ANSWER B.**

E5D13 How much power is consumed in a circuit consisting of a 100 ohm resistor in series with a 100 ohm inductive reactance drawing 1 ampere?

A. 70.7 Watts.
B. 100 Watts.
C. 141.4 Watts.
D. 200 Watts.

The 100 ohms inductive reactance in the coil will not be a factor in this problem, as a pure inductance will not consume power in this circuit. Ohm's Law, $P = I^2 \times R$, will result in $1 \times 1 = 1$, times R of 100 ohms resistance, equal to ***100 watts*** of power consumed. **ANSWER B.**

$P = I^2 \times R$
$P = (1 \times 1) \times 100$
$P = 1 \times 100$
$P = 100$ watts

Where: P = Power
I = Amps
R = Resistance in ohms

E5D14 What is reactive power?

A. Wattless, nonproductive power.
B. Power consumed in wire resistance in an inductor.
C. Power lost because of capacitor leakage.
D. Power consumed in circuit Q.

The letters AC are found in the word ***reactive*** (meaning out-of-phase). It indicates ***non-productive power*** produced in circuits containing inductors and capacitors. Since out-of-phase power in inductors will tend to cancel out-of-phase power in capacitors, reactive power is ***wattless***. It is not converted to heat and dissipated. It is energy being stored temporarily in a field and then returned to the circuit. It circulates back and forth in a coil's magnetic field and/or a capacitor's electrostatic field. **ANSWER A.**

Good To Know:

Working Backwards

You need to be able to work every electronics formula we've discussed forwards, backwards, and upside down. What this means is that you need to know $E = I \times R$, $R = E/I$, and $I = E/R$. You always need to be able to figure the unknown based on the two givens. This not only applies to Ohm's Law, which we gave as an example, but to every *formula you've learned. For example, we know that power factor is equal to the COSINE of the phase angle between voltage and current. PF = COS ß. But do you know how to figure out the phase angle if you're given the power factor?*

To do this, we need to take the inverse *function. So, let's say we're given a power factor of 0.5 and asked to find the phase angle. In this case we need to take the ARC COSINE of 0.5. On a calculator, it's simple. Type in 0.5, press the INVERSE function key, and then hit the COS^{-1} key, which will give you 60 degrees.*

These types of questions do come up in FCC examinations, but more importantly, they come up in real world electronics problems! Very seldom does a real world troubleshooting problem come as "neatly packaged" as the exam questions you've practiced!

Working problems backwards is a great way to check your work, and is also a good way to be sure you're using your calculator correctly! If you take the Log_{10} of 14, you'll get 1.146 followed by 28 more digits (depending on the precision of your calculator). You should be able to take that same monstrous number, hit 10^x and come back with exactly 14!

Practice using your scientific calculator on problems where you KNOW the answer. For example, not all calculators handle the memories the same way, and the time to find out how YOURS does this is not in the heat of an exam! Likewise, the PRIORITY OF OPERATION is not *the same on every calculator! When in doubt, ALWAYS use parentheses when combining addition and multiplication, to FORCE the order of operation you expect.*

E5D15 What is the power factor of an R-L circuit having a 45 degree phase angle between the voltage and the current?

A. 0.866.
B. 1.0.
C. 0.5.
D. 0.707.

No matter how many decades we've been playing with a scientific calculator, we always visualize these problems using rectangular coordinates. Likewise, you'll get a far greater understanding of these concepts by actually graphing them than by simply punching in numbers on a calculator. However, since you are allowed to use your calculator, the sequence here is simply ***45 COS = 0.707***. By now that 0.707 should be painfully familiar to you! It is 1 over the square root of 2. Because it is so familiar, it is often used as a "wrong answer distractor" on a lot of questions. Don't be led astray by this trick. (Except in cases like this when it actually *is* the right answer!) **ANSWER D.**

E5D16 What is the power factor of an R-L circuit having a 30 degree phase angle between the voltage and the current?

A. 1.73.
B. 0.5.
C. 0.866.
D. 0.577.

The power factor of a circuit is equal to the cos ϕ where ϕ is the phase angle, in this case 30 degrees. At ***30 degrees***, the ***power factor is 0.866***. If you don't pass your Extra Class exam, you could be 86'd out of the room! **ANSWER C.**

E5D17 How many watts are consumed in a circuit having a power factor of 0.6 if the input is 200VAC at 5 amperes?

A. 200 watts.
B. 1000 watts.
C. 1600 watts.
D. 600 watts.

Here are two simple problems that are based on the formula:

True power = E (volts) × I (amps) × PF (power factor).

True power = ***200 volts X 5 amps X 0.6 = 600 watts***. Easy, huh? **ANSWER D.**

E5D18 How many watts are consumed in a circuit having a power factor of 0.71 if the apparent power is 500VA?

A. 704 W.
B. 355 W.
C. 252 W.
D. 1.42 mW.

Consumed power = Apparent Power (Volt Amps) x Power Factor. Here they have already calculated apparent power at 500 VA (Volt Amps). ***Multiply apparent power by the given power factor 500 × 0.71 = 355 watts*** consumed in the circuit. **ANSWER B.**

E4E04 How can conducted and radiated noise caused by an automobile alternator be suppressed?

A. By installing filter capacitors in series with the DC power lead and a blocking capacitor in the field lead.
B. By installing a noise suppression resistor and a blocking capacitor in both leads.
C. By installing a high-pass filter in series with the radio's power lead and a low-pass filter in parallel with the field lead.
D. By connecting the radio's power leads directly to the battery and by installing coaxial capacitors in line with the alternator leads.

Connect your radio's red and black leads ***directly to your battery's*** positive and negative terminals using the shortest possible run. Be sure you put a fuse right next to the battery's positive pick-off point for safety. Also, it is recommended that you ***install coaxial capacitors in the alternator leads***. **ANSWER D.**

E4E05 How can noise from an electric motor be suppressed?

A. By installing a high pass filter in series with the motor's power leads.
B. By installing a brute-force AC-line filter in series with the motor leads.
C. By installing a bypass capacitor in series with the motor leads.
D. By using a ground-fault current interrupter in the circuit used to power the motor.

One way of reducing noise coming up the line from an electric motor is to ***install an AC-line filter in series with the motor leads***. The incorrect answers have the right components, but the wrong usage. RF beads are used to keep RF off of specific circuits, and bypass capacitors in parallel are used to bypass noise. As for the ground-fault interrupter, this won't help with motor noise. Go with the ***brute-force AC-line filter in series***. **ANSWER B.**

E6D06 What core material property determines the inductance of a toroidal inductor?

A. Thermal impedance.
B. Resistance.
C. Reactivity.
D. Permeability.

The key words in the question are "***core material property.***" The material that determines the ***inductance of a toroid is the permeability*** of the iron donut. **ANSWER D.**

E6D07 What is the usable frequency range of inductors that use toroidal cores, assuming a correct selection of core material for the frequency being used?

A. From a few kHz to no more than 30 MHz.
B. From less than 20 Hz to approximately 300 MHz.
C. From approximately 10 Hz to no more than 3000 kHz.
D. From about 100 kHz to at least 1000 GHz.

If you carefully select the core material, the ***frequency range*** of the inductor will span approximately ***20 Hz to 300 MHz***. Refer to the charts of core size and core specs to determine which popular toroidal cores to use. **ANSWER B.**

E6D08 What is one reason for using powdered-iron cores rather than ferrite cores in an inductor?

A. Powdered-iron cores generally have greater initial permeability.
B. Powdered-iron cores generally maintain their characteristics at higher currents.
C. Powdered-iron cores generally require fewer turns to produce a given inductance.
D. Powdered-iron cores use smaller diameter wire for the same inductance.

When designing high current circuits, such as those in linear amplifiers, ***powdered-iron toroids offer the greatest high current capabilities*** while maintaining a stable inductance and good temperature stability. **ANSWER B.**

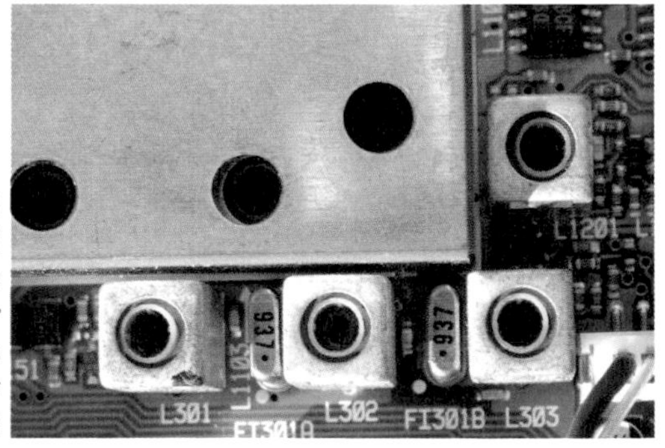

Don't be tempted to adjust any coil core unless you have the shop manual for the radio on hand. Technical shop manuals are sold separately from your radio. Unless you have all the scopes and test equipment on hand, stay away from doing any coil turning!

E6D05 What is one reason for using ferrite cores rather than powdered-iron in an inductor?

A. Ferrite toroids generally have lower initial permeability.
B. Ferrite toroids generally have better temperature stability.
C. Ferrite toroids generally require fewer turns to produce a given inductance value.
D. Ferrite toroids are easier to use with surface mount technology.

A sound reason to use ferrite toroids rather than powdered-iron toroids is the fact that ***ferrite toroids require fewer turns of wire to produce a given inductance*** value. **ANSWER C.**

E6D10 What is a primary advantage of using a toroidal core instead of a solenoidal core in an inductor?

A. Toroidal cores confine most of the magnetic field within the core material.
B. Toroidal cores make it easier to couple the magnetic energy into other components.
C. Toroidal cores exhibit greater hysteresis.
D. Toroidal cores have lower Q characteristics.

The ***toroidal core inductor concentrates the magnetic field within the core material***, protecting other nearby components from stray magnetic fields from it. **ANSWER A.**

E6D01 How many turns will be required to produce a 5-microhenry inductor using a powdered-iron toroidal core that has an inductance index (A_L) value of 40 microhenries/100 turns?

A. 35 turns. C. 79 turns.
B. 13 turns. D. 141 turns.

It used to be said that radio is nothing more than winding coils. If you do any amount of homebrew radio construction, this is still largely true! The powdered iron or ferrite toroidal core has made coil winding a lot less tedious and more precise than winding hundreds of feet of wire around an oatmeal box as in days of yore. The manufacturer of your powdered iron or ferrite core will supply some base permeability data, which will tell you how much inductance your inductor will have for a given number of turns. For powdered iron cores, the value will normally be specified in microhenries per 100 turns. But here's a little snag: the inductance of any inductor, regardless of core, goes up as the square of the number of turns. So, you'll end up using this formula:

$$N = 100\sqrt{\frac{L}{A_L}}$$

where: N is number of turns
L is inductance in **microhenries**
A_L (A sub L) is inductance index in **μH per 100 turns**

$$N = 100\sqrt{\frac{5}{40}}$$

$$N = 100\sqrt{0.125} = 100\sqrt{12.5 \times 10^{-2}}$$

$$N = 3.53 \times 10^{-1} \times 10^{2} = 3.53 \times 10^{1} = 35.3$$

To work out the problem with the details from this question, we take the desired value: 5 microhenries. We know the value of A_L is 40. (Don't divide by 100. The A_L given has already had that done.) So 5/40 = 0.125. Take the square root of 0.125, which gives us 0.3535. Finally, multiply the whole thing by 100, which gives us 35.35, rounded down to ***35 turns***. (Whew!) Don't fret too much about the fraction of a turn, in most cases you can round up or down to the nearest full turn. **ANSWER A.**

A Hypersil transformer, using a tape wound iron core is used in a great number of larger power supplies. This construction is efficient and quiet.

E6D11 How many turns will be required to produce a 1-mH inductor using a core that has an inductance index (A_L) value of 523 millihenries/1000 turns?

A. 2 turns.
B. 4 turns.
C. 43 turns.
D. 229 turns.

Here is the formula to calculate the number of turns required to produce a specific inductance when you know the inductance index, A_L, of the material. A_L is determined and published by the core manufacturer and is an index that is the microhenries of inductance per 100 turns or millihenries of inductance per 1000 turns of single layer wire on the core.

$$N = 1000\sqrt{\frac{L}{A_L}}$$

where: N is number of turns needed
L is inductance in **millihenries** of core inductor
A_L (A sub L) is inductance index in **mH per 1000 turns**

$$N = 1000\sqrt{\frac{1}{523}}$$

$$N = 1000\sqrt{0.001912} = 1000\sqrt{19.12 \times 10^{-4}}$$

$$N = 1 \times 10^{3} \times 4.37 \times 10^{-4}$$

$$N = 4.37 \times 10^{1} = 43.7$$

Here's how to work it out on your calculator. Got a calculator, right? Hit the clear button several times. Then enter ***1 ÷ 523 =*** The result is ***0.001912. Press √*** and the result is ***0.0437264. Press × 1000 = N for the answer 43.7264***. Your closest correct answer is 43. **ANSWER C**

You must use the correct toroidal core mix to obtain the proper amount of inductance for your circuit.

CD 5 TRACK 2

Components in Your New Rig

E6A01 In what application is gallium arsenide used as a semiconductor material in preference to germanium or silicon?

A. In high-current rectifier circuits.
B. In high-power audio circuits.
C. In microwave circuits.
D. In very low frequency RF circuits.

Here's a new word for your amateur radio vocabulary: GaAsFET. The GaAs stands for ***gallium arsenide***, and FET stands for field effect transistor. We use the ***GaAsFET*** transistor in VHF, UHF, and ***microwave receivers***. This transistor offers excellent signal gain with an extremely-low noise figure. **ANSWER C.**

E6A02 Which of the following semiconductor materials contains excess free electrons?

A. N-type.
B. P-type.
C. Bipolar.
D. Insulated gate.

There are ***more free electrons*** orbiting the nucleus in an N-type material. **ANSWER A.**

You will have one question from this group about semiconductor materials. Remember this phrase, "ANFREE" for an N-type material where antimony atoms or arsenic atoms have been added, and there are more free electrons in an N-type semiconductor. "PIUMH" (pronounced like sneezing) is a P-type material produced by adding either galIIUM or indIUM, producing more "Potholes" than free electrons. Also, anything that PROVIDES is a donor, and an Acceptor impurity will Add holes. Got it? Good reading about semiconductors is found in the latest edition of the American Radio Relay League Handbook for Radio Amateurs.

E6A04 What is the name given to an impurity atom that adds holes to a semiconductor crystal structure?

A. Insulator impurity.
B. N-type impurity.
C. Acceptor impurity.
D. Donor impurity.

The ***impurity atom that adds holes*** to a semiconductor crystal structure is called an ***acceptor impurity***. It's an acceptable impurity. **ANSWER C.**

E6A05 What is the alpha of a bipolar junction transistor?

A. The change of collector current with respect to base current.
B. The change of base current with respect to collector current.
C. The change of collector current with respect to emitter current.
D. The change of collector current with respect to gate current.

The ***alpha*** of a bipolar junction transistor ***is the ratio of a change of collector current to the change of emitter current*** in a common base configuration. It tells you the current gain of the device only; in an actual amplifier the current gain can be considerably less, so alpha is actually a best-case scenario. Your overall amplifier can never have more current gain than the alpha of your transistor. **ANSWER C.**

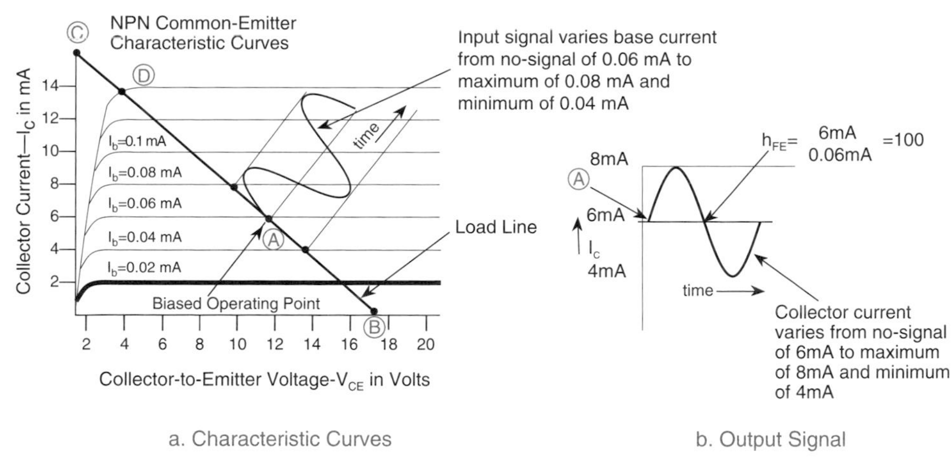

Biasing sets "No-Signal" operating point

E6A06 What is the beta of a bipolar junction transistor?

A. The frequency at which the current gain is reduced to 1.
B. The change in collector current with respect to base current.
C. The breakdown voltage of the base to collector junction.
D. The switching speed of the transistor.

Another similar figure is the ***beta*** of the transistor, which ***is the ratio of collector current change to the change in base current***, in the common emitter configuration. While the common emitter or *grounded emitter* configuration is more familiar to most hams, in the early days of transistors the common base configuration was more common, so the Alpha of a transistor was deemed more "basic." However, for most voltage amplifiers the common emitter is more prevalent, so the beta figure is more useful. In either case, alpha or beta, it is a measurement of the effective current gain of the transistor, sometimes loosely defined as "amplification factor." **ANSWER B.**

Bipolar Transistor Basics – PNP and NPN

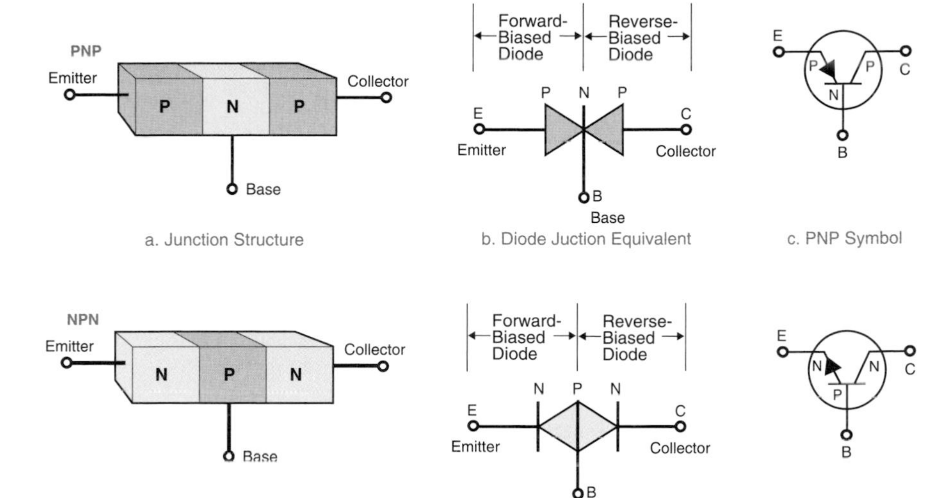

a. Junction Structure b. Diode Juction Equivalent c. PNP Symbol

d. Junction Structure e. Diode Junction Equivalent f. NPN Symbol

There are two types of transistors, bipolar and field-effect. A bipolar transistor is a combination of two junctions of semiconductor material built into a semiconductor chip (usually silicon). There are two types – PNP and NPN. Their junction structures are shown in Figure a and d, respectively. For a transistor that produces gain, the emitter-base junction is a forward-biased diode, and the collector-base junction is a reverse-biased diode. The diode equivalents of PNP and NPN transistors are shown in Figure b and e, respectively. However, unlike a reverse-biased diode that does not conduct current (except for a small leakage current), the reverse-biased collector-base junction of a bipolar transistor conducts collector current that is controlled by the current into the base at the base-emitter junction. The normal active device operation is shown below in Figure g and h.

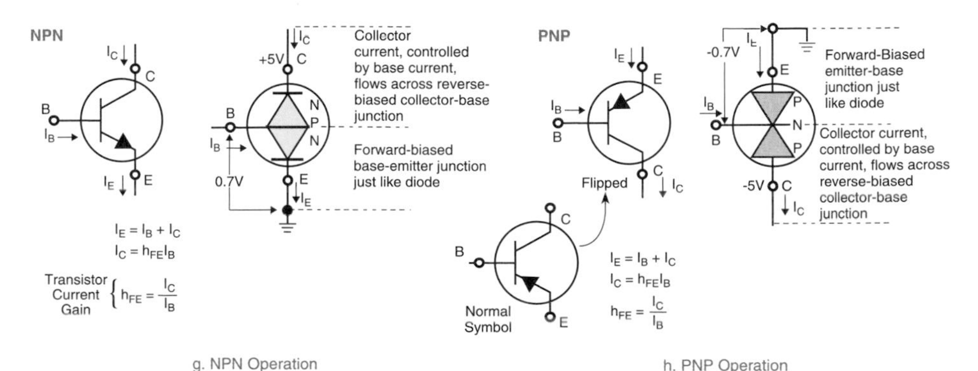

g. NPN Operation h. PNP Operation

An NPN silicon transistor P base is 0.7 V more positive that its N emitter, and the N collector is several more volts more positive than the emitter. The emitter current is the sum of the base and collector current. The current gain under any DC operating condition is hFE, the ratio of I_C to I_B, and current gains of 50 to 200 are common in modern-day silicon transistors. hFE is actually called "the common-emitter" current gain because the emitter is common in the circuit.

A PNP silicon transistor N base is 0.7 V negative with respect to its emitter in order to have the P emitter more positive than the N base. The P collector is several volts negative from the emitter to keep the collector-base junction reverse biased. As shown in Figure h, the same current equations apply, and the current gain, hFE, is the same. The major difference is in the polarity of the voltages for operation. For NPN common-emitter operation the base and collector voltages are positive with respect to the emitter; while for the PNP the voltages are negative.

Source: *Basic Communications Electronics*, Hudson & Luecke, © 1999 Master Publishing, Inc., Niles, IL

E6A07 Which of the following indicates that a silicon NPN junction transistor is biased on?

A. Base-to-emitter resistance of approximately 6 to 7 ohms.
B. Base-to-emitter resistance of approximately 0.6 to 0.7 ohms.
C. Base-to-emitter voltage of approximately 6 to 7 volts.
D. Base-to-emitter voltage of approximately 0.6 to 0.7 volts.

When an NPN transistor is biased "on," the voltage on the base is positive with respect to the emitter. Free electrons can flow from the emitter to the base. A characteristic of a ***forward-biased NPN junction*** is that the ***voltage drop*** across it is always approximately ***0.6 to 0.7V***. You don't need to remove the transistor from the circuit for this easy transistor check. **ANSWER D.**

E6A08 What term indicates the frequency at which the grounded-base current gain of a transistor has decreased to 0.7 of the gain obtainable at 1 kHz?

A. Corner frequency.
B. Alpha rejection frequency.
C. Beta cutoff frequency.
D. Alpha cutoff frequency.

Grounded-base is the same as a common base configuration. Common base means alpha. Since ***1 kHz*** is mentioned in this question, they are referring to the ***alpha cutoff frequency*** in a transistor. **ANSWER D.**

E6A09 What is a depletion-mode FET?

A. An FET that exhibits a current flow between source and drain when no gate voltage is applied.
B. An FET that has no current flow between source and drain when no gate voltage is applied.
C. Any FET without a channel.
D. Any FET for which holes are the majority carriers.

Your handheld battery may drain unexpectedly if ***depletion-mode FETs*** are used in many of the circuits. Even though there is no gate voltage applied, ***current still continues to flow*** from these components. **ANSWER A.**

E6A10 In Figure E6-2, what is the schematic symbol for an N-channel dual-gate MOSFET?

A. 2.
B. 4.
C. 5.
D. 6.

Note that the ***arrow is pointing iN***, and there are now ***two gates***. **ANSWER B.**

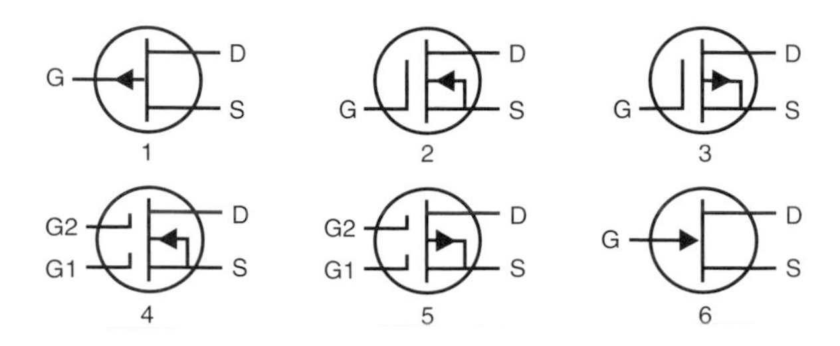

Figure E6-2

E6A11 In Figure E6-2, what is the schematic symbol for a P-channel junction FET?

A. 1.
B. 2.
C. 3.
D. 6.

In the ***P-channel junction FET***, the ***arrow is Pointing out***. **ANSWER A.**

E6A12 Why do many MOSFET devices have internally connected Zener diodes on the gates?

A. To provide a voltage reference for the correct amount of reverse-bias gate voltage.
B. To protect the substrate from excessive voltages.
C. To keep the gate voltage within specifications and prevent the device from overheating.
D. To reduce the chance of the gate insulation being punctured by static discharges or excessive voltages.

The "front end" transistors of the modern ham transceiver must sometimes sustain major amounts of incoming voltage as a result of signals from nearby transmitters, or a nearby lightning strike. This amount of ***voltage could destroy a MOSFET*** if it weren't for the built-in ***gate-protective Zener diodes***. **ANSWER D.**

E6A14 How does DC input impedance at the gate of a field-effect transistor compare with the DC input impedance of a bipolar transistor?

A. They are both low impedance.
B. An FET has low input impedance; a bipolar transistor has high input impedance.
C. An FET has high input impedance; a bipolar transistor has low input impedance.
D. They are both high impedance.

A field effect transistor, ***FET,*** is much easier to work with in circuits than the simple bipolar transistor. Its ***higher input impedance*** is less likely to load down a circuit to which it is attached, and the FET can handle big swings in signal levels. **ANSWER C.**

E6A16 What are the majority charge carriers in N-type semiconductor material?

A. Holes.
B. Free electrons.
C. Free protons.
D. Free neutrons.

In an ***N-type*** semiconductor material, there are more ***free electrons*** than holes. **ANSWER B.**

E6A15 Which semiconductor material contains excess holes in the outer shell of electrons?

A. N-type.
B. P-type.
C. Superconductor-type.
D. Bipolar-type.

If there are ***more holes*** than electrons, it is a ***P-type*** semiconductor. **ANSWER B.**

E6A17 What are the names of the three terminals of a field-effect transistor?

A. Gate 1, gate 2, drain.
B. Emitter, base, collector.
C. Emitter, base 1, base 2.
D. Gate, drain, source.

Gosh Darn Semiconductors! ***Gate, drain, source***. Gosh darn field effect transistor, a semiconductor! **ANSWER D.**

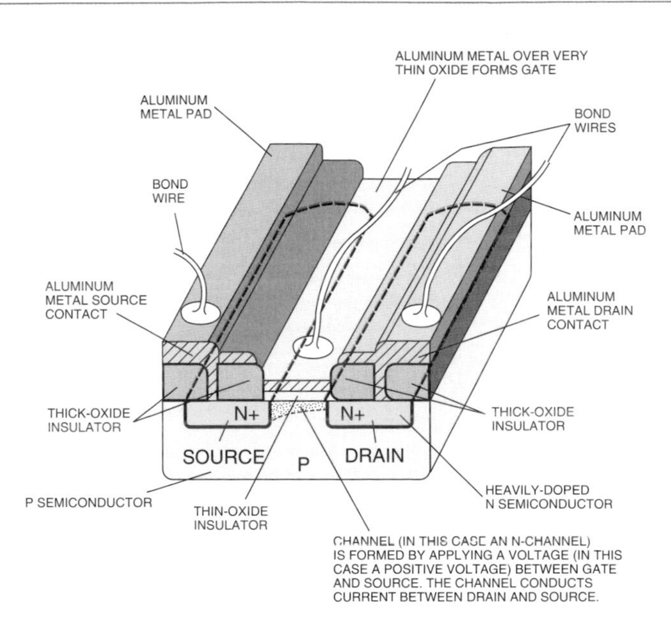

Pictorial of FET Construction (N-Channel Enhancement)

Unlike bipolar transistors that depend on current into the base to control collector current, field-effect transistor current between source and drain is controlled by a voltage on a gate. The basic structure of an N-channel MOSFET is shown here. Heavily-doped N semiconductor material forms source and drain regions in a P semiconductor material substrate. The region between the source and drain is the gate region, where a thin layer of oxide insulates the P semiconductor substrate underneath from a metal plate that is deposited over the thin oxide. A thick oxide layer over the source and drain regions insulates metal connection pads from the substrate. Holes in this thick oxide layer allow the metal pads to contact the source and drain. There are four common types of MOSFETs: P-channel depletion and enhancement mode devices; and N-channel depletion and enhancement mode devices.

Source: *Basic Communications Electronics*, Hudson & Luecke, © 1999 Master Publishing, Inc., Niles, IL

E6B01 What is the most useful characteristic of a Zener diode?

A. A constant current drop under conditions of varying voltage.
B. A constant voltage drop under conditions of varying current.
C. A negative resistance region.
D. An internal capacitance that varies with the applied voltage.

Always remember that a ***Zener diode*** is for voltage regulation which gives you ***constant voltage*** even though current changes. **ANSWER B.**

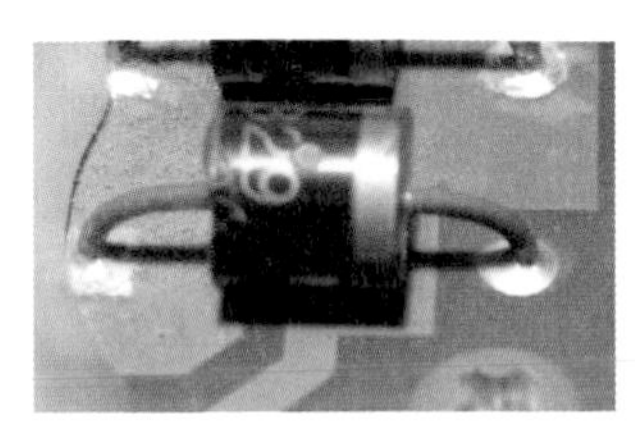

Zener Diode

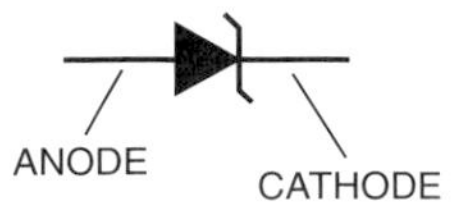

Here is the schematic symbol of a Zener diode. Since a diode only passes energy in one direction, look for that one-way arrow, plus a "Z" indicating it is a Zener diode. Doesn't that vertical line look like a tiny "Z?"

E6B02 What is an important characteristic of a Schottky diode as compared to an ordinary silicon diode when used as a power supply rectifier?

A. Much higher reverse voltage breakdown.
B. Controlled reverse avalanche voltage.
C. Enhanced carrier retention time.
D. Less forward voltage drop.

The Schottky diode is regularly found in solar power and power supply circuits because of its lower forward resistance and a slightly higher peak inverse voltage (PIV) rating. They are slightly more expensive than traditional power supply diodes, but they are used regularly in mobile and home solar systems because the ***Schottky diode offers minimal voltage drop*** while charging the sealed battery system, and they prevent current from flowing out of the battery system and back into the solar panel at night. **ANSWER D.**

E6B03 What special type of diode is capable of both amplification and oscillation?

A. Point contact.
B. Zener.
C. Tunnel.
D. Junction.

The ***tunnel diode*** is commonly used in both ***amplifier and oscillator*** circuits. The negative resistance region is particularly useful in oscillators. **ANSWER C.**

E6B04 What type of semiconductor device is designed for use as a voltage controlled capacitor?

A. Varactor diode.
B. Tunnel diode.
C. Silicon-controlled rectifier.
D. Zener diode.

The ***varactor diode*** is used to tune VHF and UHF circuits by ***varying the voltage*** applied to it. **ANSWER A.**

E6B06 Which of the following is a common use of a hot-carrier diode?

A. As balanced mixers in FM generation.
B. As a variable capacitance in an automatic frequency control circuit.
C. As a constant voltage reference in a power supply.
D. As a VHF/UHF mixer or detector.

When you get involved with ***VHF and UHF*** equipment, you'll find ***hot-carrier diodes*** commonly used as ***mixers or detectors*** because of their low noise figure characteristics. Hot-carrier diodes also have a lower forward voltage drop, making them more sensitive in mixer circuits and even in crystal radio detectors. **ANSWER D.**

E6B07 What is the failure mechanism when a junction diode fails due to excessive current?

A. Excessive inverse voltage.
B. Excessive junction temperature.
C. Insufficient forward voltage.
D. Charge carrier depletion.

Guess what will kill most electronic components? High temperature! The limit of the maximum forward current in a junction diode is the junction temperature. This is why you'll find these diodes mounted to the chassis of your equipment. The chassis acts as a heat sink to keep the ***junction temperature*** from ***exceeding its maximum limit***. **ANSWER B.**

E6B08 Which of the following describes a type of semiconductor diode?

A. Metal-semiconductor junction.
B. Electrolytic rectifier.
C. CMOS-field effect.
D. Thermionic emission diode.

The two main categories of ***semiconductor diodes*** are PN junction and ***metal-semiconductor junction***. The metal-semiconductor diode might be found in transistors, power supply rectifier diodes, and photovoltaic panels. **ANSWER A.**

E6B09 What is a common use for point contact diodes?

A. As a constant current source.
B. As a constant voltage source.
C. As an RF detector.
D. As a high voltage rectifier.

Because it only conducts in one direction, the little ***point contact diode*** may be used as an ***RF detector*** in some VHF and UHF equipment. **ANSWER C.**

E6B12 What is one common use for PIN diodes?

A. As a constant current source.
B. As a constant voltage source.
C. As an RF switch.
D. As a high voltage rectifier.

PIN diodes may be used in small, handheld transceivers for ***RF switching***. This gets away from mechanical relays for switching. As convenient as PIN diodes are, they do have limitations. They usually do not work well in the lower HF frequency range. And at high power levels, the DC control circuit must supply considerable power to turn the diode fully ON. **ANSWER C.**

E6A03 Why does a PN-junction diode not conduct current when reverse biased?

A. Only P-type semiconductor material can conduct current.
B. Only N-type semiconductor material can conduct current.
C. Holes in P-type material and electrons in the N-type material are separated by the applied voltage, widening the depletion region.
D. Excess holes in P-type material combine with the electrons in N-type material, converting the entire diode into an insulator.

A depletion region is the area of a transistor junction where charge carriers are very few to non-existent. The ***greater the electric field*** applied to this junction, the ***greater the depletion region***. **ANSWER C.**

E6B05 What characteristic of a PIN diode makes it useful as an RF switch or attenuator?

A. Extremely high reverse breakdown voltage.
B. Ability to dissipate large amounts of power.
C. Reverse bias controls its forward voltage drop.
D. A large region of intrinsic material.

The PIN diode operating at forward bias will act as a closed switch, passing microwave radio frequency signals. If you don't have transmit/receive relays in that new rig, chances are the silent T/R circuitry is a PIN diode. The ***PIN diode*** has a ***layer of intrinsic, un-doped semiconductor material*** between the P and N type materials. Forward bias can turn it into a completed circuit switch, yet a decrease of forward bias may turn it into a variable resistance to RF signals passing through. **ANSWER D.**

E6B11 What is used to control the attenuation of RF signals by a PIN diode?

A. Forward DC bias current.
B. A sub-harmonic pump signal.
C. Reverse voltage larger than the RF signal.
D. Capacitance of an RF coupling capacitor.

The ***PIN diode*** can not only work as an RF switch, but also as an ***RF attenuator*** if the ***forward DC bias current*** is reduced. With just the right amount of forward bias, the PIN diode will act as a variable resistance to RF signals. **ANSWER A.**

E6F07 What is a solid state relay?

A. A relay using transistors to drive the relay coil.
B. A device that uses semiconductors to implement the functions of an electromechanical relay.
C. A mechanical relay that latches in the on or off state each time it is pulsed.
D. A passive delay line.

The ***PIN diode,*** sometimes called a ***solid state relay***, is capable of passing microwave and RF signals when forward biased. The PIN diode is not capable of replacing the high voltage and high current handling properties of mechanical relays in automatic antenna tuners. Electromechanical relays in tuners will stick around for some time to come. **ANSWER B.**

E6B10 In Figure E6-3, what is the schematic symbol for a light-emitting diode?

A. 1. C. 6.
B. 5. D. 7.

The light-emitting diode (LED) is a giveaway on the schematic diagram! Look for the ***two arrows indicating an illuminating device***. **ANSWER B.**

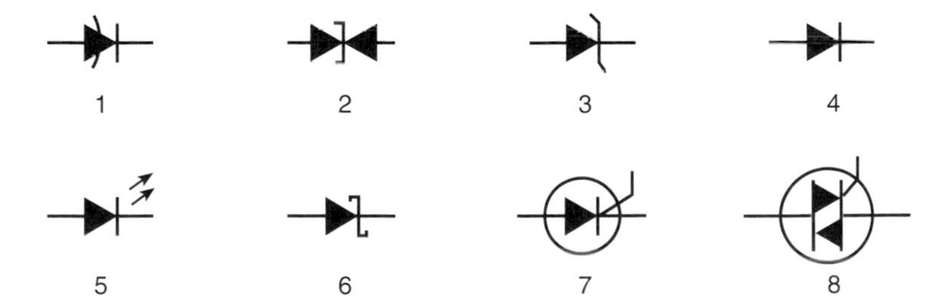

Figure E6-3

E6B13 What type of bias is required for an LED to emit light?

A. Reverse bias. C. Zero bias.
B. Forward bias. D. Inductive bias.

The ***LED*** requires ***forward bias*** in order to ***illuminate***. As indicated in symbol 5 of Figure E6-3, the LED is a diode and must be forward biased to give off light. **ANSWER B.**

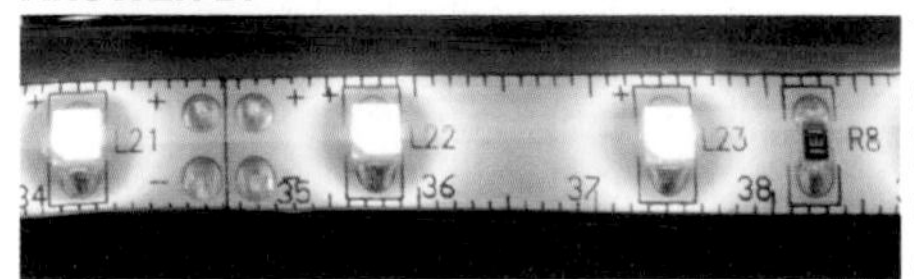

The popular 12 volt LED flexible light strip for shack illumination. Remember, if you use a pulse width modulated dimmer you'll have lots of HF noise to contend with in the dimmed mode.

E6A13 What do the initials CMOS stand for?

A. Common Mode Oscillating System.
B. Complementary Mica-Oxide Silicon.
C. Complementary Metal-Oxide Semiconductor.
D. Common Mode Organic Silicon.

When we talk about transistorized devices, they are in fact semi-conductors. A *CMOS* is made from layers of *metal-oxide semiconductor* material. **ANSWER C.**

A pair of large scale integrated (LSI) circuit chips.

E6C05 What is an advantage of CMOS logic devices over TTL devices?

A. Differential output capability.
B. Lower distortion.
C. Immune to damage from static discharge.
D. Lower power consumption.

The Complementary Metal Oxide Semiconductor (CMOS) contains both N-channel and P-channel MOSFETs within a single chip. The greatest advantage of the *CMOS* is its *extremely low power consumption* in that new handheld you just bought for the kids. **ANSWER D.**

E6C06 Why do CMOS digital integrated circuits have high immunity to noise on the input signal or power supply?

A. Larger bypass capacitors are used in CMOS circuit design.
B. The input switching threshold is about two times the power supply voltage.
C. The input switching threshold is about one-half the power supply voltage.
D. Input signals are stronger.

The *CMOS* device is relatively *immune to noise* because its *switching threshold is one-half the power supply voltage*. As a result, much larger power supply noise transients or input transients will not cause a transition in the input state when compared to other logic circuit types. **ANSWER C.**

E6E04 Which is the most common input and output impedance of circuits that use MMICs?

A. 50 ohms.
B. 300 ohms.
C. 450 ohms.
D. 10 ohms.

The beauty of the *MMIC* (Monolithic Microwave Integrated Circuit) is that it has a *characteristic impedance of 50 ohms*: just what the microwavers love. **ANSWER A.**

E6E03 Which of the following materials is likely to provide the highest frequency of operation when used in MMICs?

A. Silicon.
B. Silicon nitride.
C. Silicon dioxide.
D. Gallium nitride.

When most people think of semiconductors, they think of silicon. While silicon has a lot going for it – it's cheap, abundant, and can handle high temperatures – it is not the world's best semiconductor. Many other elements and compounds are much faster than silicon or have other highly desirable traits, such as exceptionally low noise. A monolithic microwave integrated circuit (*MMIC*) usually uses something other than silicon to do its job; one of the best performers being *gallium nitride*. **ANSWER D.**

E6E07 Which of the following is typically used to construct a MMIC-based microwave amplifier?

A. Ground-plane construction.
B. Microstrip construction.
C. Point-to-point construction.
D. Wave-soldering construction.

Up at microwave frequencies, we use ***microstrip*** construction, which is ideal for the MMIC. **ANSWER B.**

Monolithic Microwave Integrated Circuit
Photo Courtesy of Hewlett Packard Co.

E6E06 What characteristics of the MMIC make it a popular choice for VHF through microwave circuits?

A. The ability to retrieve information from a single signal even in the presence of other strong signals.
B. Plate current that is controlled by a control grid.
C. Nearly infinite gain, very high input impedance, and very low output impedance.
D. Controlled gain, low noise figure, and constant input and output impedance over the specified frequency range.

A monolithic microwave integrated circuit (MMIC) is a radio frequency "chip" that is designed for high gain, wide bandwidth, and good stability. The MMIC usually only has an input terminal, an output terminal and a power terminal, so it often resembles a simple "pill" style RF transistor. However MMICs are often incorporated into blocks or modules with SMA or other convenient RF connectors. MMICs can be cascaded to provide extremely ***high values of RF gain, and provide close to 50 ohm impedance matching at the input and output*** ports across their normal operating range. A complete receiving system can be built by plugging MMICs together with a minimum of external components. **ANSWER D.**

E6E08 How is voltage from a power supply normally furnished to the most common type of monolithic microwave integrated circuit (MMIC)?

A. Through a resistor and/or RF choke connected to the amplifier output lead.
B. MMICs require no operating bias.
C. Through a capacitor and RF choke connected to the amplifier input lead.
D. Directly to the bias voltage (VCC IN) lead.

The operating bias voltage for the MMIC is developed through a ***resistor and/or RF choke connected to the amplifier output lead***. **ANSWER A.**

E6C01 What is the function of hysteresis in a comparator?

A. To prevent input noise from causing unstable output signals.
B. To allow the comparator to be used with AC input signals.
C. To cause the output to change states continually.
D. To increase the sensitivity.

Hysteresis can be thought of as electrical "stickiness" or friction. It can be designed into a circuit to require a little extra "nudge" to make it change state, thus ***making the circuit less susceptible to random noise*** or other signals. **ANSWER A.**

E6C02 What happens when the level of a comparator's input signal crosses the threshold?

A. The IC input can be damaged.
B. The comparator changes its output state.
C. The comparator enters latch-up.
D. The feedback loop becomes unstable.

As the name suggests, a comparator compares the input at one terminal to the input at another terminal, and makes a concrete decision as to which input is larger. ***When the level of a comparator's input signal crosses the threshold, it changes its output state.*** A comparator is the simplest analog-to-digital converter, sometimes called a "1 bit" converter (which usually costs less than two bits). **ANSWER B.**

E6E10 What is the packaging technique in which leadless components are soldered directly to circuit boards?

A. Direct soldering.
B. Virtual lead mounting.
C. Stripped lead.
D. Surface mount.

It's probably safe to say that ***surface mount*** construction is the predominant construction method in all modern electronics. What once was all done with through-hole leads and tedious two-sided work, is now done completely on one side of a printed circuit board. While at first a bit daunting, repairing and building SMD circuits can be learned in a few hours. If an old geezer like Eric can learn to do it, so can you! **ANSWER D.**

Lead-less components require surface mount soldering skills if you plan to remove and replace a defective component.

E6E12 Why are high-power RF amplifier ICs and transistors sometimes mounted in ceramic packages?

A. High-voltage insulating ability.
B. Better dissipation of heat.
C. Enhanced sensitivity to light.
D. To provide a low-pass frequency response.

Very few materials known to man have high thermal conductivity as well as high electrical resistivity. Most of the time high electrical conductivity and high thermal conductivity are inseparable. Certain ceramics such as Beryllium Oxide have the unusual combination of properties necessary for good heat sink construction; that is, high thermal conductivity while acting as a near perfect electrical insulator. (Diamond is another such substance, but for obvious reasons we don't see a lot of it in amateur radio equipment!) High power RF integrated circuits are certainly more efficient than vacuum tube amplifiers, but they still must dissipate significant amounts of heat. Mounting ***a ceramic package device*** to a large heat sink ***will help keep things cool***. **ANSWER B.**

Power output transistors and transistor modules get warm, especially in the digital modes. The transistor on the right has an external heat sensor to self-protect against melt down.

E6E09 Which of the following component package types would be most suitable for use at frequencies above the HF range?

A. TO-220.
B. Axial lead.
C. Radial lead.
D. Surface mount.

Surface mount devices can be made extremely small, sometimes bordering on the microscopic. As such, they have very small parasitic inductance, making them suitable for VHF and microwave applications. ***Surface mount*** soldering is a skill worth learning. It just takes some practice and a good magnifying glass. **ANSWER D.**

E6E02 Which of the following device package is a through-hole type?

A. DIP.
B. PLCC.
C. Ball grid array.
D. SOT.

DIP (dual in-line package) construction is the traditional form of integrated circuit. The DIP device is mounted on top of a printed circuit board (PCB) with its leads passing through holes to be soldered on the trace side of the board. Individual transistors and other "old school" components are generally ***through-hole construction***. Unfortunately, many parts are becoming more difficult to find as surface mount devices are becoming the standard. It's a good idea to learn how to do surface mount soldering if you plan on doing much homebrewing or repairing modern radio gear. **ANSWER A.**

E6E11 What is a characteristic of DIP packaging used for integrated circuits?

A. Package mounts in a direct inverted position.
B. Low leakage doubly insulated package.
C. Two chips in each package (Dual In Package).
D. A total of two rows of connecting pins placed on opposite sides of the package (Dual In-line Package).

Back before everything became microscopic, most of us semi-old-timers used SIP (Single In-Line Package) or DIP (Dual In-Line Package) through-hole components for most of our construction projects. These components were named not for the chips they contained, as that could be many, but rather for the ***number of rows of connecting pins*** radiating from the package. DIPs have a row of pins on each side (Dual). SIP and DIP chips are still available, but becoming less standard, as SMD takes over the landscape. (Eric's done everything from point-to-point, pre-circuit-board vacuum tube wiring, to the latest SMD construction. He says it's a good idea to learn it all!) **ANSWER D.**

A gaggle of 5% tolerance resistors on a PCB. Notice the gold forth stripe, indicating their relative precision.

Logically Speaking of Counters

E6C08 In Figure E6-5, what is the schematic symbol for a NAND gate?

A. 1. C. 3.
B. 2. D. 4.

Since the ***NAND*** gate has a ***logic "0"*** at its output, spot the little ***tiny "o" on the nose of the symbol that looks like the letter D***. It looks the same as the AND gate, except for the small circle at its output. **ANSWER B.**

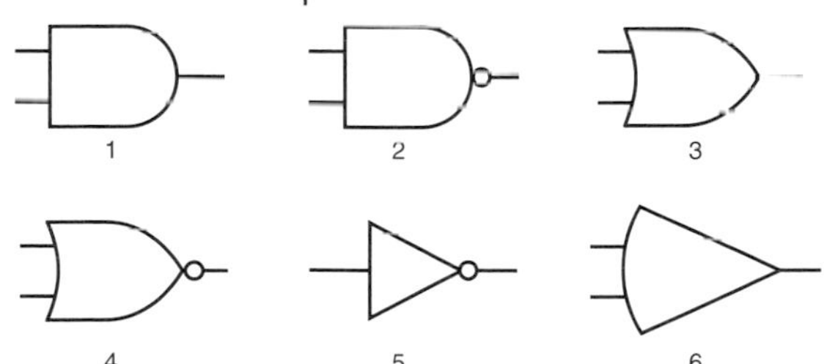

Figure E6-5

E6C10 In Figure E6-5, what is the schematic symbol for a NOR gate?

A. 1. C. 3.
B. 2. D. 4.

Since the ***NOR gate*** will give us a logic "0" at its output if any inputs are logic "1," look for the little "o" on the nose and a concave input. It looks just like an OR gate, except for the ***small circle on its nose***. **ANSWER D.**

E6C11 In Figure E6-5, what is the schematic symbol for the NOT operation (inverter)?

A. 2. C. 5.
B. 4. D. 6.

The ***NOT gate is triangular*** in shape and has a ***little "o" on its nose***. It looks altogether different from any of the other logic gates. You'll probably have one question based on gate symbols. Make up flashcards, and see if you can easily call out which is which. **ANSWER C.**

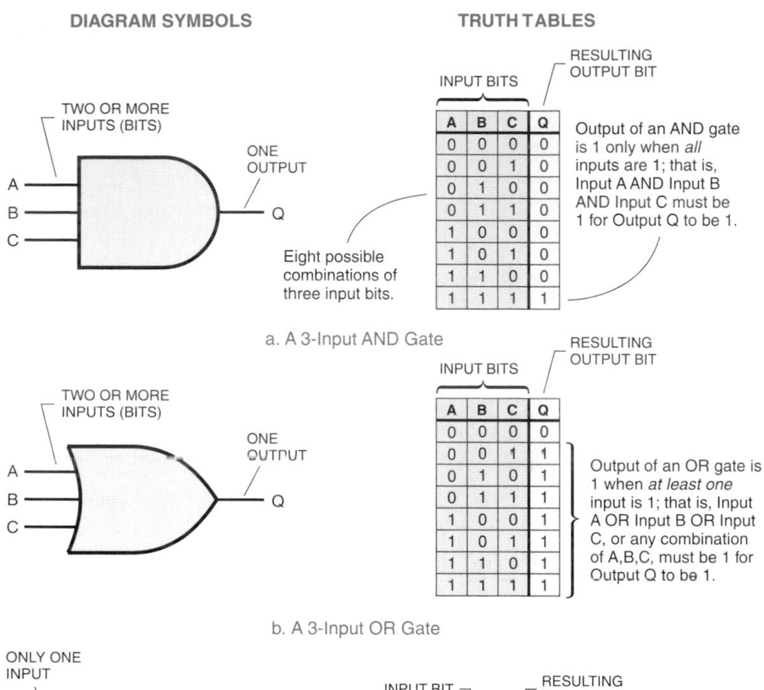

A	B	C	Q
0	0	0	0
0	0	1	0
0	1	0	0
0	1	1	0
1	0	0	0
1	0	1	0
1	1	0	0
1	1	1	1

a. A 3-Input AND Gate

A	B	C	Q
0	0	0	0
0	0	1	1
0	1	0	1
0	1	1	1
1	0	0	1
1	0	1	1
1	1	0	1
1	1	1	1

b. A 3-Input OR Gate

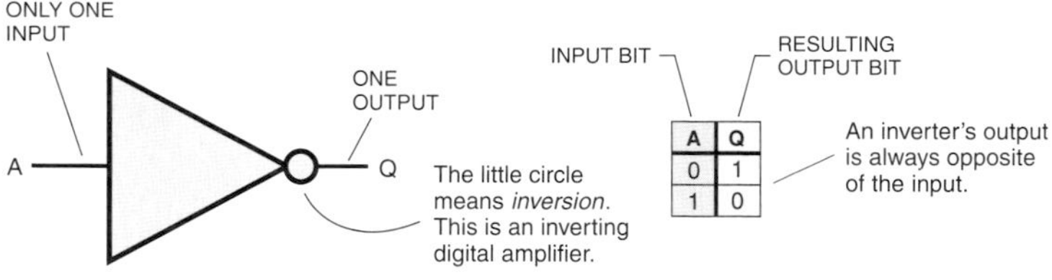

A	Q
0	1
1	0

c. A NOT Gate Inverter

AND, OR, and NOT Truth Tables

Source: *Basic Electronics* © 1994, Master Publishing, Inc., Niles, Illinois

E7A07 What logical operation does a NAND gate perform?

A. It produces logic "0" at its output only when all inputs are logic "0".
B. It produces logic "1" at its output only when all inputs are logic "1".
C. It produces logic "0" at its output if some but not all inputs are logic "1".
D. It produces logic "0" at its output only when all inputs are logic "1".

The word ***NAND*** reminds us of the word "naw", meaning no or nothing. The NAND gate ***produces a logic "0" at its output only when all inputs are logic "1"***. **ANSWER D.**

E7A08 What logical operation does an OR gate perform?

A. It produces logic "1" at its output if any or all inputs are logic "1".
B. It produces logic "0" at its output if all inputs are logic "1".
C. It only produces logic "0" at its output when all inputs are logic "1".
D. It produces logic "1" at its output if all inputs are logic "0".

The ***OR*** gate will produce a ***logic "1"*** at its output ***if any one or all inputs are a logic "1"***. **ANSWER A.**

E7A09 What logical operation is performed by an exclusive NOR gate?

A. It produces logic "0" at its output only if all inputs are logic "0".
B. It produces logic "1" at its output only if all inputs are logic "1".
C. It produces logic "0" at its output if any single input is logic "1".
D. It produces logic "1" at its output if any single input is logic "1".

The ***NOR gate*** will produce a ***logic "0" if any or all inputs are a logic "1"***. **ANSWER C.**

E7A10 What is a truth table?

A. A table of logic symbols that indicate the high logic states of an op-amp.
B. A diagram showing logic states when the digital device output is true.
C. A list of inputs and corresponding outputs for a digital device.
D. A table of logic symbols that indicate the logic states of an op-amp.

We use a ***truth table*** in digital circuitry to characterize a digital device's function. It ***lists the inputs and outputs*** for a digital device. If you buy the owner's technical manual on a piece of Amateur Radio gear, you will usually find a page describing the truth table of digital devices used in the equipment. **ANSWER C.**

E7A11 What type of logic defines "1" as a high voltage?

A. Reverse Logic.
B. Assertive Logic.
C. Negative logic.
D. Positive Logic.

In a ***positive logic*** circuit, the ***logic "1" is a high level***, or the most positive level. Think of the four letters in high and the four letters in plus (for positive). The actual voltage value represented by a 1 will depend on the logic family, such as TTL, ECL, CMOS, etc. **ANSWER D.**

E7A12 What type of logic defines "0" as a high voltage?

A. Reverse Logic.
B. Assertive Logic.
C. Negative logic.
D. Positive Logic.

The NOT gate inverter with an input bit of "1" will result in an opposite output, a ***"0" as a high voltage***. The letter N is a good way to remember ***NOT and negative logic***. **ANSWER C.**

E7A01 Which is a bi-stable circuit?

A. An AND gate.
B. An OR gate.
C. A flip-flop.
D. A clock.

In electronics, a ***flip-flop*** circuit has two stable states and can be used to store state information. Bi-stable stands for ***two stable states***. **ANSWER C.**

E7A03 Which of the following can divide the frequency of a pulse train by 2?

A. An XOR gate.
B. A flip-flop.
C. An OR gate.
D. A multiplexer.

A ***flip-flop*** can be used to ***divide a pulse train*** (the frequency of an AC signal). **ANSWER B.**

E7A04 How many flip-flops are required to divide a signal frequency by 4?

A. 1.
B. 2.
C. 4.
D. 8.

Since a flip-flop has two stable states, you will need ***two flip-flops to divide a signal frequency by four***. **ANSWER B.**

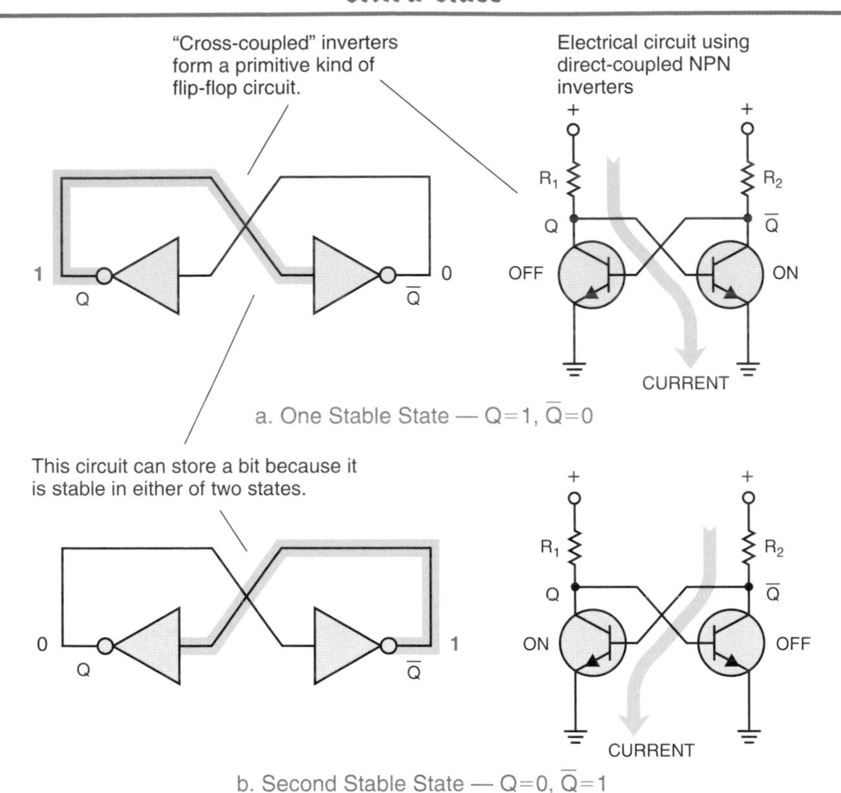

Basic concept of the flip-flop, also called a bistable element or a static memory element.
Source: *Basic Electronics* © 1994, Master Publishing, Inc., Niles, Illinois

E7A05 Which of the following is a circuit that continuously alternates between two states without an external clock?

A. Monostable multivibrator.
B. J-K flip-flop.
C. T flip-flop.
D. Astable multivibrator.

An ***astable multivibrator*** continuously switches back and forth between ***two unstable states without an external clock***. **ANSWER D.**

E7A06 What is a characteristic of a monostable multivibrator?

A. It switches momentarily to the opposite binary state and then returns to its original state after a set time.
B. It produces a continuous square wave oscillating between 1 and 0.
C. It stores one bit of data in either a 0 or 1 state.
D. It maintains a constant output voltage, regardless of variations in the input voltage.

This type of ***multivibrator may momentarily be monostable*** (MO MO): it stays in an original state until triggered to its other state, where it remains for a time usually determined by external components, after which it ***returns to the original state***. This is sometimes called a "one-shot" multivibrator. It is often used to create a precise time delay. The extremely popular 555 IC is capable of operating as a one-shot multivibrator. **ANSWER A.**

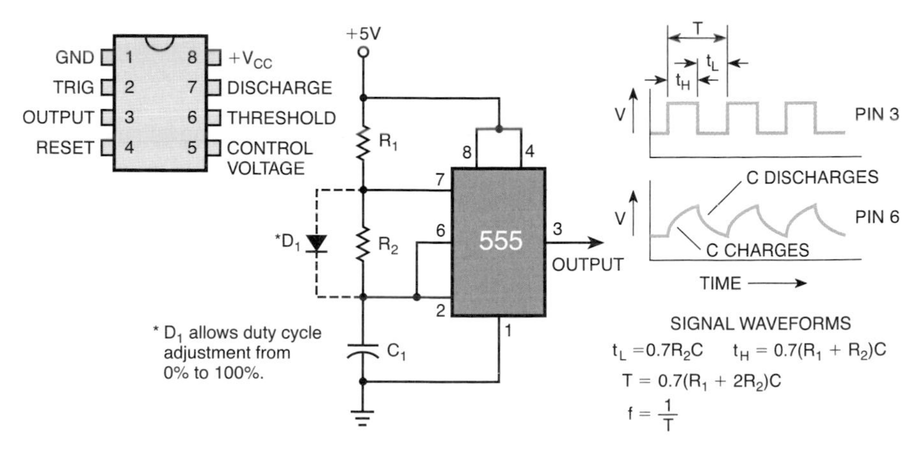

An IC 555 timer can be used as a bistable, astable, or monostable multivibrator. This schematic shows it wired as an astable multivibrator. The resulting astable voltage waveforms at pins 3 and 6 are shown.
Source: *Basic Digital Electronics,* Evans, ©1996, Master Publishing, Inc., Niles, IL

E8A09 How many levels can an analog-to-digital converter with 8 bit resolution encode?

A. 8.
B. 8 multiplied by the gain of the input amplifier.
C. 256 divided by the gain of the input amplifier.
D. 256.

The number of values that any digital number can represent is calculated by 2^N, where N is the number of bits. ***2^8 is 256***. Keep in mind that the first number is 0! So the actual numbers represented are 0-255. **ANSWER D.**

E6C07 What best describes a pull-up or pull-down resistor?

A. A resistor in a keying circuit used to reduce key clicks.
B. A resistor connected to the positive or negative supply line used to establish a voltage when an input or output is an open circuit.
C. A resistor that insures that an oscillator frequency does not drive lower over time.
D. A resistor connected to an op-amp output that only functions when the logic output is false.

Digital logic gates don't generally like inputs "flapping in the breeze." Such a condition can lead to some "interesting" effects, to say the least. Depending on the type of gate, a high resistance ***pull-up or pull-down resistor may be connected to an input gate*** to "tie" that gate to a specific voltage when not being toggled, thus ***preventing false outputs***. Sometimes the pull-up or pull-down resistor is built into the chip, sometimes not. Always check your data sheets! **ANSWER B.**

E6C09 What is a Programmable Logic Device (PLD)?

A. A device to control industrial equipment.
B. A programmable collection of logic gates and circuits in a single integrated circuit.
C. Programmable equipment used for testing digital logic integrated circuits.
D. An algorithm for simulating logic functions during circuit design.

Logic devices can be categorized as "hard-wired" or programmable. A ***PLD*** is a medium or large scale ***integrated circuit that can be programmed*** by commands from the outside world. It need not be as elaborate as a complete microprocessor, but is generally much more complex than any hard-wired logic circuit you are likely to put together. The Field Programmable Gate Array (FPGA) is one popular type of PLD. **ANSWER B.**

This is a UV erasable EEPROM, which makes it possible for a manufacturer to upgrade your rig's receiver to take in new bands, such as those proposed below 500 kHz.

E6C14 What is the primary advantage of using a Programmable Gate Array (PGA) in a logic circuit?

A. Many similar gates are less expensive than a mixture of gate types.
B. Complex logic functions can be created in a single integrated circuit.
C. A PGA contains its own internal power supply.
D. All of these choices are correct.

Cracking out the soldering iron is the traditional way of engineering and modifying circuits, digital or otherwise, but sometimes it's a lot more convenient to be able to reconfigure a circuit on the fly. A ***Programmable Gate Array allows you to quickly modify the characteristics of a logic circuit*** through software commands. **ANSWER B.**

E8A10 What is the purpose of a low pass filter used in conjunction with a digital-to-analog converter?

A. Lower the input bandwidth to increase the effective resolution.
B. Improve accuracy by removing out of sequence codes from the input.
C. Remove harmonics from the output caused by the discrete analog levels generated.
D. All of these choices are correct.

The problem with the ***D-to-A*** converter is ***harmonics***, which are usually easier to deal with using a ***low pass filter***. You don't need the same "brick wall" anti-aliasing filter required in the A-to-D device. **ANSWER C.**

E6C03 What is tri-state logic?

A. Logic devices with 0, 1, and high impedance output states.
B. Logic devices that utilize ternary math.
C. Low power logic devices designed to operate at 3 volts.
D. Proprietary logic devices manufactured by Tri-State Devices.

Tri-state logic devices can cut down on the number of individual OR gates needed to route data in a bus wire where there may be many "listeners." The tri-state device is identified by one additional control lead. The truth table looks like this for the tri-gate device:

A	Control	B
0	on	0
1	on	1
0	off	high impedance
1	off	high impedance

Tri-state logic devices offer three levels of output: "0," "1," and high impedance.
ANSWER A.

E6C04 What is the primary advantage of tri-state logic?

A. Low power consumption.
B. Ability to connect many device outputs to a common bus.
C. High speed operation.
D. More efficient arithmetic operations.

In today's electronics, multiple signals may travel on a pair of wires called a bus with multiple sources listening on that bus for a specific command. *Tri-state logic gates are used to cut down on the number of components necessary along the bus* to accommodate all these signals. **ANSWER B.**

E6C12 What is BiCMOS logic?

A. A logic device with two CMOS circuits per package.
B. A FET logic family based on bimetallic semiconductors.
C. A logic family based on bismuth CMOS devices.
D. An integrated circuit logic family using both bipolar and CMOS transistors

Relatively new to the complementary metal oxide semiconductor is *BiCMOS* logic, *combining MOS* logic *with* increased current capabilities of a *bipolar transistor*. The BiCMOS inverter also offers increased speed over a stand-alone CMOS gate on a large capacitive load. This increases the current multiplying capability of the bipolar output transistor. The downside to the BiCMOS is its higher cost due to its intricate construction when compared to individual components that do the same job. **ANSWER D.**

E6C13 Which of the following is an advantage of BiCMOS logic?

A. Its simplicity results in much less expensive devices than standard CMOS.
B. It is totally immune to electrostatic damage.
C. It has the high input impedance of CMOS and the low output impedance of bipolar transistors.
D. All of these choices are correct.

The *BiCMOS* device offers a *high input impedance* associated with CMOS, yet a *low output impedance* that you would find with a bipolar transistor. It also provides increased gain with low power dissipation in heat. The BiCMOS inverter can be used as a buffer to drive large load capacitances offering increased speed in critical design networks. **ANSWER C.**

It's Only Logical!

Since in recent years "technology" and "digital" are nearly synonymous, it's easy to overlook just how recent these concepts are. Searching through the ARRL periodical archives, I discovered that the very first mention of digital logic was in the November, 1962 QST article, "Logic for Amateurs," which described some simple decision making circuits using vacuum tubes. *Actually, the article never used the term "digital"...that was to appear about six years later in that same venerable publication. Shucks, I was already in* high school *and building my own stereo systems by that time, which shows us just how* recent *the concept is...at least in Amateur Radio circles!*

How...or why....did digital *technology (and thought processes) come to dominate our "electronics" thinking? After all, we'd been doing radio just fine, thank you, for nearly two thirds of a century before digital and logic were familiar terms.*

Well, not exactly. Things like light switches and telegraphs keys and rotary phone dials were around at least as long as radio, and these are all logical devices. If you have a room with a pair of "three way" light switches, you actually have an "exclusive OR" function. If Switch A is up, the light lights; if switch B is up, the light lights; and if BOTH are up, the light stays dark. This is simple combinational logic.

If you have ever used a dial telephone (gasp!) you have worked with sequential logic. *The first digit you dial operates one set of 10 contact rotary relays at the phone office, which then transfers you to the next set of 10 contact rotary relays, which transfers you to the next set of 10 contact relays, and so on. What the system does with each digit you dial is dependent on the previous digit you dialed...in other words, you need some kind of memory...in this case, the physical position of each rotary relay is* stored. *(By the way, if you've never seen mechanical phone relays in action, it's something to behold. There are some great videos of this ancient technology on YouTube!)*

Digital truth tables are nothing more than maps that allow you to follow through a maze of open or closed switches to get an electrical voltage to the final destination. Of course, we don't often use mechanical switches for digital logic, as these have been largely replaced by transistors, or other more exotic semiconductors.

Transistors are a lot smaller, quieter, and faster than mechanical switches, but they perform the same task. MOST digital logic gates are of the inverting *type, NAND, NOR, because common emitter transistors invert the input and output voltages. This is no mystery, as we've already studied those, right? Transistor switches can be placed in* series *to form an AND (or NAND) function, while transistor switches in* parallel *form an OR (or NOR) function. It's that simple! Just plumbing and pipes.*

If you have enough time, space, and patience, you can build most digital devices with discrete transistors, and it wasn't too long ago that most electronics students did just that. In high school, I built a 10:1 counter using a pile of "bare" transistors and it actually worked! Nowadays, of course, digital integrated circuits can have thousands or millions of transistors, but they're still nothing but collections of switches.

E4A14 What is the purpose of the prescaler function on a frequency counter?

A. It amplifies low level signals for more accurate counting.
B. It multiplies a higher frequency signal so a low-frequency counter can display the operating frequency.
C. It prevents oscillation in a low-frequency counter circuit.
D. It divides a higher frequency signal so a low-frequency counter can display the input frequency.

You will find a ***prescaler*** in frequency counters where it is used to ***divide down HF and VHF signals*** so they can be counted and ***displayed on a low-frequency counter***. **ANSWER D.**

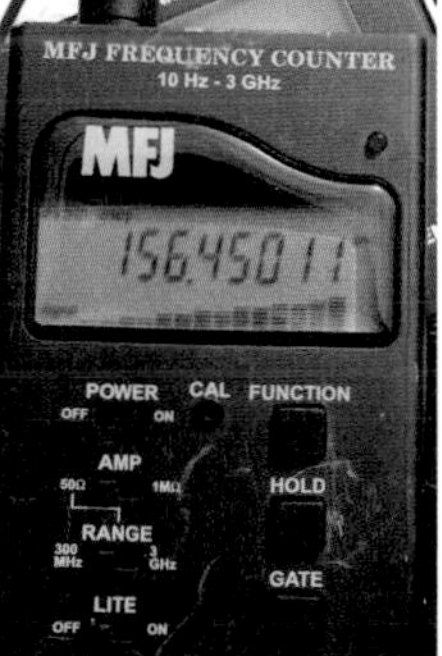

The portable frequency counter is a good way to see what frequency is found on an un-marked land mobile radio that you plan to convert to 2 meters. The counter is reading marine VHF channel 9.

E7A02 What is the function of a decade counter digital IC?

A. It produces one output pulse for every ten input pulses.
B. It decodes a decimal number for display on a seven segment LED display.
C. It produces ten output pulses for every input pulse.
D. It adds two decimal numbers together.

A ***decade counter*** digital IC gives ***one output pulse for every 10 input pulses***. **ANSWER A.**

E4B01 Which of the following factors most affects the accuracy of a frequency counter?

A. Input attenuator accuracy.
B. Time base accuracy.
C. Decade divider accuracy.
D. Temperature coefficient of the logic.

When shopping for a ***frequency counter***, check out the frequency ***time base accuracy***. The stability of the counter will depend on the internal crystal reference. You should regularly check your crystal reference when calibrating transmit equipment for digital operation with no frequency error greater than 5 Hz. **ANSWER B.**

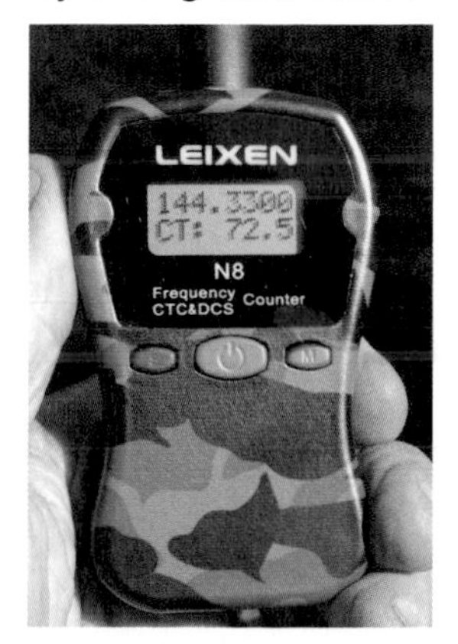

These inexpensive counters will give you a ball park reading of on-channel or not. But only a frequency counter with a temperature compensated crystal time base is accurate for digital work.

E4B05 If a frequency counter with a specified accuracy of +/- 10 ppm reads 146,520,000 Hz, what is the most the actual frequency being measured could differ from the reading?

A. 146.52 Hz.
B. 10 Hz.
C. 146.52 kHz.
D. 1465.20 Hz.

Keystroke the following on your calculator: Clear ***146.52 X 10 = 1465.2 Hz. You are multiplying the frequency in MHz by the time base accuracy of 10 ppm*****. ANSWER D.**

E4B03 If a frequency counter with a specified accuracy of +/- 1.0 ppm reads 146,520,000 Hz, what is the most the actual frequency being measured could differ from the reading?

A. 165.2 Hz.
B. 14.652 kHz.
C. 146.52 Hz.
D. 1.4652 MHz.

You may have a question on your test regarding the error readout on a frequency counter with a specific part per million (ppm) time base. It's easy to solve these problems with a simple calculator. Since they're asking parts per million, first convert the frequency given in Hz to MHz. You do this simply by moving the decimal point 6 places to the left. Now multiply the frequency in MHz times the parts per million time base accuracy (in this case 1 ppm). This will give you the error in Hz they are looking for in the answer. Try this on your calculator:
Clear ***146.52 × 1 = 146.52 Hz*****. ANSWER C**

Counter Readout Error
Readout Error = f × a

f = Frequency in **MHz** being measured
a = Counter accuracy in **parts per million**

E4B04 If a frequency counter with a specified accuracy of +/- 0.1 ppm reads 146,520,000 Hz, what is the most the actual frequency being measured could differ from the reading?

A. 14.652 Hz.
B. 0.1 MHz.
C. 1.4652 Hz.
D. 1.4652 kHz.

Hertz to MHz will give you a frequency of 146.52. Multiply that by 0.1 parts per million, and see your calculator display 14.652 in Hz. The keystrokes are: Clear ***146.52 × 0.1 = 14.652*****. ANSWER A.**

E4A15 What is an advantage of a period-measuring frequency counter over a direct-count type?

A. It can run on battery power for remote measurements.
B. It does not require an expensive high-precision time base.
C. It provides improved resolution of low-frequency signals within a comparable time period.
D. It can directly measure the modulation index of an FM transmitter.

A common frequency counter calculates the number of cycles per second (Hertz) of the near-field frequency it is measuring. These low-cost counters are amazingly precise as they average out the readout from internal counters. In the ***period-measuring frequency counter***, the RF wave is divided into 360 degrees for a ***more precise time measurement***. Phase difference comparisons take place in the reference oscillator, leading to improved signal frequency resolution within a comparable time period. This takes place inside many GPS receivers where signals are measured down to 10 to the minus 15th second, the femto second! **ANSWER C.**

CD 5 TRACK 4

Optos and OpAmps Plus Solar

E7G12 What is an integrated circuit operational amplifier?

A. A high-gain, direct-coupled differential amplifier with very high input impedance and very low output impedance.
B. A digital audio amplifier whose characteristics are determined by components external to the amplifier.
C. An amplifier used to increase the average output of frequency modulated amateur signals to the legal limit.
D. A RF amplifier used in the UHF and microwave regions.

The ***operational amplifier (op-amp)*** is a direct-coupled ***differential amplifier*** that offers high gain and high input impedance. The "differential" wording refers to the op-amp input design where the output is determined by the difference in voltage between the two inputs. Different op-amp circuits and characteristics are determined by the external components connected to the amplifier. This integrated circuit op-amp offers high-gain with very ***high input*** impedance and very ***low output impedance***. The gain and frequency response of an op-amp are almost entirely determined by external feedback components, primarily resistors and capacitors.
ANSWER A.

E7G02 What is the effect of ringing in a filter?

A. An echo caused by a long time delay.
B. A reduction in high frequency response.
C. Partial cancellation of the signal over a range of frequencies.
D. Undesired oscillations added to the desired signal.

Ringing in a filter may contribute to ***unwanted oscillations*** that could actually ride along with the desired signal, making it more difficult to copy the signal you want to hear. **ANSWER D.**

E7G05 How can unwanted ringing and audio instability be prevented in a multi-section op-amp RC audio filter circuit?

A. Restrict both gain and Q.
B. Restrict gain but increase Q.
C. Restrict Q but increase gain.
D. Increase both gain and Q.

In order to keep an ***op-amp*** from going into oscillation, ***both the gain and the Q are restricted*** and set by the feedback circuit. **ANSWER A.**

E7G06 Which of the following is the most appropriate use of an op-amp active filter?

A. As a high-pass filter used to block RFI at the input to receivers.
B. As a low-pass filter used between a transmitter and a transmission line.
C. For smoothing power supply output.
D. As an audio filter in a receiver.

You will find the ***op-amp RC active filter*** in the ***audio section*** of your new transceiver's receiver. The number of filter options with more filter schemes usually increases as the price of an HF transceiver increases. **ANSWER D.**

E7G08 How does the gain of an ideal operational amplifier vary with frequency?

A. It increases linearly with increasing frequency.
B. It decreases linearly with increasing frequency.
C. It decreases logarithmically with increasing frequency.
D. It does not vary with frequency.

The ***gain*** on an ideal op-amp ***does not vary with frequency***. Op-amp design is primarily based on the ideal op-amp, where the characteristics are determined only by the feedback components. However, all real world op-amps have an open loop gain-bandwidth product (GBP) that limits the ultimate performance of an op-amp. **ANSWER D.**

E7G07 What magnitude of voltage gain can be expected from the circuit in Figure E7-4 when R1 is 10 ohms and RF is 470 ohms?

A. 0.21.
B. 94.
C. 47.
D. 24.

Take a look at the operational amplifier basics chart on page 218 and remember the gain formula of an inverting IC op-am

E ▶ $G = -R_F \div R_1 = R_F/R_1$.

The minus sign means the output is out of phase with the input. R_F = 470 ohms; R_1 = 10 ohms. ***Dividing 10 into 470, yields 47***. On a calculator, the keystrokes are: Clear, 470 ÷ 10 = 47. Not really a brain-buster, was it? **ANSWER C.**

Op-Amp devices.

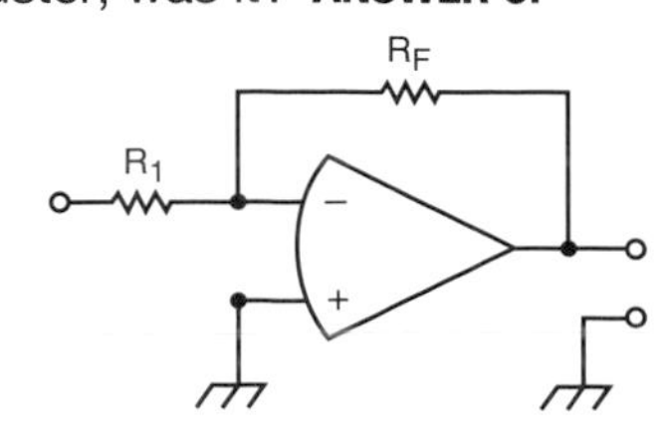

Figure E7-4

E7G09 What will be the output voltage of the circuit shown in Figure E7-4 if R_1 is 1000 ohms, RF is 10,000 ohms, and 0.23 volts DC is applied to the input?

A. 0.23 volts.
B. 2.3 volts.
C. -0.23 volts.
D. -2.3 volts.

This time we work the formula: $V_{OUT}/V_{IN} = -R_F/R_1$, when $R_1 = 1000$, and $R_F = 10{,}000$, the gain of the op-amp is $V_{OUT}/V_{IN} = -10$. As a result, the ***output = -10 × V_{IN} = -10 × 0.23 = -2.3 volts*****.** The gain of an op-amp is easy to calculate because it is determined entirely by the external resistors: being the simple quotient of the feedback resistance over the input resistance. Since the output voltage is 180 degrees out of phase with the input voltage, op-amps will be operated in the *inverting* mode for stability. A major clue that you're using an inverting amplifier is the minus sign in the terminal into which you're putting the signal! In this case, the gain is 10,000/1000 or 10:1. So any voltage we apply will be amplified by 10. But since it's an *inverting* amplifier, we have to use the minus sign. So our signal is ultimately multiplied by -10X the input signal. Nothing to it! **ANSWER D.**

E7G10 What absolute voltage gain can be expected from the circuit in Figure E7-4 when R1 is 1800 ohms and RF is 68 kilohms?

A. 1.
B. 0.03.
C. 38.
D. 76.

Now we're back to the gain of an inverting IC operational amplifier. ***Simply divide 1,800 ohms into 68,000 ohms***. Remember, the gain is the quotient of the feedback resistance over the input resistance. Your calculator keystrokes are: clear, 68000 ÷ 1800 = 37.777, rounded off to your ***correct answer of 38***. Simple, huh? **ANSWER C.**

E7G11 What absolute voltage gain can be expected from the circuit in Figure E7-4 when R1 is 3300 ohms and RF is 47 kilohms?

A. 28.
B. 14.
C. 7.
D. 0.07.

One more time, an easy one – ***divide*** the small number ***(3,300) into*** the large number ***(47,000)***, and you end up with 14.24, rounded to the correct answer of ***14***. **ANSWER B.**

E7G04 What is meant by the term op-amp input offset voltage?

A. The output voltage of the op-amp minus its input voltage.
B. The difference between the output voltage of the op-amp and the input voltage required in the immediately following stage.
C. The differential input voltage needed to bring the open loop output voltage to zero.
D. The potential between the amplifier input terminals of the op-amp in an open loop condition.

If the op-amp has been constructed properly, you should see almost ***no voltage between the amplifier input terminals*** when the feedback loop is closed. **ANSWER C.**

E7G03 What is the typical input impedance of an integrated circuit op-amp?

A. 100 ohms.
B. 1000 ohms.
C. Very low.
D. Very high.

Remember that the ***input impedance*** to the ideal ***op-amp*** is always ***Hi iN***. **ANSWER D.**

Operational Amplifier Basics

Ideal Operational Amplifier

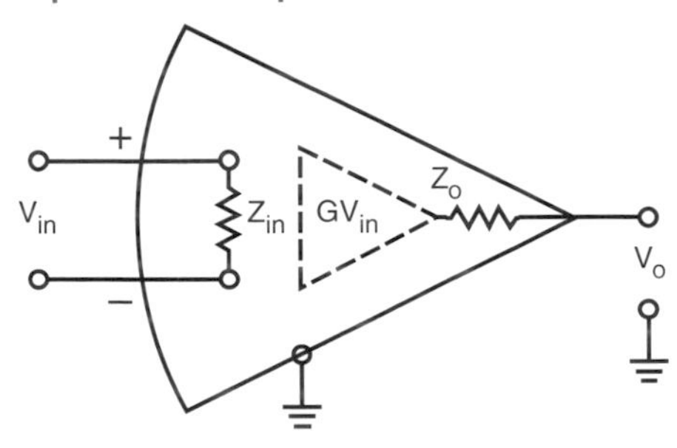

Z_{in} = Infinity

G = Gain = Infinity

Z_o = Zero

Bandwith = Infinity

V_o has no offset
(V_o=0 when V_{in}=0)

IC Operational Ampliifer

Even though IC operational amplifiers do not meet all the ideal specifications, G and Z_{in} are very large. Because G is very large, any small input voltage would drive the output into saturation (for practical supply voltages); therefore, normal operational amplifier operation is with feedback to set the gain. Here's an example for an inverting amplifier:

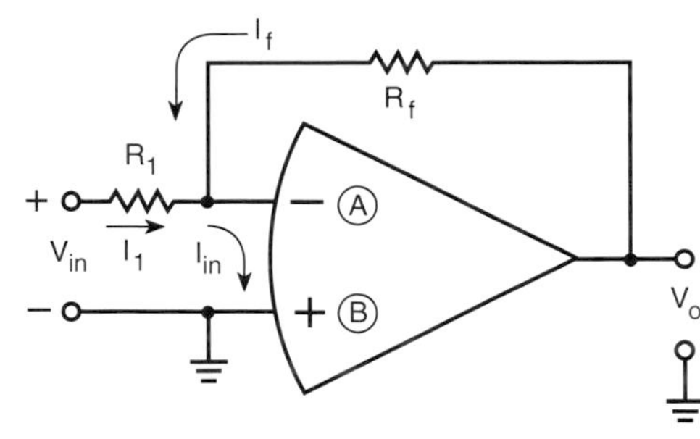

Since Z_{in} is very large, $I_{in} = 0$

$$\therefore I_1 + I_f = 0 \qquad \therefore \frac{V_{in}}{R_1} + \frac{V_o}{R_f} = 0$$

$$I_1 = \frac{V_{in}}{R_1} \qquad \frac{V_o}{R_f} = -\frac{V_{in}}{R_1}$$

$$I_f = \frac{V_o}{R_f} \qquad \therefore \frac{V_o}{V_{in}} = -\frac{R_f}{R_1}$$

Ⓐ Inverting Input — An input voltage that is more positive on this input will cause the output voltage to be less positive.

Ⓑ Non-Inverting Input — An input voltage that is more positive on this input will cause the output voltage to be more positive.

The gain of an *inverting IC operational amplifier* is:

E ▶ $$G = -\frac{R_f}{R_1}$$

The minus sign means the output is out of phase with the input.

E7G01 What is the typical output impedance of an integrated circuit op-amp?

A. Very low.
B. Very high.
C. 100 ohms.
D. 1000 ohms.

Remember that the ***output impedance*** of the ideal op-amp is always ***low out*****.**
ANSWER A.

E6F01 What is photoconductivity?

A. The conversion of photon energy to electromotive energy.
B. The increased conductivity of an illuminated semiconductor.
C. The conversion of electromotive energy to photon energy.
D. The decreased conductivity of an illuminated semiconductor.

The ***photocell*** has high resistance when no light shines on it and a ***varying resistance when light shines on it***. It can be used in an ON or OFF state as a simple beam alarm across a doorway, or may be found in almost all modern SLR cameras as a sensitive light meter to judge the amount of light present. **ANSWER B.**

E6F02 What happens to the conductivity of a photoconductive material when light shines on it?

A. It increases.
B. It decreases.
C. It stays the same.
D. It becomes unstable.

In the dark, a photocell has high resistance. When you ***shine light on it***, the ***conductivity increases*** as the resistance decreases. **ANSWER A.**

E6F03 What is the most common configuration of an optoisolator or optocoupler?

A. A lens and a photomultiplier.
B. A frequency modulated helium-neon laser.
C. An amplitude modulated helium-neon laser.
D. An LED and a phototransistor.

Optocouplers are found in many modern, high-frequency transceivers. When you spin the dial to tune, you are not turning a giant capacitor, but rather interrupting an ***LED*** light beam on a ***phototransistor***. **ANSWER D.**

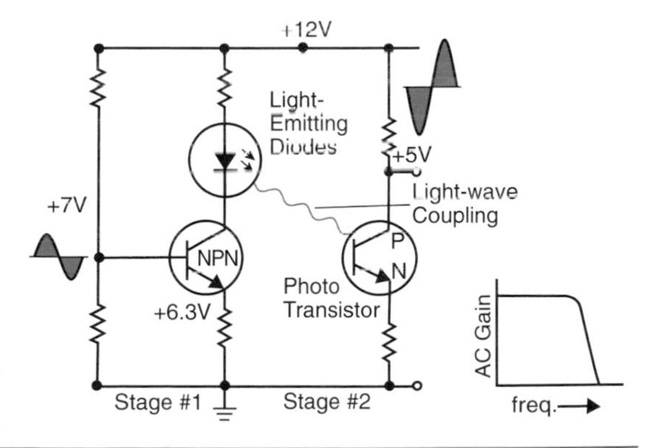

Optical Coupling
Source: *Basic Communications Electronics*, Hudson & Luecke, © 1999 Master Publishing, Inc., Niles, IL

E6F08 Why are optoisolators often used in conjunction with solid state circuits when switching 120VAC?

A. Optoisolators provide a low impedance link between a control circuit and a power circuit.
B. Optoisolators provide impedance matching between the control circuit and power circuit.
C. Optoisolators provide a very high degree of electrical isolation between a control circuit and the circuit being switched.
D. Optoisolators eliminate the effects of reflected light in the control circuit.

The chief advantage of the ***optoisolator*** at household voltages is to provide a ***very high degree of electrical isolation between*** the control circuit and the house power circuit. Look for the answer ***"isolation"*** to agree with the optoisolator device. **ANSWER C.**

E6F05 Which describes an optical shaft encoder?

A. A device which detects rotation of a control by interrupting a light source with a patterned wheel.

B. A device which measures the strength of a beam of light using analog to digital conversion.

C. A digital encryption device often used to encrypt spacecraft control signals.

D. A device for generating RTTY signals by means of a rotating light source.

When you turn the big VFO knob of that new high-frequency transceiver you awarded to yourself for passing the Extra Class exam, an ***LED light source shining through*** clear and black ***patterns on a wheel*** will mimic rotation of a control while a logic circuit counts the number of light pulses making it through to the ***optical shaft encoder***. **ANSWER A.**

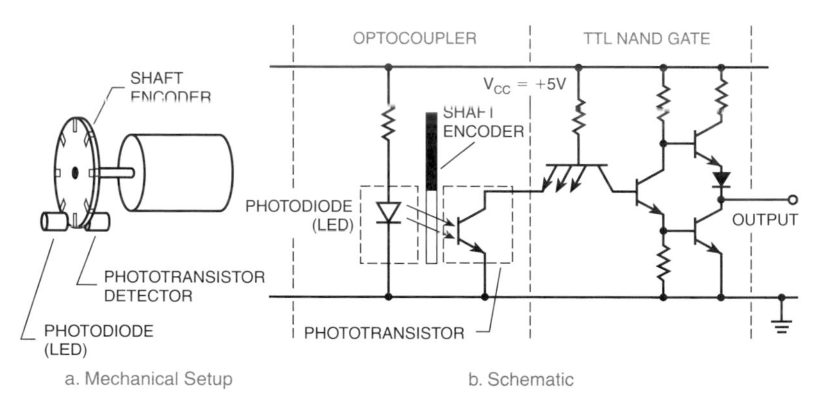

Optocoupler Used for Shaft Encoder

E6F04 What is the photovoltaic effect?

A. The conversion of voltage to current when exposed to light.

B. The conversion of light to electrical energy.

C. The conversion of electrical energy to mechanical energy.

D. The tendency of a battery to discharge when used outside.

The term ***"photovoltaic"*** refers to ***light photons being converted to electrical energy*** such as in solar panels. **ANSWER B.**

In full sunlight these four solar panels, which are wired in parallel, can produce better than 10 amps 12 VDC charging power.

E6F09 What is the efficiency of a photovoltaic cell?

A. The output RF power divided by the input DC power.

B. The effective payback period.

C. The open-circuit voltage divided by the short-circuit current under full illumination.

D. The relative fraction of light that is converted to current.

Lead sulfide offers good efficiency in a photovoltaic cell – lead takes in the infrared and cadmium takes in visible light. ***Solar cell efficiency is the relative fraction of light that ultimately gets converted to current***. **ANSWER D.**

E6F06 Which of these materials is affected the most by photoconductivity?

A. A crystalline semiconductor.
B. An ordinary metal.
C. A heavy metal.
D. A liquid semiconductor.

The ***crystalline semiconductor*** is found inside the optical shaft encoder mechanism inside your transceiver. It might also be found outside your ham shack as part of your solar-charging system. **ANSWER A.**

Resistance Variation of a Photodiode

Light Incidence	Current I	Resistance	Test Conditions
Dark	5 nA	2000 MΩ	$V_R = 10$ V $^*E_e = 0$
Light	15 µA	666.7 kΩ	$V_R = 10$ V $E_e = 250$ µW/cm² at 940 nm

*Irradiance (E_e) is the radiant power per unit area incident on surface

---

E6F11 What is the approximate open-circuit voltage produced by a fully-illuminated silicon photovoltaic cell?

A. 0.1 V.
B. 0.5 V.
C. 1.5 V.
D. 12 V.

While that new solar panel you used at Field Day has an output of 17 volts with no load, a ***single cell*** of that panel ***is rated at 0.5 volts***. Now you know why that panel is so big! **ANSWER B.**

E6F12 What absorbs the energy from light falling on a photovoltaic cell?

A. Protons.
B. Photons.
C. Electrons.
D. Holes.

Photon energy from sunlight is absorbed by the electrons in the photovoltaic cell thus creating voltage output from sunlight. When sunlight photons strike the panel, an ***electron will absorb the energy*** and develop a new energy level, ***creating the photovoltaic process***. **ANSWER C.**

E6F10 What is the most common type of photovoltaic cell used for electrical power generation?

A. Selenium.
B. Silicon.
C. Cadmium Sulfide.
D. Copper oxide.

The photovoltaic material of a ***solar cell*** is made up of ***silicon wafers*** which offer relatively high voltage output and durability for the many years it will be out in the sunlight. **ANSWER B.**

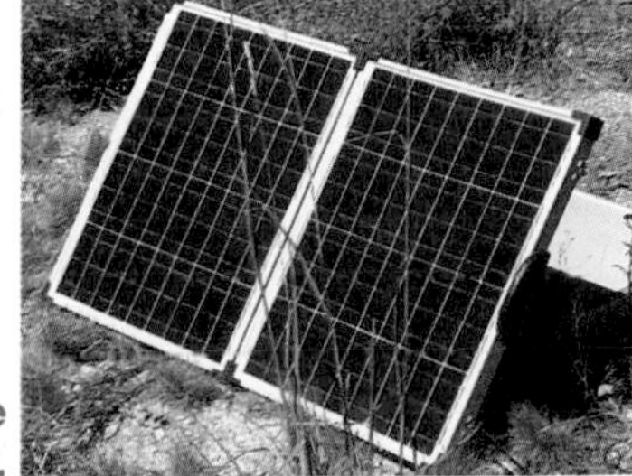

This portable solar panel will produce about 4 amps in bright sunlight.

E7D09 What is the main reason to use a charge controller with a solar power system?

A. Prevention of battery undercharge.
B. Control of electrolyte levels during battery discharge.
C. Prevention of battery damage due to overcharge.
D. Matching of day and night charge rates.

A bank of solar cells puts out voltage in proportion to the amount of light falling on it, which may be more than the batteries can handle during bright sunlight. A ***charge controller equalizes the voltage to the batteries*** during the hours when the cells are charging, thus ***preventing battery damage***. **ANSWER C.**

Solar panels that put out more than 2 amps should always have a regulator in series to insure that you don't cook your batteries during full sunlight.

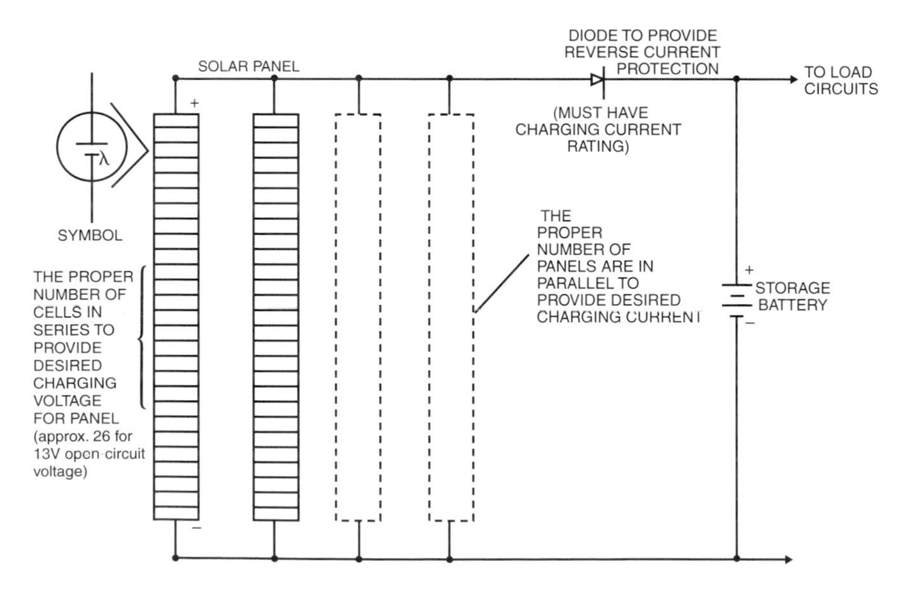

Schematic of Solar Panel for Charging Storage Battery.

CD 5 TRACK 5

Test Gear, Testing, Testing 1, 2, 3

E4A02 Which of the following parameters would a spectrum analyzer display on the vertical and horizontal axes?

A. RF amplitude and time.
B. RF amplitude and frequency.
C. SWR and frequency.
D. SWR and time.

A spectrum analyzer resembles an oscilloscope, but instead of the ***horizontal axis*** representing time, it ***represents frequency***, while the ***vertical axis represents RF amplitude***. A spectrum analyzer can take several forms, ranging from a simple "panadapter" used for checking band activity, to a precise measuring instrument for alignment and troubleshooting of radio equipment of all kinds. The waterfall display of a soundcard digital software package is an audio frequency spectrum analyzer. **ANSWER B.**

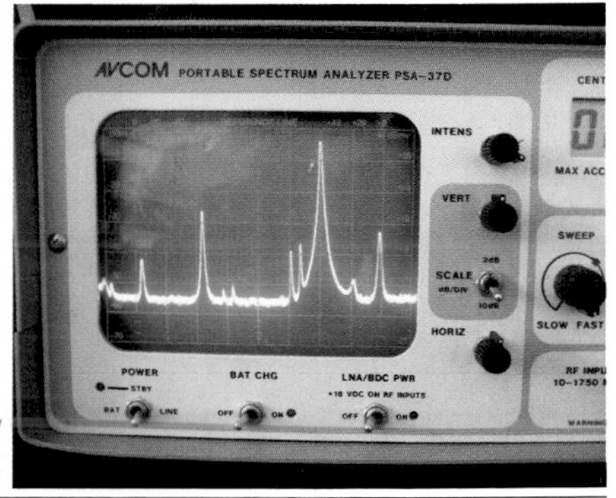

This analog CRT spectrum analyzer is battery operated, but the CRT really eats up the juice!

E4A03 Which of the following test instruments is used to display spurious signals and/or intermodulation distortion products in an SSB transmitter?

A. A wattmeter.
B. A spectrum analyzer.
C. A logic analyzer.
D. A time-domain reflectometer.

A ***spectrum analyzer*** is the most convenient way to check your transmitter's RF emissions. Though not inexpensive, a well-equipped radio shop will have a spectrum analyzer. But it's not the only way to check your on-the-air signal. A general coverage receiver can be a valuable and much less expensive piece of test equipment. No ham shack should be without one. **ANSWER B.**

E4B10 Which of the following describes a method to measure intermodulation distortion in an SSB transmitter?

A. Modulate the transmitter with two non-harmonically related radio frequencies and observe the RF output with a spectrum analyzer.
B. Modulate the transmitter with two non-harmonically related audio frequencies and observe the RF output with a spectrum analyzer.
C. Modulate the transmitter with two harmonically related audio frequencies and observe the RF output with a peak reading wattmeter.
D. Modulate the transmitter with two harmonically related audio frequencies and observe the RF output with a logic analyzer.

When looking at your signal on a spectrum analyzer, a whistle into the microphone is not a good test for transmitter signal distortion and intermodulation. Using *two non-harmonically related audio tones* from a two tone generator is a much better way to observe the characteristics of the signal on your spectrum analyzer. **ANSWER B.**

E4A12 Which of the following procedures is an important precaution to follow when connecting a spectrum analyzer to a transmitter output?

A. Use high quality double shielded coaxial cables to reduce signal losses.
B. Attenuate the transmitter output going to the spectrum analyzer.
C. Match the antenna to the load.
D. All of these choices are correct.

Never transmit directly into a *spectrum analyzer*. You'll blow it up! Your *transmitter signal must be attenuated* to just a fraction of a watt to protect the spectrum analyzer. Make it a practice to NEVER directly couple to a spectrum analyzer. Rather, use a little pigtail and "sniff" your transmit RF. **ANSWER B.**

E4B02 What is an advantage of using a bridge circuit to measure impedance?

A. It provides an excellent match under all conditions.
B. It is relatively immune to drift in the signal generator source.
C. It is very precise in obtaining a signal null.
D. It can display results directly in Smith chart format.

Multiband, high-frequency antenna traps, containing both coil inductive reactance and outer sleeve capacitive reactance, can be tuned to a specific frequency that will stop the RF signal from going further out on the element. A portable impedance-measuring bridge circuit will show a much more precise dip, or null, in voltage, as opposed to many other methods of impedance measurement. Try to loosely couple the impedance *bridge circuit* so the *pronounced dip* is as *precisely on frequency* as possible. **ANSWER C.**

E4B14 What happens if a dip meter is too tightly coupled to a tuned circuit being checked?

A. Harmonics are generated.
B. A less accurate reading results.
C. Cross modulation occurs.
D. Intermodulation distortion occurs.

If you put your *dip meter right on top of a coil*, you will have a *less accurate reading*. You want to only get close enough to see a smooth dip on your readout. On the SWR analyzer type of dip meters, everything is done for you automatically inside the box. All you need to do is screw on the antenna cable and sweep the frequency dial to see where the antenna is resonant. **ANSWER B.**

E4B06 How much power is being absorbed by the load when a directional power meter connected between a transmitter and a terminating load reads 100 watts forward power and 25 watts reflected power?

A. 100 watts.
C. 25 watts.
B. 125 watts.
D. 75 watts.

A directional wattmeter is a valuable instrument for every radio amateur. While functionally similar to an SWR meter, the directional wattmeter gives you actual forward and reflected power and is more revealing in many cases. The actual power delivered to a load (which may be an antenna or some intermediate device, such as an amplifier) is always the forward power minus the reflected power; in the case of this question, ***75 watts***. While it's usually best to adjust an antenna for zero reflected power, it's not always convenient or possible. So, one of the best investments you can make is in a good low-loss antenna tuner in conjunction with low-loss transmission line. By means of conjugate matching the antenna tuner causes any reflected power to be re-reflected and added back to the original forward power, assuring that all the power is eventually delivered to the antenna. Required reading for every radio amateur is Maxell's ***Reflections II***. **ANSWER D.**

E4A11 Which of the following is good practice when using an oscilloscope probe?

A. Keep the signal ground connection of the probe as short as possible.
B. Never use a high impedance probe to measure a low impedance circuit.
C. Never use a DC-coupled probe to measure an AC circuit.
D. All of these choices are correct.

When working on radio equipment using an ***oscilloscope probe***, the ***probe's*** signal ground lead should be ***kept as short as possible*** and clipped to the ground chassis near where you are testing. The short ground lead will minimize stray pickup from other stages of the radio. **ANSWER A.**

When working on sensitive component equipment, make sure YOU have a grounded body connection, and that the scope ground lead is tied into a close-by ground bus, with the shortest lead length possible.

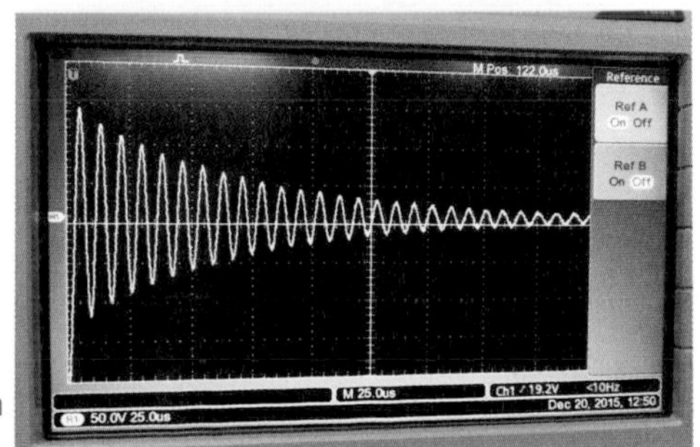

Damped oscillation seen on a digital oscilloscope.

E4A13 How is the compensation of an oscilloscope probe typically adjusted?

A. A square wave is displayed and the probe is adjusted until the horizontal portions of the displayed wave are as nearly flat as possible.
B. A high frequency sine wave is displayed and the probe is adjusted for maximum amplitude.
C. A frequency standard is displayed and the probe is adjusted until the deflection time is accurate.
D. A DC voltage standard is displayed and the probe is adjusted until the displayed voltage is accurate.

If you just purchased a new probe for the oscilloscope you regularly use when working on your ham radio equipment, use your ***square wave*** generator and adjust the horizontal portion of the displayed wave ***to be as flat as possible***. **ANSWER A.**

E4A05 What might be an advantage of a digital vs an analog oscilloscope?

A. Automatic amplitude and frequency numerical readout.
B. Storage of traces for future reference.
C. Manipulation of time base after trace capture.
D. All of these choices are correct.

Digital vs. analog is a consideration for all test equipment. Both have advantages. In the case of an oscilloscope, a Digital Storage Oscilloscope (DSO) is an amazing piece of test equipment. ***Automatic readout of amplitude and frequency*** help with accuracy. The unit ***stores waveforms*** in a convenient form for archiving, collaborating, and sharing. The memory capacity of the unit also allows for local ***manipulation of the stored data***. The DSO is not only an amazing piece of test equipment, it's becoming more versatile and inexpensive all the time. You almost can't afford not to have one in your ham shack! ***All of these choices*** are correct. **ANSWER D.**

While the analog CRT oscilloscope has served us well on the test bench, a modern digital scope offers many advantages when testing radio equipment.

E4A01 Which of the following parameters determines the bandwidth of a digital or computer-based oscilloscope?

A. Input capacitance.
B. Input impedance.
C. Sampling rate.
D. Sample resolution.

If you've spent a lot of time working with analog oscilloscopes, you may have to recalibrate your thinking and terminology a bit when working with a new Digital Storage Oscilloscope (DSO). While input capacitance and impedance affect bandwidth, the ***sampling rate*** is generally the limiting factor of the frequencies that can be measured. The Nyquist sampling theorem says that you need to sample the signal at twice the rate of its highest frequency component, so for a given DSO with a fixed sampling rate, the frequency bandwidth is very clearly defined. **ANSWER C.**

E4A04 What determines the upper frequency limit for a computer soundcard-based oscilloscope program?

A. Analog-to-digital conversion speed of the soundcard.
B. Amount of memory on the soundcard.
C. Q of the interface of the interface circuit.
D. All of these choices are correct.

A soundcard based oscilloscope program is an inexpensive alternative to a stand-alone Digital Storage Oscilloscope (DSO). The soundcard performs the same analog-to-digital sampling that is performed in the DSO, with the same requirements and restrictions. As in the DSO, bandwidth is determined by the sampling rate or ***Analog-to-Digital (A-to-D) conversion speed***, as dictated by the Nyquist Sampling Theorem. **ANSWER A.**

E4A06 What is the effect of aliasing in a digital or computer-based oscilloscope?

A. False signals are displayed.
B. All signals will have a DC offset.
C. Calibration of the vertical scale is no longer valid.
D. False triggering occurs.

Any digital sampling system is capable of generating "alias" signals, and the Digital Storage Oscilloscope is no exception. You can blame Nyquist's Sampling Theorem for the inevitable existence of aliases. There are a number of ways to identify and avoid ***false alias signals***. The operator's manual for your DSO will show combinations of settings and frequencies to avoid to side-step aliasing. **ANSWER A.**

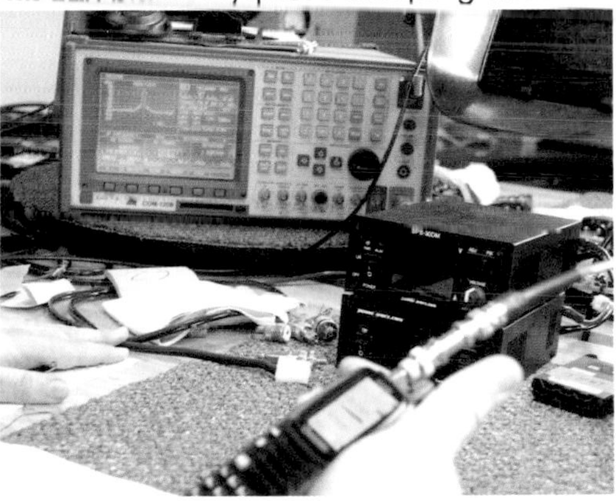

When upgrading to a digital scope, be aware that you will need to read the manual and learn to recognize tiny "alias" signals that are not part of the real signal you are measuring.

E4B08 Which of the following is a characteristic of a good DC voltmeter?

A. High reluctance input.
B. Low reluctance input.
C. High impedance input.
D. Low impedance input.

Your modern high-frequency or VHF/UHF transceiver has circuits so sensitive that they will quit oscillating if you attempt to measure ***DC voltage*** with an inexpensive, low-impedance multimeter. ***High impedance input*** – usually greater than 25 k ohms per volt – will help prevent loading that changes the characteristics of the circuit. **ANSWER C.**

Digital Multitester
Source: *Using Your Meter* © 2012, Master Publishing, Inc., Niles, Illinois

E4B09 What is indicated if the current reading on an RF ammeter placed in series with the antenna feed line of a transmitter increases as the transmitter is tuned to resonance?

A. There is possibly a short to ground in the feed line.
B. The transmitter is not properly neutralized.
C. There is an impedance mismatch between the antenna and feed line.
D. There is more power going into the antenna.

When Gordo is working aboard boats testing the output of a long-wire antenna tuner, he inserts a series thermocouple RF ammeter to monitor RF output levels. This helps him confirm that the tuner is operating properly when he changes bands, and see ***RF power increase*** as much as an amp ***going into the long-wire antenna***. **ANSWER D.**

The R.F. ammeter is a severely underutilized instrument in present day amateur radio. In the early days of ham radio, it was the only piece of test equipment most hams had at their disposal. It can tell you more about what's really happening in an antenna system than just about any other piece of test equipment even to this day.

E4B12 What is the significance of voltmeter sensitivity expressed in ohms per volt?

A. The full scale reading of the voltmeter multiplied by its ohms per volt rating will indicate the input impedance of the voltmeter.
B. When used as a galvanometer, the reading in volts multiplied by the ohms per volt rating will determine the power drawn by the device under test.
C. When used as an ohmmeter, the reading in ohms divided by the ohms per volt rating will determine the voltage applied to the circuit.
D. When used as an ammeter, the full scale reading in amps divided by ohms per volt rating will determine the size of shunt needed.

When measuring anything in a circuit, it is important that the measuring instrument cause as little change to the circuit as possible. Since a voltmeter is placed across a circuit (connected between two points in that circuit), it acts as a resistance in parallel with that part of the circuit. In order to cause the least change, the meter should be as high a resistance as possible. Further, ***by multiplying the ohms per volt rating times the maximum voltage on a given scale of the meter, you can determine the equivalent resistance your meter will look like to the circuit***. **ANSWER A.**

E4A10 Which of the following displays multiple digital signal states simultaneously?

A. Network analyzer.
B. Bit error rate tester.
C. Modulation monitor.
D. Logic analyzer.

At its core, a ***logic analyzer*** is nothing more than an oscilloscope with a lot of input channels. Modern logic analyzers have all the features of a Digital Storage Oscilloscope (DSO) and then some. A "mixed mode" DSO may include "normal" analog input channels in addition to logic analyzer inputs. A logic analyzer normally uses "1-bit" A-to-D conversion, which can be extremely fast. **ANSWER D.**

E4B07 What do the subscripts of S parameters represent?

A. The port or ports at which measurements are made.
B. The relative time between measurements.
C. Relative quality of the data.
D. Frequency order of the measurements.

S parameter notation can seem like Klingon if you're used to "normal" (lumped constant) circuit analysis, but is extremely revealing, as well as a whole lot of fun, once you get used to it. In S parameter notation, the first subscript ***identifies the port*** we're looking at for a response, and the second subscript is the port through which the "stimulus" is transmitted. Now that more hams are using network analyzers, S parameter notation is becoming more common. **ANSWER A.**

E4B13 Which S parameter is equivalent to forward gain?

A. S11. C. S21.
B. S12. D. S22.

In S parameter notation, the first subscript identifies the port we're looking at for some response, and the second subscript is the "stimulus" which causes the response. So to translate ***S21*** into English: Let's look at port 2 and see what happens here when we do something at port 1. This sounds like an *amplifier*, doesn't it? What does the output of an amplifier (port 2) look like when we put a signal into the input (Port 1)? **ANSWER C.**

E4B16 Which S parameter represents return loss or SWR?

A. S11. C. S21.
B. S12. D. S22.

In S parameter notation, ***S11 describes a one-port network***. We're looking at the same port for both a stimulus and a response. In a typical ***SWR*** measurement, we are looking at a reflected wave at the same port that we're applying a forward wave to. A network analyzer is capable of dealing with both one-port and two-port networks. **ANSWER A.**

E4B17 What three test loads are used to calibrate a standard RF vector network analyzer?

A. 50 ohms, 75 ohms, and 90 ohms.
B. Short circuit, open circuit, and 50 ohms.
C. Short circuit, open circuit, and resonant circuit.
D. 50 ohms through 1/8 wavelength, 1/4 wavelength, and 1/2 wavelength of coaxial cable.

One of the neat things about a vector network analyzer (VNA) is that it will tell you the impedance at the far end of a transmission line of any arbitrary length, provided you perform the proper calibration procedure. This is extremely useful if you don't care to lug a network analyzer up the tower with you! You can make all the measurements on the ground. ***The VNA has three calibration points: shorted, open, and terminated (50 ohms).*** Perform these calibrations at the load end of the transmission line you intend to use. It will automatically compensate for the length and loss of the transmission line. Then connect the transmission line to your antenna and take the readings from your VNA in the comfort of your shack. They will be accurate! **ANSWER B.**

E9A14 What is meant by the radiation resistance of an antenna?

A. The combined losses of the antenna elements and feed line.
B. The specific impedance of the antenna.
C. The value of a resistance that would dissipate the same amount of power as that radiated from an antenna.
D. The resistance in the atmosphere that an antenna must overcome to be able to radiate a signal.

Radiation resistance is an important term because it compares the power radiated from an antenna to the equivalent resistance that would dissipate the same amount of power as heat. The higher the radiation resistance of an antenna, the better its performance. Don't confuse radiation resistance with simple DC resistance – they are quite different. Radiation resistance is ***the equivalent resistance that would dissipate the same amount of power as that radiated from the antenna***. On the other hand, ohmic resistance is a real resistance which converts radio frequency currents directly into heat. The ratio of radiation resistance to ohomic resistance determines the efficiency of the antenna. **ANSWER C.**

E9A05 What is included in the total resistance of an antenna system?

A. Radiation resistance plus space impedance.
B. Radiation resistance plus transmission resistance.
C. Transmission-line resistance plus radiation resistance.
D. Radiation resistance plus ohmic resistance.

To calculate the ***total resistance*** of an antenna system, ***add radiation resistance plus all ohmic resistances*** within the coils and conductors. **ANSWER D.**

E9A04 Which of the following factors may affect the feed point impedance of an antenna?

A. Transmission-line length.
B. Antenna height, conductor length/diameter ratio and location of nearby conductive objects.
C. The settings of an antenna tuner at the transmiiter.
D. Sunspot activity and time of day.

You're thinking about building a new beam. Be sure to consider the three largest factors that will affect the ***feed point impedance*** of that antenna. These include getting it more than a half wavelength in ***height above the ground***, the ***element length to diameter ratio***, and ***how close it is to a nearby conductive object*** (like another band beam below it). **ANSWER B.**

Multiple antennas on the same mast can influence the impedance of all the antennas in the stack.

E9A03 Why would one need to know the feed point impedance of an antenna?

A. To match impedances in order to minimize standing wave ratio on the transmission line.
B. To measure the near-field radiation density from a transmitting antenna.
C. To calculate the front-to-side ratio of the antenna.
D. To calculate the front-to-back ratio of the antenna.

When the antenna (the load) ***has the same characteristic impedance as the feed line, maximum power will be transferred to the load***. A mismatch may result in high SWR, transmit power feedback, and potential transmitter heat-up. Most coaxial cable presents a 50 ohm impedance when terminated in a non-reactive 50 ohm resistance**.** **ANSWER A.**

This hex beam wire antenna was flat on 10, 15, 20 and 40 meters, with no external tuner. What a signal at the Quartzfest gathering in Arizona!

E9A09 How is antenna efficiency calculated?

A. (radiation resistance / transmission resistance) x 100 percent.
B. (radiation resistance / total resistance) x 100 percent.
C. (total resistance / radiation resistance) x 100 percent.
D. (effective radiated power / transmitter output) x 100 percent.

The efficiency of your ***antenna*** is an important consideration when you plan to upgrade your antenna system. ***Efficiency equals the radiation resistance divided by the total resistance, multiplied by 100 percent***. You want to keep your total (ohmic) resistance as low as possible to make your antenna more efficient. Generally, bigger antenna elements have greater radiation resistance, lower ohmic resistance, higher efficiency, and increased bandwidth. Bigger is always better! **ANSWER B.**

E9A10 Which of the following choices is a way to improve the efficiency of a ground-mounted quarter-wave vertical antenna?

A. Install a good radial system.
B. Isolate the coax shield from ground.
C. Shorten the radiating element.
D. Reduce the diameter of the radiating element.

Most ***vertical antennas*** are one-quarter wavelength long and use a mirror image counterpoise to compare to that of a halfwave antenna. A good ***ground radial system*** will give your signal more punch, lower the noise floor, lower your take-off angle, and give you good DX. **ANSWER A.**

As soon as Jim Ford, N6JF, ran 4 radials per band, ERP was much improved for some fun QRP contacts. Yes, a wet lawn will help, too.

E9A11 Which of the following factors determines ground losses for a ground-mounted vertical antenna operating in the 3 MHz to 30 MHz range?

A. The standing wave ratio.
B. Distance from the transmitter.
C. Soil conductivity.
D. Take-off angle.

Dirt is ground, right? Likely not – dry ***soil*** is not nearly as effective a ***conductor*** as 4 quarter-wavelength radials per band on that new HF vertical antenna in your back yard. Looking to really increase your 5-band trap vertical performance on HF? Replace those quarter wavelength ground wires with 3-inch wide copper foil strips, 4 per band, each a quarter wavelength long. **ANSWER C.**

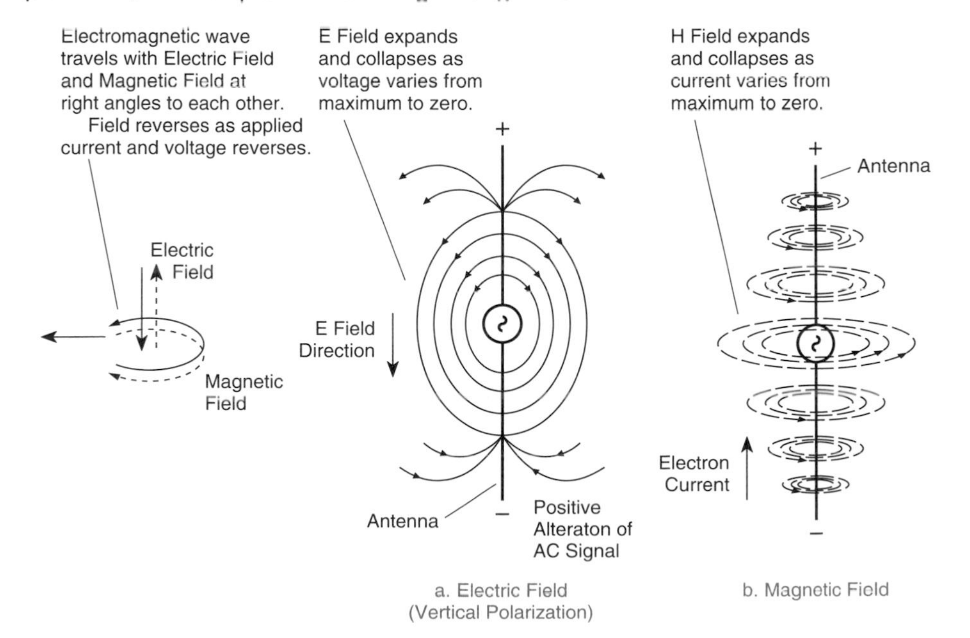

Propagating Electromagnetic Waves
Source: *Antennas – Selection and Installation*, A.J. Evans, Copyright ©1986 Master Publishing, Inc., Niles, Illinois

E9C11 How is the far-field elevation pattern of a vertically polarized antenna affected by being mounted over seawater versus rocky ground?

A. The low-angle radiation decreases.
B. The high-angle radiation increases.
C. Both the high-angle and low-angle radiation decrease.
D. The low-angle radiation increases.

Here we go – mount it over ***salt water*** and achieve a ***low-angle radiation increase***. This gives you longer-range skywave signals. **ANSWER D.**

E9C13 What is the main effect of placing a vertical antenna over an imperfect ground?

A. It causes increased SWR.
B. It changes the impedance angle of the matching network.
C. It reduces low-angle radiation.
D. It reduces losses in the radiating portion of the antenna.

When Gordo outfits boaters with their high frequency antenna systems, he always couples the antenna ground system to sea water. Nothing beats a direct contact to conductive sea water for the antenna ground, lowering the takeoff angle for maximum DX. ***Mounting a vertical antenna on your roof*** and relying only on the metal vent pipe as the ground plane ***will produce multiple near-vertical radiation angles and can reduce the DX low-angle radiation*** capability you were hoping to achieve. A vertical antenna needs a major ground system below it if you plan to work more than the lower 48! **ANSWER C.**

E9D11 Which of the following types of conductors would be best for minimizing losses in a station's RF ground system?

A. A resistive wire, such as spark plug wire.
B. A wide flat copper strap.
C. A cable with six or seven 18 gauge conductors in parallel.
D. A single 12 gauge or 10 gauge stainless steel wire.

What happens when you pass RF through a coil? The coil has inductive reactance and the impedance will be high at certain frequencies. For ground systems to be effective, minimize any wire runs that will look like a coil and create unwanted impedance. ***Copper ground foil is our choice for good grounding*** techniques. It is available in 25-foot, 50-foot, and custom-length rolls from your local ham radio dealer or a copper foil manufacturer. Three inch wide foil is Gordo's favorite for minimum inductive reactance. **ANSWER B.**

Copper ground foil minimizes inductive reactance and improves grounding to RF frequencies on High Frequency.

E9D12 Which of the following would provide the best RF ground for your station?

A. A 50 ohm resistor connected to ground.
B. An electrically short connection to a metal water pipe.
C. An electrically short connection to 3 or 4 interconnected ground rods driven into the Earth.
D. An electrically short connection to 3 or 4 interconnected ground rods via a series RF choke.

A good earth ground for your new Extra Class-reward transceiver should be ***copper foil connected to three or four ground rods, driven into the earth***. Follow manufacturer suggestions on how to ground your tower, too. **ANSWER C.**

E0A01 What is the primary function of an external earth connection or ground rod?

A. Reduce received noise.
B. Lightning protection.
C. Reduce RF current flow between pieces of equipment.
D. Reduce RFI to telephones and home entertainment systems.

While there isn't much you can do if a lightning bolt has your number on it, you can reduce the possibility of it doing too much damage if you direct the charge to ground outside your radio shack. There is no silver bullet when it comes to effectively grounding your station; you may have to decide which is more important for your situation, ***lightning protection***, RF suppression, or electrostatic discharge protection. **ANSWER B.**

E9C01 What is the radiation pattern of two 1/4-wavelength vertical antennas spaced 1/2-wavelength apart and fed 180 degrees out of phase?

A. Cardioid.
B. Omni-directional.
C. A figure-8 broadside to the axis of the array.
D. A figure-8 oriented along the axis of the array.

An array of two quarter-wave vertical antennas spaced ***one-half wavelength*** apart and ***fed 180 degrees out of phase*** results in a ***figure-8 radiation pattern*** with best radiation ***in line with the antennas***. **ANSWER D.**

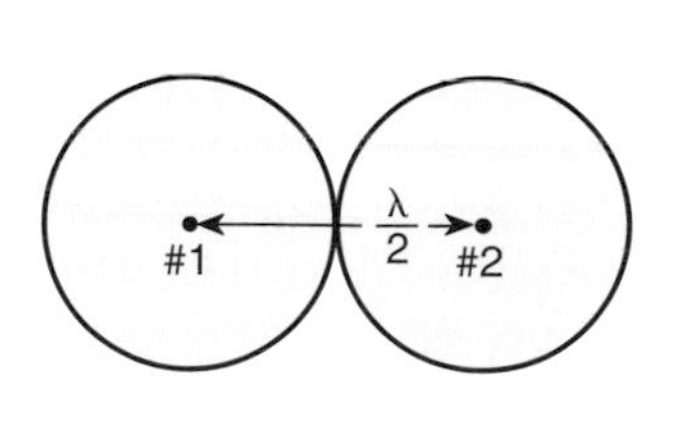

a. Two λ/4 Verticals Fed 180° Out of Phase (Figure 8 End-Fire In-Line)

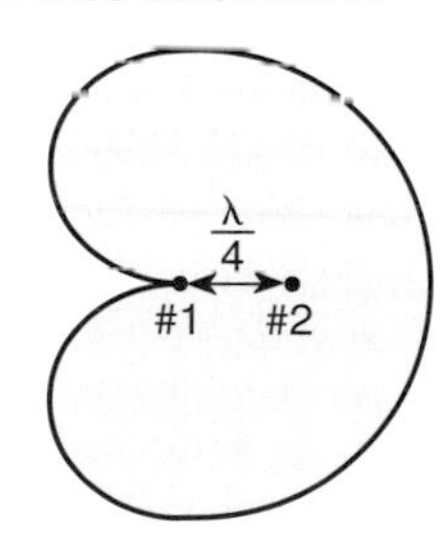

b. Two λ/4 Verticals Fed 90° Out of Phase (Cardioid)

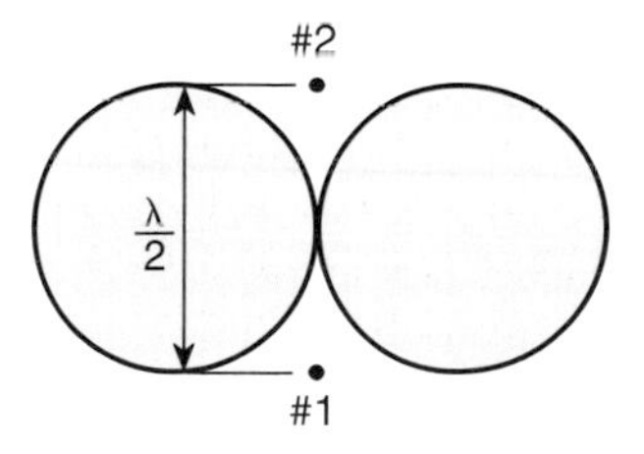

c. Two λ/4 Verticals Fed In Phase (Figure 8 Broadside)

Antenna Radiation Patterns

E9C02 What is the radiation pattern of two 1/4 wavelength vertical antennas spaced 1/4 wavelength apart and fed 90 degrees out of phase?

A. Cardioid.
B. A figure-8 end-fire along the axis of the array.
C. A figure-8 broadside to the axis of the array.
D. Omni-directional.

An array of ***two quarter-wave vertical antennas*** spaced ***one-quarter wavelength apart*** and ***fed 90 degrees out of phase*** results in a ***cardioid pattern***. This cardioid or unidirectional pattern is needed for direction finding. **ANSWER A.**

E9C03 What is the radiation pattern of two 1/4 wavelength vertical antennas spaced 1/2 wavelength apart and fed in phase?

A. Omni-directional.
B. Cardioid.
C. A Figure-8 broadside to the axis of the array.
D. A Figure-8 end-fire along the axis of the array.

An array of two quarter-wave vertical antennas spaced ***one-half wavelength apart*** and ***fed in phase*** will also result in a ***figure-8 pattern***, but maximum gain will be ***broadside to the antennas***. **ANSWER C.**

E9C04 What happens to the radiation pattern of an unterminated long wire antenna as the wire length is increased?

A. The lobes become more perpendicular to the wire.
B. The lobes align more in the direction of the wire.
C. The vertical angle increases.
D. The front-to-back ratio decreases.

If you have an infinite amount of property, a long wire antenna can give you an infinite amount of gain! However, the point of diminishing returns is usually reached once a long wire antenna exceeds six or seven wavelengths. The major lobe of the radiation pattern of a long wire antenna tends to follow the direction of the wire. The longer the wire is (in terms of wavelength), the closer the ***major lobes are to the direction of the wire*** itself. Combinations of long-wire antennas in the form of Rhombics or V-beams achieve very high gain by aligning the major lobes of the individual long-wire portions. **ANSWER B.**

E9C08 What is a folded dipole antenna?

A. A dipole one-quarter wavelength long.
B. A type of ground-plane antenna.
C. A dipole consisting of one wavelength of wire forming a very thin loop.
D. A dipole configured to provide forward gain.

The folded dipole is a fun one to build out of television twin-lead cable. The twin-lead ***folded dipole is constructed by using a half-wavelength piece of twin lead to form a loop that is one full wavelength long***. Scrape the insulation from the wire ends and twist the wires together on each end. In the center of one of the leads in the twin-lead, cut the lead and scrape some of the insulation off to expose bare wires on the end of each quarter-wave piece. Feed these wires with standard 300 ohm twin-lead, or connect a balun transformer to the bare wires to permit 50 ohm or 75 ohm coax cable to feed the folded dipole. **ANSWER C.**

E9C07 What is the approximate feed point impedance at the center of a two-wire folded dipole antenna?

A. 300 ohms.
B. 72 ohms.
C. 50 ohms.
D. 450 ohms.

The ***impedance at the center of a folded dipole*** may be a whopping ***300 ohms***. You'll need an impedance-matching transformer to feed it with 50 ohm coax; however, ordinary TV twin-lead will match very well. **ANSWER A.**

E9C05 What is an OCFD antenna?

A. A dipole fed approximately 1/3 the way from one end with a 4:1 balun to provide multiband operation.
B. A remotely tunable dipole antenna using orthogonally controlled frequency diversity.
C. An eight band dipole antenna using octophase filters.
D. A multiband dipole antenna using one-way circular polarization for frequency diversity.

A center-fed dipole will usually work just fine on odd harmonics. For example, a 40 meter dipole will show a near perfect match on 15 meters. However, this is not the case on the even harmonics. A center-fed 40 meter dipole will have a horrendous mismatch on 20 meters; this is because on 20 meters the mid-point is a voltage loop, which has extremely high impedance – many thousands of ohms. In the 1920's Loren Windom, 8GZ/8ZG, discovered that if you feed a dipole antenna about 14% off-center the impedance will be reasonably uniform across both the even and odd harmonic bands, and on the order of 600 ohms. This presented a pretty good match to the single-wire feedline that was quite common at the time. The original Windom fell into disfavor when hams began using coaxial cable shortly after WWII. Recently, the basic principles of the Windom have been resurrected in the form of the ***off-center fed dipole (OCFD), which uses a 4:1 balun/step up transformer at the feed point to provide a decent match across most HF amateur bands***. While the OCFD is often called a Windom, it is not. A true Windom uses a single-wire feed-line. **ANSWER A.**

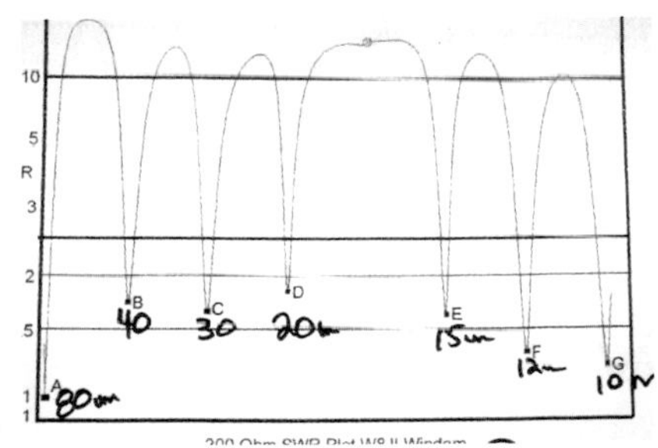

This off center fed antenna has a 4:1 balun at the feed point, spaced away from the metal mast. The dipole offers resonance on all bands provided you work with it long enough when pruning each end.

E9C09 What is a G5RV antenna?

A. A multi-band dipole antenna fed with coax and a balun through a selected length of open wire transmission line.
B. A multi-band trap antenna.
C. A phased array antenna consisting of multiple loops.
D. A wide band dipole using shorted coaxial cable for the radiating elements and fed with a 4:1 balun.

There are times when having a high SWR isn't such a bad thing; in fact it might be absolutely necessary. The ***open wire transmission line section of a G5RV*** antenna is just such a case. This line section acts as an impedance matching transformer and tuning "stub." While there is a high SWR on the open wire line, the bottom end of the line is close to 50 ohms, providing a reasonable match to the 50 ohm coax preceding it. A balun/transformer at the bottom end assures that the line radiation on the coaxial segment is low. **ANSWER A.**

E9C10 Which of the following describes a Zepp antenna?

A. A dipole constructed from zip cord.
B. An end fed dipole antenna.
C. An omni-directional antenna commonly used for satellite communications.
D. A vertical array capable of quickly changing the direction of maximum radiation by changing phasing lines.

The ***Zepp*** antenna, so named for the Zeppelin airships on which it was commonly deployed, ***is an end fed halfwave dipole.*** A quarter-wave ladder line "Q-section" matches the transmitter's low output impedance to the extremely high impedance at the end of the Zepp dipole. Incidentally, the popular J-pole vertical antenna is, for all practical purposes, a "stretched out" Zepp. **ANSWER B.**

E9C12 Which of the following describes an extended double Zepp antenna?

A. A wideband vertical antenna constructed from precisely tapered aluminum tubing.
B. A portable antenna erected using two push support poles.
C. A center fed 1.25 wavelength antenna (two 5/8 wave elements in phase).
D. An end fed folded dipole antenna.

The ***Extended Double Zepp (EDZ) antenna*** has the greatest possible broadside gain for a single wire antenna, about 3dB of gain over a dipole at typical heights. (The same is true for its half-size vertical variation, the ***5/8 wave*** vertical or ground plane). The EDZ, like the 5/8 wave vertical, is not self-resonant. Its feed point consists of some capacitive reactance, so it must have inductive loading of some kind. In addition, the resistive component is somewhat higher than 50 ohms, so some form of matching network is usually necessary. **ANSWER C.**

E9A01 What describes an isotropic antenna?

A. A grounded antenna used to measure earth conductivity.
B. A horizontally polarized antenna used to compare Yagi antennas.
C. A theoretical antenna used as a reference for antenna gain.
D. A spacecraft antenna used to direct signals toward the earth.

An ***isotropic radiator*** is simply a theoretical antenna; and in every April 1st issue of many magazines, advertisers try to sell "isotropic radiators" to the unsuspecting ham! We use it as a ***theoretical reference to calculate the gain*** or loss of real antenna systems. **ANSWER C.**

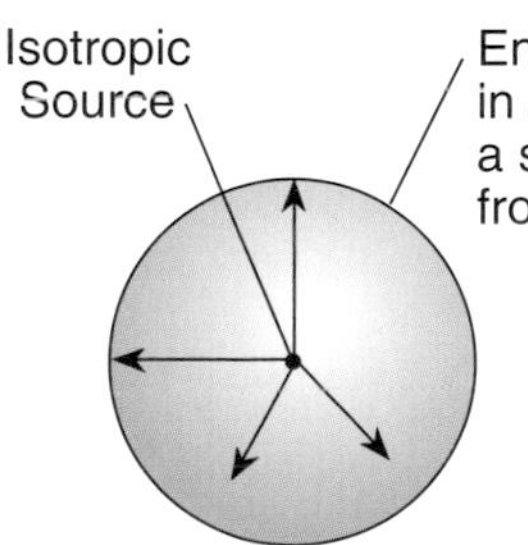

Isotropic Radiator Pattern

E9A02 What antenna has no gain in any direction?

A. Quarter-wave vertical.
B. Yagi.
C. Half-wave dipole.
D. Isotropic antenna.

On paper, it is the theoretical ***isotropic radiator*** that has equal signal dispersion in all directions and ***no gain in any direction***. **ANSWER D.**

E9B07 How does the total amount of radiation emitted by a directional gain antenna compare with the total amount of radiation emitted from an isotropic antenna, assuming each is driven by the same amount of power?

A. The total amount of radiation from the directional antenna is increased by the gain of the antenna.
B. The total amount of radiation from the directional antenna is stronger by its front-to-back ratio.
C. They are the same.
D. The radiation from the isotropic antenna is 2.15 dB stronger than that from the directional antenna.

This is a good question to get you thinking about the isotropic antenna – an antenna that radiates equally well in all directions. If you were to pump 100 watts into this theoretical antenna, you would get 100 watts out, unity gain, in every direction, including up and down. Now you switch to the beam antenna and presto, those 100 watts are focused in one direction, with minimum up and minimum down radiation. By taking energy in all directions and concentrating it in just one direction with a beam, you have improved your signal on transmit and receive. But the total amount of radiation emitted is that same 100 watts at the feed point, so there is no magical new power created in the beam, and thus ***there is no difference between the two antennas***. A ***100 watt*** light bulb hanging by a wire and a 100 watt bulb with a polished reflector ***is still 100 watts***, right? Sure, but one bulb looks brighter, with the same amount of power, thanks to the reflector! **ANSWER C.**

E9A07 What is meant by antenna gain?

A. The ratio of the radiated signal strength of an antenna in the direction of maximum radiation to that of a reference antenna.
B. The ratio of the signal in the forward direction to that in the opposite direction.
C. The ratio of the amount of power radiated by an antenna compared to the transmitter output power.
D. The final amplifier gain minus the transmission line losses.

We normally rate ***antenna gain*** figures in dB. This is the ***numerical ratio relating the radiated signal strength of your antenna to that of a reference antenna.*** Generally, the ham with the higher gain antenna will get better signal reports, but not always. **ANSWER A.**

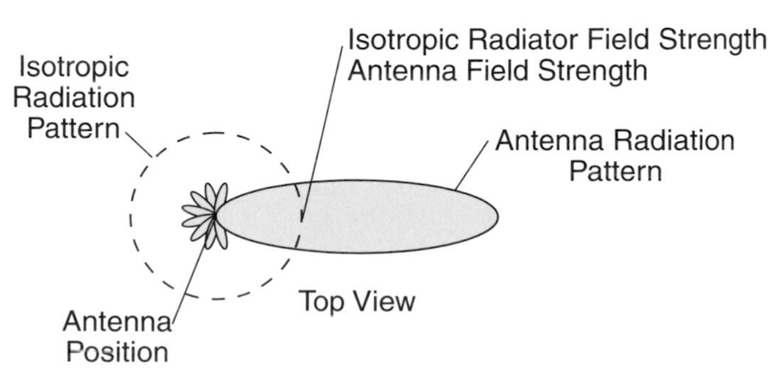

"Gain" of an Antenna

E9A12 How much gain does an antenna have compared to a 1/2-wavelength dipole when it has 6 dB gain over an isotropic antenna?

A. 3.85 dB. C. 8.15 dB.
B. 6.0 dB. D. 2.79 dB.

The halfwave dipole has about 2.15 dB gain over an isotropic radiator. If an antenna under measurement has 6 dB gain over an isotropic radiator, it has ***6 dB - 2.15 dB = about 3.85 dB gain over the dipole*.** **ANSWER A.**

E9A13 How much gain does an antenna have compared to a 1/2-wavelength dipole when it has 12 dB gain over an isotropic antenna?

A. 6.17 dB. C. 12.5 dB.
B. 9.85 dB. D. 14.15 dB.

Here is a bigger antenna that has 12 dB gain over an isotropic radiator, so it has ***12 dB gain - 2.15 dB = about 9.85 dB over the dipole*.** **ANSWER B.**

E9D08 What happens as the Q of an antenna increases?

A. SWR bandwidth increases.
B. SWR bandwidth decreases.
C. Gain is reduced.
D. More common-mode current is present on the feed line.

As with any other resonant circuit, the Q of an antenna is inversely proportional to the bandwidth (***as Q increases, bandwidth decreases***). In the calculations for an antenna, the R used in the equation Q = X/R is *radiation resistance*. Antennas with higher radiation resistance tend to be more broad banded and have a lower Q. **ANSWER B.**

E9D09 What is the function of a loading coil used as part of an HF mobile antenna?

A. To increase the SWR bandwidth.
B. To lower the losses.
C. To lower the Q.
D. To cancel capacitive reactance.

Since X_L (inductive reactance) must equal X_C (capacitive reactance) in order for a whip antenna to achieve resonance, we need to ***add a loading coil*** on an antenna for lower frequency operations to ***tune out capacitive reactance*.** **ANSWER D.**

A high Q mobile coil (left) with the contactor disc seen in the upper 2/3 of the antenna coil. These motorized high Q antennas really improve performance on 40 and 75 meters.

E9A08 What is meant by antenna bandwidth?

A. Antenna length divided by the number of elements.
B. The frequency range over which an antenna satisfies a performance requirement.
C. The angle between the half-power radiation points.
D. The angle formed between two imaginary lines drawn through the element ends.

If you operate on the lower bands, such as 40 meters, antenna bandwidth is an important consideration. ***Bandwidth is the frequency range over which you can expect your antenna to perform well*.** Small antennas on 40 meters don't have a very wide bandwidth. While they will perform on the 40 meter band, you may have only a narrow window of useable frequencies. Monster antennas for 40 meters have good bandwidth giving you more usable frequencies. **ANSWER B.**

E9D06 What happens to the bandwidth of an antenna as it is shortened through the use of loading coils?

A. It is increased.
B. It is decreased.
C. No change occurs.
D. It becomes flat.

When you ***shorten a vertical antenna with a loading coil, bandwidth decreases*** and losses increase. **ANSWER B.**

E9D04 Why should an HF mobile antenna loading coil have a high ratio of reactance to resistance?

A. To swamp out harmonics.
B. To maximize losses.
C. To minimize losses.
D. To minimize the Q.

Those giant open-air coils halfway up the mast on a mobile whip antenna are ***high Q coils that minimize losses***. The bigger the coil, the lower the losses. **ANSWER C.**

E9D03 Where should a high Q loading coil be placed to minimize losses in a shortened vertical antenna?

A. Near the center of the vertical radiator.
B. As low as possible on the vertical radiator.
C. As close to the transmitter as possible.
D. At a voltage node.

To minimize coil losses and increase performance of a loaded vertical antenna, the ***loading coil*** should be near the ***center of the vertical radiator***, not at the base. **ANSWER A.** ☞ **www.hy-gain.com**

See the bird cage capacity loading hat above the coil? It increases performance on the lower bands, like 40 meters.

E9D07 What is an advantage of using top loading in a shortened HF vertical antenna?

A. Lower Q.
B. Greater structural strength.
C. Higher losses.
D. Improved radiation efficiency.

Top loading and helical loading are used on a shortened HF vertical antenna to ***improve radiation efficiency***. With this in mind, it's easy to understand why a little 96-inch CB whip antenna with an automatic antenna tuner will not perform as well as a resonant center-loaded, helical-loaded, or top-loaded vertical antenna. It doesn't matter how much power you put into an antenna by fooling the radio with the tuner; if the antenna is not resonant great amounts of power are lost. **ANSWER D.**

E9D10 What happens to feed point impedance at the base of a fixed length HF mobile antenna as the frequency of operation is lowered?

A. The radiation resistance decreases and the capacitive reactance decreases.
B. The radiation resistance decreases and the capacitive reactance increases.
C. The radiation resistance increases and the capacitive reactance decreases.
D. The radiation resistance increases and the capacitive reactance increases.

As we ***operate lower on the bands***, the base ***feed point resistance decreases*** (lower decreases), and the ***capacitive reactance (X_C) increases***. You can visualize the correct answer by thinking: the lower you operate in frequency, the more coil (inductive reactance, X_L) you will need to offset and equal the increased capacitive reactance. **ANSWER B.**

E9D05 What is a disadvantage of using a multiband trapped antenna?

A. It might radiate harmonics.
B. It radiates the harmonics and fundamental equally well.
C. It is too sharply directional at lower frequencies.
D. It must be neutralized.

Trap antennas have one big disadvantage – on older equipment, they can radiate harmonics. Since ham bands may be harmonically related, it's easy to see that the second harmonic of 7 MHz can radiate from a trap antenna quite nicely because the antenna also has perfect resonance on 14 MHz. So watch out for ***harmonics on a trap antenna***. **ANSWER A.**

E9B08 How can the approximate beamwidth in a given plane of a directional antenna be determined?

A. Note the two points where the signal strength of the antenna is 3 dB less than maximum and compute the angular difference.
B. Measure the ratio of the signal strengths of the radiated power lobes from the front and rear of the antenna.
C. Draw two imaginary lines through the ends of the elements and measure the angle between the lines.
D. Measure the ratio of the signal strengths of the radiated power lobes from the front and side of the antenna.

On directional antennas, beamwidth is an important consideration. ***Beamwidth is*** always described by ***the angle between the two points where the signal strength is down 3 dB from the maximum signal point***. The "tighter" your main lobe pattern, the narrower the beamwidth. **ANSWER A.**

E9B09 What type of computer program technique is commonly used for modeling antennas?

A. Graphical analysis.
B. Method of Moments.
C. Mutual impedance analysis.
D. Calculus differentiation with respect to physical properties.

"Method of Moments," is a computer program ***for modeling a proposed antenna*** on the computer screen. **ANSWER B.**

E9B10 What is the principle of a Method of Moments analysis?

A. A wire is modeled as a series of segments, each having a uniform value of current.
B. A wire is modeled as a single sine-wave current generator.
C. A wire is modeled as a series of points, each having a distinct location in space.
D. A wire is modeled as a series of segments, each having a distinct value of voltage across it.

Method of Moments (MOM) software models a wire as a series of segments where ***each segment has a uniform value of antenna current***. **ANSWER A.**

E9B13 What does the abbreviation NEC stand for when applied to antenna modeling programs?

A. Next Element Comparison.
B. Numerical Electromagnetic Code.
C. National Electrical Code.
D. Numeric Electrical Computation.

NEC-based antenna modeling stands for Numerical Electromagnetics Code. NEC simulates the electromagnetic response of antennas and metal structures. Under contract to the U.S. Navy, Jerry Burke and A. Poggio wrote the NEC/MOM family of programs at Lawrence Livermore Labs in 1981. NEC2 was later released to the public and is now available on most computing platforms. **ANSWER B.**
To learn more about NEC, ☞ **www.nec2.org**.

E9B11 What is a disadvantage of decreasing the number of wire segments in an antenna model below the guideline of 10 segments per half wavelength?

A. Ground conductivity will not be accurately modeled.
B. The resulting design will favor radiation of harmonic energy.
C. The computed feed point impedance may be incorrect.
D. The antenna will become mechanically unstable.

When modeling antennas on a computer, we always stay above 10 segments per half wavelength so we don't cause the *feed point impedance to be miscalculated using less than* the recommended *10 segments per half wavelength*. **ANSWER C.**

E9B14 What type of information can be obtained by submitting the details of a proposed new antenna to a modeling program?

A. SWR vs frequency charts.
B. Polar plots of the far field elevation and azimuth patterns.
C. Antenna gain.
D. All of these choices are correct.

Modern antenna programs will show *SWR* plots at a certain frequency, *polar plots, antenna gain, and feed point impedance*. In some programs, you can even tweak your proposed antenna by adding in the diameter of those cool-looking polished copper tube elements! **ANSWER D.**

E9H01 When constructing a Beverage antenna, which of the following factors should be included in the design to achieve good performance at the desired frequency?

A. Its overall length must not exceed 1/4 wavelength.
B. It must be mounted more than 1 wavelength above ground.
C. It should be configured as a four-sided loop.
D. It should be one or more wavelengths long.

The *Beverage antenna* is a tremendous signal-grabber. For best performance, it really *needs to be L O N G*! The longer the better for a Beverage antenna.
ANSWER D.

E9H09 Which of the following describes the construction of a receiving loop antenna?

A. A large circularly polarized antenna.
B. A small coil of wire tightly wound around a toroidal ferrite core.
C. One or more turns of wire wound in the shape of a large open coil.
D. A vertical antenna coupled to a feed line through an inductive loop of wire.

A receiving loop antenna, for ham radio receive purposes, is usually many turns of wire wound in the shape of a large open coil or square loop. The more turns you add, the greater the gain of the receiving loop antenna. Most modern, high-frequency transceivers offer a separate antenna port for your homebrewed "one or more turns of wire" ***large open-coil antenna***. **ANSWER C.**

Home brew tunable loop with the giant capacitor under the white hood.

E9H10 How can the output voltage of a multiple turn receiving loop antenna be increased?

A. By reducing the permeability of the loop shield.
B. By increasing the number of wire turns in the loop and reducing the area of the loop structure.
C. By winding adjacent turns in opposing directions.
D. By increasing either the number of wire turns in the loop or the area of the loop structure or both.

To make a loop antenna more powerful, ***either increase the number of wire turns*** in the loop ***or make the loop physically larger***. **ANSWER D.**

The more turns you add to a loop the greater the receive output voltage.

E9H11 What characteristic of a cardioid pattern antenna is useful for direction finding?

A. A very sharp peak.
B. A very sharp single null.
C. Broad band response.
D. High-radiation angle.

A ***cardioid pattern*** is shaped like a heart. There is a broad reception pattern in one direction and a ***deep null*** in the other. When the button is pushed to "sense" the direction of the incoming signal on a direction-finding system that has a cardioid antenna pattern, the deep null in the pattern is normally 180 degrees away from the incoming signal. It's easier to use the null point to sense the direction of the incoming signal than it is the broad pattern – look for the minimum signal, not the maximum. See the illustration at E9C01, page 235. **ANSWER B.**

E9H04 What is an advantage of using a shielded loop antenna for direction finding?

A. It automatically cancels ignition noise in mobile installations.
B. It is electro-statically balanced against ground, giving better nulls.
C. It eliminates tracking errors caused by strong out-of-band signals.
D. It allows stations to communicate without giving away their position.

Transmitter hunting down on 75 meters is an international sport. Most operators use a shielded loop antenna for T-hunting because the *shielding offers better directivity*. **ANSWER B.**

E9H05 What is the main drawback of a wire-loop antenna for direction finding?

A. It has a bidirectional pattern.
B. It is non-rotatable.
C. It receives equally well in all directions.
D. It is practical for use only on VHF bands.

Wire loop direction finding antennas will pick up maximum signal strength when the loop is broadside to the signal source. It is bi-directional – it will *pick up signal sources on either side of the loop*. **ANSWER A.** ☞ **www.lwca.org**

E9H07 Why is it advisable to use an RF attenuator on a receiver being used for direction finding?

A. It narrows the bandwidth of the received signal to improve signal to noise ratio.
B. It compensates for the effects of an isotropic antenna, thereby improving directivity.
C. It reduces loss of received signals caused by antenna pattern nulls, thereby increasing sensitivity.
D. It prevents receiver overload which could make it difficult to determine peaks or nulls.

When using a receiver for direction finding, a step adjustable or variable adjustable attenuator is necessary on the input when you get extremely close to the signal source. You *increase attenuation* in order *to keep your receiver from overloading* as you maintain maximum sensitivity for signal variations as the antenna direction is changed. **ANSWER D.**

E9H08 What is the function of a sense antenna?

A. It modifies the pattern of a DF antenna array to provide a null in one direction.
B. It increases the sensitivity of a DF antenna array.
C. It allows DF antennas to receive signals at different vertical angles.
D. It provides diversity reception that cancels multipath signals.

On high frequency automatic radio direction finder systems, the double loop may employ an additional rigid whip to act as a *sense antenna*. Push the ADF sense button and the fixed double loop DF (direction finding) antenna *will null out* ambiguity as to which direction the incoming signal is arriving from at that double loop. The sense antenna must be rigid, never a flexing whip. **ANSWER A.**

The telescopic whip on this direction finding loop extends up to provide a sense capability. It lets the operator know which end of the loop is pointing toward the transmitting station as the loop is turned slightly.

E9H06 What is the triangulation method of direction finding?

A. The geometric angles of sky waves from the source are used to determine its position.
B. A fixed receiving station plots three headings to the signal source.
C. Antenna headings from several different receiving locations are used to locate the signal source.
D. A fixed receiving station uses three different antennas to plot the location of the signal source.

Amateur radio operators participating in a joint "T-hunt" will swing their beams and record a magnetic bearing to where signal strength is the greatest. ***Through triangulation, they can*** then ***determine the approximate location of the transmitting station***. All participating stations should agree to use either magnetic north or true north for their heading directions. **ANSWER C.**

E9B01 In the antenna radiation pattern shown in Figure E9-1, what is the 3 dB beamwidth?

A. 75 degrees.
B. 50 degrees.
C. 25 degrees.
D. 30 degrees.

Looking carefully at the diagram in Figure E9-1, notice that the main lobe of the pattern intersects the 3 dB points at approximately plus and minus 25 degrees. Hence, the ***3-dB beamwidth is approximately 50 degrees***. **ANSWER B.**

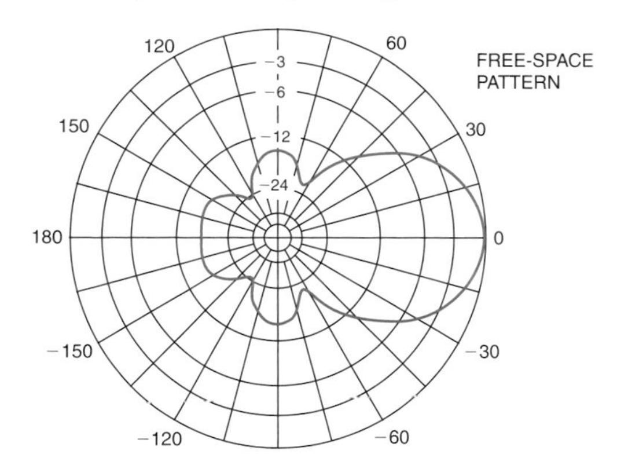

Figure E9-1

E9B02 In the antenna radiation pattern shown in Figure E9-1, what is the front-to-back ratio?

A. 36 dB.
B. 18 dB.
C. 24 dB.
D. 14 dB.

If you look carefully at Figure E9-1, ***the rear lobe is about -18 dB down from the front lobe***. **ANSWER B.**

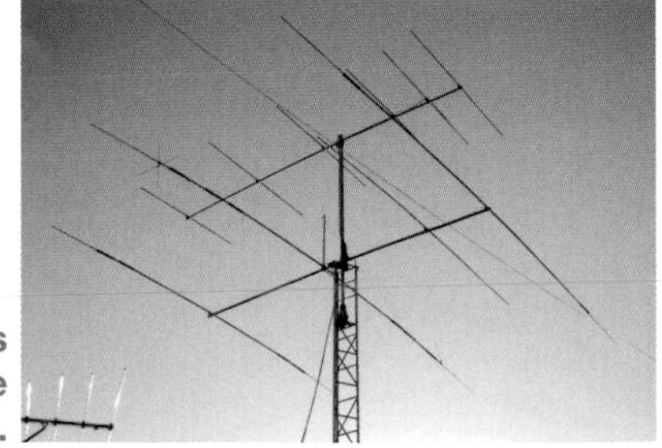

Gordo's 6 meter beam is on the top, and his four-band HF beam is on the bottom. The tower is aluminum from Aluma Tower.

E9B03 In the antenna radiation pattern shown in Figure E9-1, what is the front-to-side ratio?

A. 12 dB.
B. 14 dB.
C. 18 dB.
D. 24 dB.

Study Figure E9-1 carefully. The front-to-side ratio is measured where the lobes perpendicular to the main lobe cross the vertical axis. Notice the scale decreases as you travel out this axis. Since the lobe crosses just a little before 12 but closer to 12 than to 24, ***14 is your best choice*** from the answers given here. **ANSWER B.**

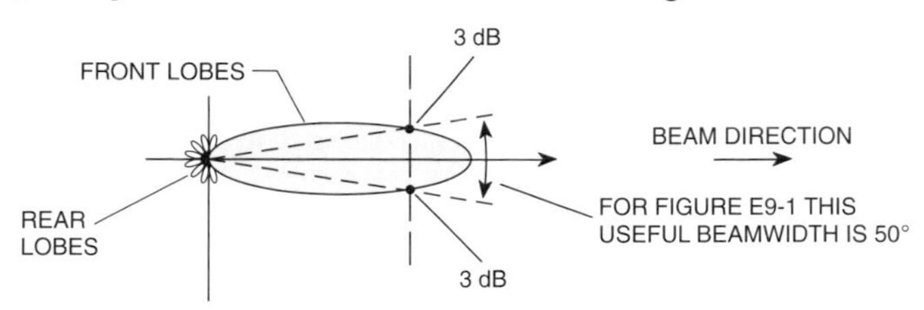

Directional Radiation Pattern of a Yagi Beam

Source: *Antennas – Selection and Installation*, A.J. Evans, Copyright ©1986 Master Publishing, Inc., Niles, Illinois

---

E9H02 Which is generally true for low band (160 meter and 80 meter) receiving antennas?

A. Atmospheric noise is so high that gain over a dipole is not important.
B. They must be erected at least 1/2 wavelength above the ground to attain good directivity.
C. Low loss coax transmission line is essential for good performance.
D. All of these choices are correct.

Many hams forget that the function of a transmitting antenna can be very different from that of a receiving antenna. Further, the performance priorities of each can be very different as well. On low bands, especially, it can be a great advantage to use a separate receiving antenna. Small, tuned receiving loops can provide a large ***advantage in signal-to-noise ratio over a full sized antenna, even if the overall gain is reduced***. **ANSWER A.**

E9B12 What is the far field of an antenna?

A. The region of the ionosphere where radiated power is not refracted.
B. The region where radiated power dissipates over a specified time period.
C. The region where radiated field strengths are obstructed by objects of reflection.
D. The region where the shape of the antenna pattern is independent of distance.

The ***far field zone*** of an antenna is where the ***antenna pattern remains constant, independent of distance***. This is generally a safe area when calculating MPE. **ANSWER D.**

E9B04 What may occur when a directional antenna is operated at different frequencies within the band for which it was designed?

A. Feed point impedance may become negative.
B. The E-field and H-field patterns may reverse.
C. Element spacing limits could be exceeded.
D. The gain may change depending on frequency.

Building your own homebrew Yagi can teach some fundamental Yagi antenna concepts:

- Close element spacing will yield narrow bandwidth response.
- Directivity sharpness depends on parasitic-element tuning for maximum gain.
- Boom length, not number of directors, influences gain the most.

A Yagi antenna with closely-spaced reflector and director elements may indeed maximize gain, but ***overall gain may have significant variations*** If your Yagi design is tuned to the bottom of the band and you begin to operate at the top of the band, several hundred kHz away from the frequency used for your homebrew design. **ANSWER D.**

This is the classic KT-34XA beam antenna. The tower climber is giving the thrust bearing a lube job.

E9D13 What usually occurs if a Yagi antenna is designed solely for maximum forward gain?

A. The front-to-back ratio increases.
B. The front-to-back ratio decreases.
C. The frequency response is widened over the whole frequency band.
D. The SWR is reduced.

On that brand new Yagi antenna you are computer modeling, the term "front to back ratio" is the ratio of the power radiated in the main lobe to the power being radiated in the opposite direction. As you work your computer antenna modeling program, you will quickly discover that ***maximum forward gain will decrease the front to back ratio***. This is why some antenna experts will sacrifice a half dB of forward gain to improve one or two nulls off the back of the beam. **ANSWER B.**

If you buy your beam, the manufacturer usually has the elements spaced for the best front-to-back ratio, slightly decreasing forward gain.

E9C15 How does the radiation pattern of a horizontally polarized 3-element beam antenna vary with its height above ground?

A. The main lobe takeoff angle increases with increasing height.
B. The main lobe takeoff angle decreases with increasing height.
C. The horizontal beamwidth increases with height.
D. The horizontal beamwidth decreases with height.

When you put up a 3-element Yagi, the high frequency antenna is usually mounted horizontally at a height above ground of at least one-half wavelength of the lowest band of operation. Most 3-element Yagis feature tri-band operation for 10, 15 and 20 meters. The lowest band, 20 meters, would like to see the antenna at one-half wavelength off the earth, or about 32 feet in the air. Anything lower than one-half wavelength from the operating frequency dramatically raises the main lobe takeoff angle. The higher the takeoff angle, the shorter the skip distance. You want to ***get your beam up at least one-half wavelength above the earth for a nice low takeoff angle*** for long-range DX. **ANSWER B.**

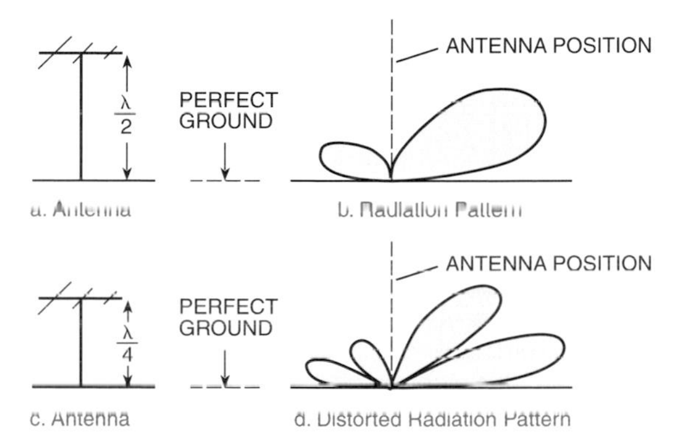

Radiation of 3-Element Yagi for Different Antenna Heights

E9C14 How does the performance of a horizontally polarized antenna mounted on the side of a hill compare with the same antenna mounted on flat ground?

A. The main lobe takeoff angle increases in the downhill direction.
B. The main lobe takeoff angle decreases in the downhill direction.
C. The horizontal beamwidth decreases in the downhill direction.
D. The horizontal beamwidth increases in the uphill direction.

If you live on the East coast, build your home and radio shack on a hill that slopes down to the ocean. Your beam will now have a main lobe whose takeoff angle decreases toward the ocean, giving you a dynamite signal to Europe. Out on the West coast, Gordo is up on some headlands with the ocean just down the way, giving him a good shot to Asia with the ***lower takeoff angle in the downhill direction***. **ANSWER B.**

E9B05 What type of antenna pattern over real ground is shown in Figure E9-2?

A. Elevation.
B. Azimuth.
C. Radiation resistance.
D. Polarization.

Figure E9-2 shows ***the takeoff elevation*** patterns of an antenna over real ground. **ANSWER A.**

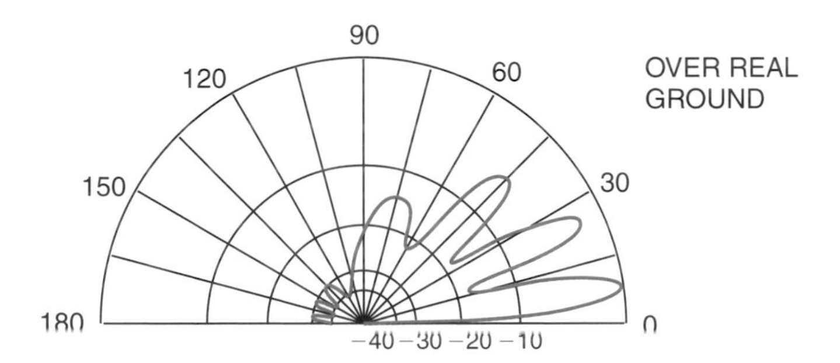

Figure E9-2

E9B06 What is the elevation angle of peak response in the antenna radiation pattern shown in Figure E9-2?

A. 45 degrees.
B. 75 degrees.
C. 7.5 degrees.
D. 25 degrees.

From the pattern in the figure we see that ***the main zero dB lobe is about 7.5 degrees up***, with two other significant lobes at higher elevations. **ANSWER C.**

E9B15 What is the front-to-back ratio of the radiation pattern shown in Figure E9-2?

A. 15 dB.
B. 28 dB.
C. 3 dB.
D. 24 dB.

The antenna ***front-to-back ratio*** in Figure E9-2 looks to be ***about 28 to 30 dB***. This is the ratio of forward "front" radiation to the slight amount of unwanted radiation in the reverse direction from the back of the antenna. This ratio is the same for transmit as well as receive. **ANSWER B.**

E9B16 How many elevation lobes appear in the forward direction of the antenna radiation pattern shown in Figure E9-2?

A. 4.
B. 3.
C. 1.
D. 7.

There are ***three big lobes*** and ***one smaller lobe in the forward direction***. **ANSWER A.**

E9D02 How can linearly polarized Yagi antennas be used to produce circular polarization?

A. Stack two Yagis fed 90 degrees out of phase to form an array with the respective elements in parallel planes.
B. Stack two Yagis fed in phase to form an array with the respective elements in parallel planes.
C. Arrange two Yagis perpendicular to each other with the driven elements at the same point on the boom fed 90 degrees out of phase.
D. Arrange two Yagis collinear to each other with the driven elements fed 180 degrees out of phase.

To get circular polarization out of a pair of Yagis, arrange the *Yagis perpendicular to each other* with the driven elements in the same plane and *fed 90 degrees out of phase*. **ANSWER C.**

E9C06 What is the effect of a terminating resistor on a rhombic antenna?

A. It reflects the standing waves on the antenna elements back to the transmitter.
B. It changes the radiation pattern from bidirectional to unidirectional.
C. It changes the radiation pattern from horizontal to vertical polarization.
D. It decreases the ground loss.

A *terminating resistor* at the end of the rhombic opposite the feed line *changes the radiation pattern from bidirectional to unidirectional*. **ANSWER B.**

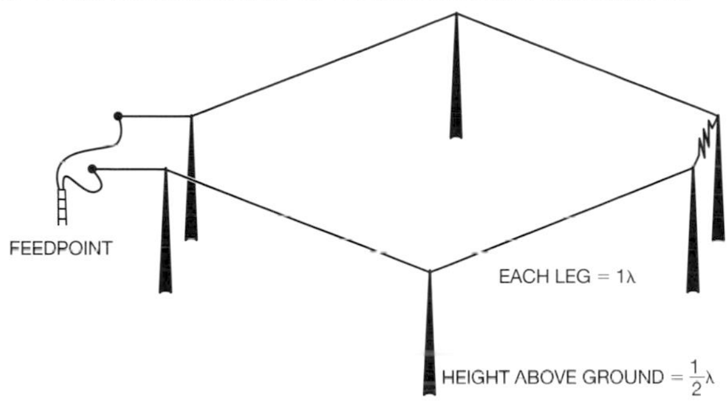

Non-Resonant Rhombic Antenna

E9D01 How does the gain of an ideal parabolic dish antenna change when the operating frequency is doubled?

A. Gain does not change.
B. Gain is multiplied by 0.707.
C. Gain increases by 6 dB.
D. Gain increases by 3 dB.

From 1270 MHz on up, you may wish to *use a parabolic dish antenna* for microwave work. If you *double the frequency, the gain of the dish increases by 6 dB, a 4 times increase*. **ANSWER C.**

E9A06 How does the beamwidth of an antenna vary as the gain is increased?

A. It increases geometrically.
B. It increases arithmetically.
C. It is essentially unaffected.
D. It decreases.

The higher the gain of an antenna the narrower the beamwidth. *Beamwidth decreases as gain increases*. **ANSWER D.**

Like a magnifying glass, the bigger the dish mechanism the smaller and more powerful the resulting radiated pattern. This 10,368.1 MHz X band microwave dish can work stations up to 500 miles away under good tropo conditions.

Feedlines and Safety

E9E01 What system matches a higher impedance transmission line to a lower impedance antenna by connecting the line to the driven element in two places spaced a fraction of a wavelength each side of element center?

A. The gamma matching system.
B. The delta matching system.
C. The omega matching system
D. The stub matching system

The ***delta*** matching system is a popular way to ***change the impedance of the transmission line to match the impedance of the antenna***. **ANSWER B.**

E9E02 What is the name of an antenna matching system that matches an unbalanced feed line to an antenna by feeding the driven element both at the center of the element and at a fraction of a wavelength to one side of center?

A. The gamma match.
B. The delta match.
C. The epsilon match.
D. The stub match.

A ***gamma match*** is used with coax cable that forms an ***unbalanced feed system*** going to a tube that acts as a capacitor in series with the connection point to change the impedance of the antenna to match that of the feed line. **ANSWER A.**

Close up of a Gamma match on a beam feed point. Look carefully behind the coaxial connector for the center insulated conductor going up and into the radiating element.

Why is Characteristic Impedance So Important?

Let's look at this Figure *to help us understand why the characteristic impedance, Z_0, of a transmission line is so important for the efficient transfer of energy.* **Figure** *a shows a transmitter coupled to an antenna with a transmission line that has a characteristic impedance of Z_0. The transmitter's output impedance is Z_T and the antenna input impedance is Z_A, the load impedance that the antenna presents to the transmission line.*

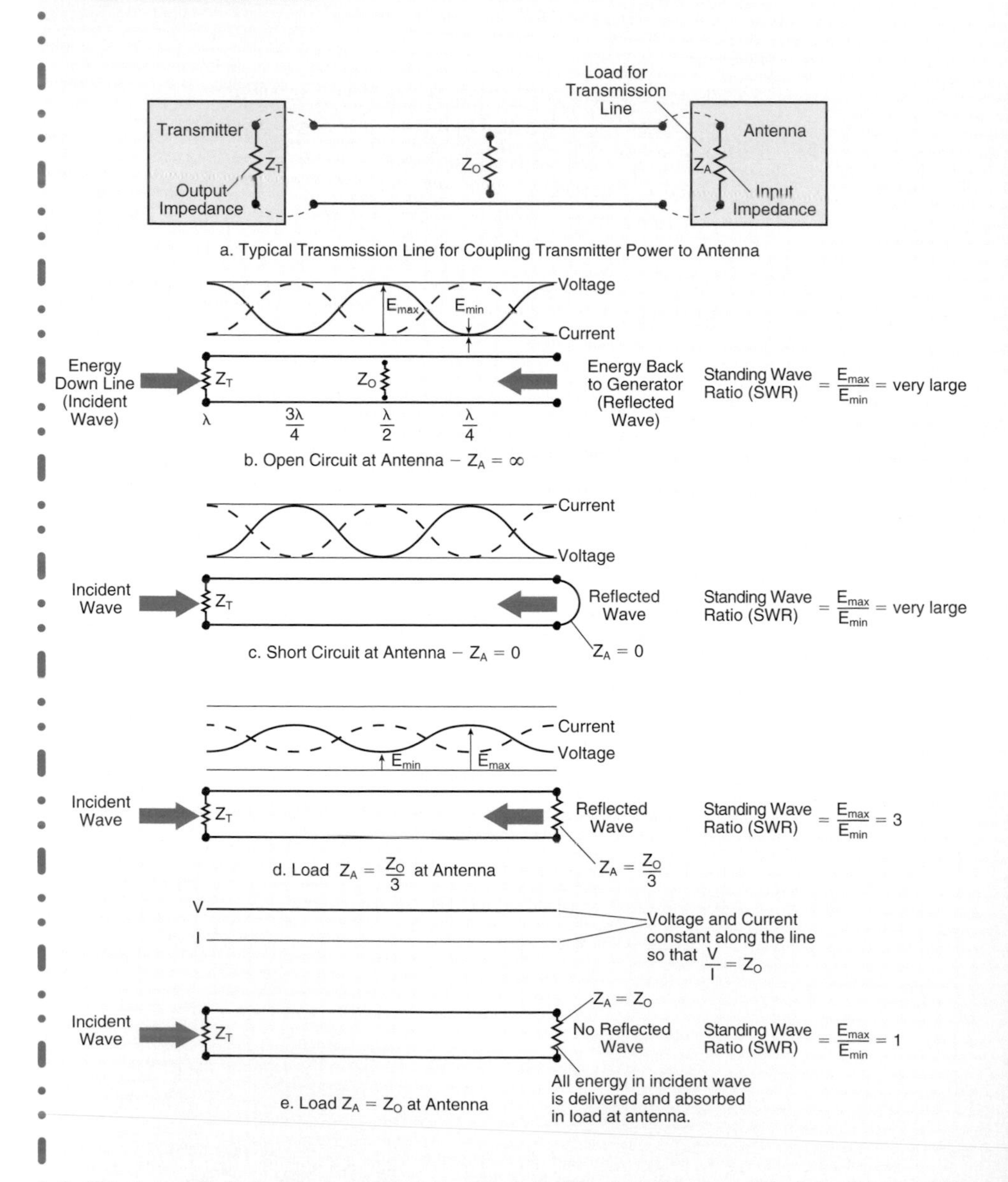

Let's assume that Z_A is infinity — in other words, the load on the transmission line at the antenna is an open circuit. Figure b *is a diagram of the interconnection and shows the voltages and currents that exist on the transmission line. Since Z_A is an open circuit, the current at the end of the line is zero and the voltage is maximum. Since the wave of energy sent from the transmitter has no load to absorb the energy, it is reflected back to the transmitter. The voltage and current waveforms shown are the combined resultant of the wave from the transmitter (called the incident wave) and the reflected wave from the load end. In this case, the total wave is reflected back and no signal is transmitted.*

Standing-Wave Ratio

The ratio of the maximum amplitude, E_{max}, of the resultant voltage waveform on the line to the minimum amplitude, E_{min}, is called the Standing-Wave Ratio, *SWR, or more correctly the* Voltage Standing-Wave Ratio, *VSWR. In the case of an open load, since there is no energy absorbed at the load and all energy delivered is absorbed as losses in the line or absorbed by the transmitter's output resistance, the VSWR is very high (in the order of 20 or greater). VSWR is a measure of how well energy is being transferred over a transmission line to the load. VSWRs below 1.5 indicate quite efficient transfer of energy (VSWR = 1 is perfect), and VSWRs above 2 begin to indicate poor matching.*

Now let's look at Figure c. *Here the Z_A is a short circuit. That means the voltage is zero at the end of the line and the current is maximum. Again, no energy is absorbed by the load. It is all reflected back down the line. The voltage and current waveforms vary in minimums and maximums every quarter wavelength, as in the open-circuit case, but are 180° out of phase from the open-circuit case. The standing-wave ratio again is very large and no energy is transferred.*

A Matched Case — $Z_A = Z_0$

Figure c *is the case where all the energy in the incident wave is absorbed by the load. There is no reflected wave because $Z_A = Z_0$. The antenna load is matched to the characteristic impedance of the transmission line and maximum transfer of power from the transmitter to the antenna occurs. The standing-wave ratio is 1 because the voltage and current in the line are constant and are such that they equal Z_0 everywhere along the line. Thus, we see how important it is to match the load impedance to the transmission line Z_0. With $Z_A = Z_0$, there will be some small losses in the transmission line, but if there is a mismatch, energy will be reflected from the load back to the transmitter and large losses on the line will be dissipated as heat.*

Source: *Basic Communications Electronics*, Hudson & Luecke, © 1999 Master Publishing, Inc., Niles, IL

E9E03 What is the name of the matching system that uses a section of transmission line connected in parallel with the feed line at or near the feed point?

A. The gamma match.
B. The delta match.
C. The omega match.
D. The stub match.

When ***a short section of transmission line*** is connected to the antenna feed line near the antenna, it is called ***stub matching***. This one is easy… think short = stub. Perhaps you once used a stub to cancel out interference in your TV receiver. **ANSWER D.**

Impedance Matching Stubs

Quarter wavelength ($\frac{\lambda}{4}$) stubs
(or multiples of odd numbers of $\frac{\lambda}{4}$ s)

Less than $\frac{1}{4}\lambda$ stub

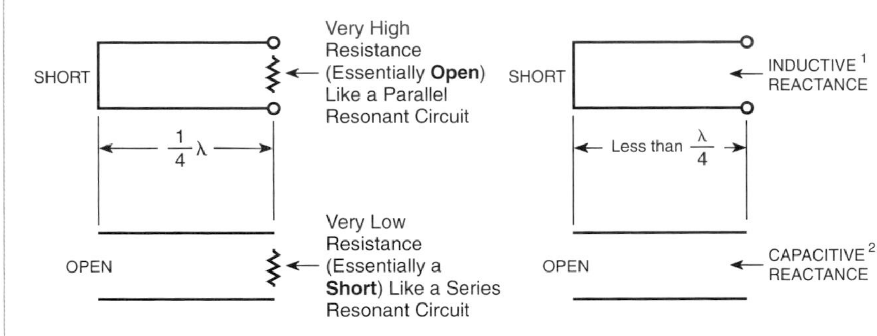

Half-wavelength ($\frac{\lambda}{2}$) stubs
(or multiples of $\frac{\lambda}{2}$ s)

Think of it as two $\frac{\lambda}{4}$ stubs in series.

Greater than $\frac{\lambda}{4}$, less than $\frac{\lambda}{2}$

Think of it as a $\frac{\lambda}{4}$ stub plus one less than $\frac{\lambda}{4}$.

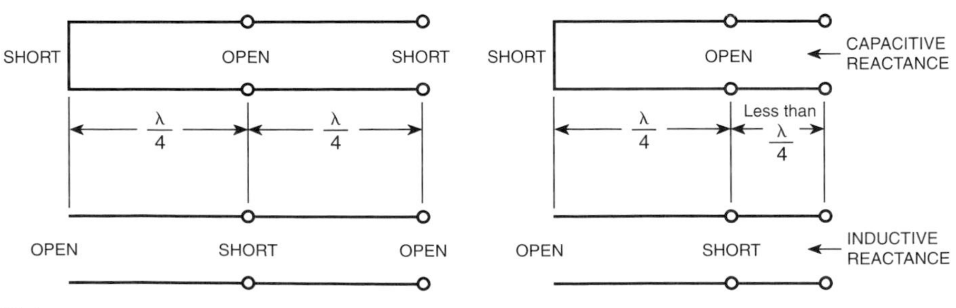

Notes

1. Since a short at end makes the diagram look like a long wire, this should remind the reader that impedance is an inductive reactance.
2. Since an open at end makes the diagram look like two parallel plates, this should remind the reader that impedance is a capacitive reactance.

E9E04 What is the purpose of the series capacitor in a gamma-type antenna matching network?

A. To provide DC isolation between the feed line and the antenna.
B. To cancel the inductive reactance of the matching network.
C. To provide a rejection notch that prevents the radiation of harmonics.
D. To transform the antenna impedance to a higher value.

In some ***gamma***-type antenna ***matching*** networks, the center conductor and PVC insulation on the center conductor of coaxial cable are used inside a tube as a series capacitance. ***Series capacitance will cancel the inductive reactance*** of the matching network and result in resonance. **ANSWER B.**

E9E05 How must the driven element in a 3-element Yagi be tuned to use a hairpin matching system?

A. The driven element reactance must be capacitive.
B. The driven element reactance must be inductive.
C. The driven element resonance must be lower than the operating frequency.
D. The driven element radiation resistance must be higher than the characteristic impedance of the transmission line.

For the ***hairpin matching system*** to work well, the ***driven element reactance must be capacitive, X_C***. **ANSWER A.**

It's always a good idea to develop a coax choke at any HF antenna feed point. This will help trap any SWR feeding back out of the antenna and prevent any SWR signal from going back down on the coax outside shield. Keep your 10 turns nice and neat!

E9E06 What is the equivalent lumped-constant network for a hairpin matching system of a 3-element Yagi?

A. Pi-network.
B. Pi-L-network.
C. A shunt inductor.
D. A series capacitor.

At VHF frequencies and higher, it is often convenient to replace lumped constant components with equivalent transmission line sections. Often such circuitry can take strange or unfamiliar physical forms, but still function the same as a "normal" electrical network. A typical 3 element Yagi has a fairly low impedance on the driven element, so it is necessary to raise the impedance to match the 50 ohm impedance of the transmission line. While a gamma match can do this, it generally has mediocre balance, which can upset the radiation pattern. This can be a serious problem when large arrays of Yagis are assembled. The ***hairpin matching system places a shunt inductance*** in parallel with the feed point. Formed by a couple of short rods, the shunt performs a function similar to that of a folded dipole. A hairpin match provides a good impedance match and a well-balanced pattern. **ANSWER C.**

E9E07 What term best describes the interactions at the load end of a mismatched transmission line?

A. Characteristic impedance.
B. Reflection coefficient.
C. Velocity factor.
D. Dielectric constant.

A ***mismatched*** transmission line will give you ***reflections***. **ANSWER B.**

E4E15 Which of the following can cause shielded cables to radiate or receive interference?

A. Low inductance ground connections at both ends of the shield.
B. Common mode currents on the shield and conductors.
C. Use of braided shielding material.
D. Tying all ground connections to a common point resulting in differential mode currents in the shield.

The best, most expensive coaxial cable in the world won't do its job properly if ***RF currents are induced or coupled on the outside shield***. A proper balun *and* transmission line orientation are usually necessary to prevent the outside of your transmission line from becoming a radiator; that is, an antenna. The degree to which feed line radiation is a problem varies from nothing to huge, depending on the application. There is no "silver balun" to solve all your transmission line problems, but understanding the principles and applying the appropriate ones will make you a happy ham. **ANSWER B.**

E4E16 What current flows equally on all conductors of an unshielded multi-conductor cable?

A. Differential-mode current.
B. Common-mode current.
C. Reactive current only.
D. Return current.

Common-mode currents are just what they sound like: currents that are equal in amplitude and phase on two or more conductors. The best way to reduce ***common-mode currents*** is to use a common-mode choke, such as passing the entire bundle through a single toroidal ferrite core or a string of ferrite beads. **ANSWER B.**

E9E08 Which of the following measurements is characteristic of a mismatched transmission line?

A. An SWR less than 1:1.
B. A reflection coefficient greater than 1.
C. A dielectric constant greater than 1.
D. An SWR greater than 1:1.

An ***SWR is greater than 1:1*** describes a ***mismatched*** transmission line. **ANSWER D.**

This modern SWR analyzer offers both an analog needle SWR meter plus a color LCD display, which is showing a bandwidth sweep.

E9E09 Which of these matching systems is an effective method of connecting a 50 ohm coaxial cable feed line to a grounded tower so it can be used as a vertical antenna?

A. Double-bazooka match.
B. Hairpin match.
C. Gamma match.
D. All of these choices are correct.

Gordo has a grounded tower that he can use as a halfwave antenna on 160 meters! The transmission line cable runs about halfway up the tower into a ***gamma match.*** The outer braid taps the metal tower leg and the center conductor is extended up and runs parallel to the same tower leg, and then makes a sharp turn and is connect to the same leg allowing for the impedance to match the transmission line. If you were to look at it with an ohm meter, it would look like a direct short. But your gamma match and a manual tuner can get you down to the band of choice. **ANSWER C.**

E9E10 Which of these choices is an effective way to match an antenna with a 100 ohm feed point impedance to a 50 ohm coaxial cable feed line?

A. Connect a 1/4-wavelength open stub of 300 ohm twin-lead in parallel with the coaxial feed line where it connects to the antenna.
B. Insert a 1/2 wavelength piece of 300 ohm twin-lead in series between the antenna terminals and the 50 ohm feed cable.
C. Insert a 1/4-wavelength piece of 75 ohm coaxial cable transmission line in series between the antenna terminals and the 50 ohm feed cable.
D. Connect a 1/2 wavelength shorted stub of 75 ohm cable in parallel with the 50 ohm cable where it attaches to the antenna.

VHF and UHF antenna stacking many times ends up with the combined antenna system presenting an impedance of 100 ohms at the feed point. This must be matched to the transmission line 50 ohms impedance. ***Calculate one quarter wavelength***, including coax velocity factor (usually 0.66) for your 75 ohm matching network, ***and insert this 75 ohm coax between the antenna terminals and the 50 ohm coax leading down to the ham shack***. **ANSWER C.**

E9E11 What is an effective way of matching a feed line to a VHF or UHF antenna when the impedances of both the antenna and feed line are unknown?

A. Use a 50 ohm 1:1 balun between the antenna and feed line.
B. Use the universal stub matching technique.
C. Connect a series-resonant LC network across the antenna feed terminals.
D. Connect a parallel-resonant LC network across the antenna feed terminals.

The ***universal stub*** takes unwanted reactance out of the driven portion of the antenna system without actually having to compute impedances up the tower. You'll need to use your trusty portable SWR analyzer as you adjust the stiff, bare wire shorting elements normally connected to the driven element of the antenna. **ANSWER B.**

E9E12 What is the primary purpose of a phasing line when used with an antenna having multiple driven elements?

A. It ensures that each driven element operates in concert with the others to create the desired antenna pattern.
B. It prevents reflected power from traveling back down the feed line and causing harmonic radiation from the transmitter.
C. It allows single-band antennas to operate on other bands.
D. It makes sure the antenna has a low-angle radiation pattern.

As an Extra Class operator, you likely will choose an exotic, multiband beam antenna to exercise your new privileges. Multiband antennas have a variety of methods to phase most elements for the desired beam pattern. "Trapless" designs may use rigid ***phasing lines to ensure each driven element operates in concert with all of the other elements*** on the antenna for the desired antenna pattern. **ANSWER A.**

E9E13 What is a use for a Wilkinson divider?

A. It divides the operating frequency of a transmitter signal so it can be used on a lower frequency band.
B. It is used to feed high-impedance antennas from a low-impedance source.
C. It is used to divide power equally between two 50 ohm loads while maintaining 50 ohm input impedance.
D. It is used to feed low-impedance loads from a high-impedance source.

With many high frequency antennas now incorporating multiple-band driven elements all off one feed line, the ***Wilkinson divider distributes the antenna input power equally*** to the multiple driven elements ***while maintaining the 50 ohm input impedance to both loads***. **ANSWER C.**

E9F08 What is the term for the ratio of the actual speed at which a signal travels through a transmission line to the speed of light in a vacuum?

A. Velocity factor.
B. Characteristic impedance.
C. Surge impedance.
D. Standing wave ratio.

Velocity factor is a ratio of how much slower radio signals travel in a conductor compared to the velocity of radio waves in free space. Radio waves in free space travel at the speed of light. In coaxial cable and along antenna conductors the radio waves travel more slowly. **ANSWER A.**

E9F01 What is the velocity factor of a transmission line?

A. The ratio of the characteristic impedance of the line to the terminating impedance.
B. The index of shielding for coaxial cable.
C. The velocity of the wave in the transmission line multiplied by the velocity of light in a vacuum.
D. The velocity of the wave in the transmission line divided by the velocity of light in a vacuum.

In free space, radio waves travel at 300 million meters per second. But in coaxial cable transmission lines and other types of feed lines, the radio waves move more slowly. ***Velocity factor*** is the ***velocity of the radio wave on the transmission line divided by the velocity of the wave in free space***. It's a ratio which always has division. **ANSWER D.**

Remember, when you are cutting coax to a specific wavelength most amateur radio coax has a velocity factor of 0.66. Always seal your feed point coax connections against rain and snow.

E9F02 Which of the following determines the velocity factor of a transmission line?

A. The termination impedance.
B. The line length.
C. Dielectric materials used in the line.
D. The center conductor resistivity.

There are some great low-loss coaxial cable types with different dielectrics separating the center conductor and the outside shield. It is the ***type of dielectric used that determines the velocity factor*** of the transmission line. **ANSWER C.**

E9F04 What is the typical velocity factor for a coaxial cable with solid polyethylene dielectric?

A. 2.70.
B. 0.66.
C. 0.30.
D. 0.10.

Most coax cable has a velocity factor of 0.66. This is not stated on the outside jacket, and may not even be known by the dealer selling the coax. Contact the manufacturer or distributor directly to find out the actual velocity factor of a particular coaxial cable. **ANSWER B.**

E9F03 Why is the physical length of a coaxial cable transmission line shorter than its electrical length?

A. Skin effect is less pronounced in the coaxial cable.
B. The characteristic impedance is higher in a parallel feed line.
C. The surge impedance is higher in a parallel feed line.
D. Electrical signals move more slowly in a coaxial cable than in air.

Radio frequency energy moves slightly *slower inside coaxial cable* than it does in free space. **ANSWER D.**

E9F05 What is the approximate physical length of a solid polyethylene dielectric coaxial transmission line that is electrically one-quarter wavelength long at 14.1 MHz?

A. 20 meters.
B. 2.3 meters.
C. 3.5 meters.
D. 0.2 meters.

When you work on antenna phasing harnesses, you will need to calculate the velocity factor for a piece of coax in order to know where to place the connectors. You must be accurate down to a fraction of an inch! You can calculate the physical length of a coax cable which is electrically one-quarter wavelength long using the formula:

$$L \text{ (in feet)} = \frac{984\lambda V}{f}$$

Where: L is antenna length in **feet**
λ is wavelength in **meters**
V is **velocity factor**
f is frequency in **MHz**

For this problem, multiply 984 × 0.25, because the coax is one-quarter wavelength. Then multiply the result by 0.66, the most common velocity factor for coax. Divide the total by 14.1, the frequency in MHz. The length comes out

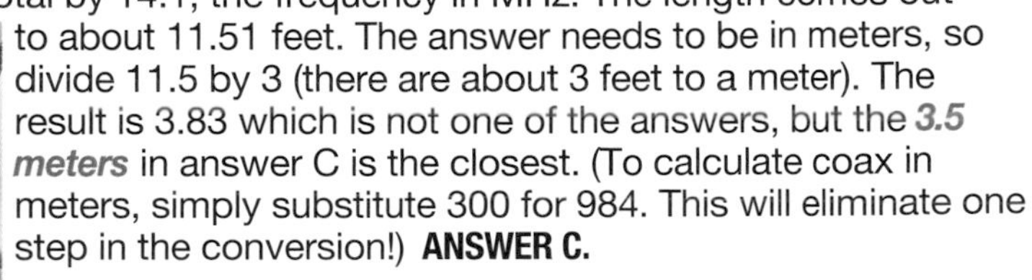

to about 11.51 feet. The answer needs to be in meters, so divide 11.5 by 3 (there are about 3 feet to a meter). The result is 3.83 which is not one of the answers, but the *3.5 meters* in answer C is the closest. (To calculate coax in meters, simply substitute 300 for 984. This will eliminate one step in the conversion!) **ANSWER C.**

Here is where coax goes bad! In wet weather, water can enter the feed point connection and get into the dielectric. Treat the connection point with flexible sealant and then wrap it with self-vulcanizing waterproof tape.

E9F06 What is the approximate physical length of an air-insulated, parallel conductor transmission line that is electrically one-half wavelength long at 14.10 MHz?

A. 15 meters.
B. 20 meters.
C. 10 meters.
D. 71 meters.

In this problem, the length of the parallel conductor feed line that will be one-half wavelength long at 14.10 MHz with a velocity factor of 0.95 is calculated by multiplying ***984 × 0.5 × 0.95*** and ***dividing by 14.10***, the frequency in MHz. ***Convert feet to meters***, and you end up with ***10 meters***. **ANSWER C.**

E9F07 How does ladder line compare to small-diameter coaxial cable such as RG-58 at 50 MHz?

A. Lower loss.
B. Higher SWR.
C. Smaller reflection coefficient.
D. Lower velocity factor.

Going to 450 ohm ladder line on the 6 meter band is one way to reduce losses over typical RG-58 coax cable. They both cost about the same which keeps your expenses to a minimum. RG-58 should be relegated to test leads, never used for antenna feed line runs. The loss is simply too high. ***450 ohm ladder line will have much lower loss at 50 MHz than small RG-58 coax***. **ANSWER A.**

E9F09 What is the approximate physical length of a solid polyethylene dielectric coaxial transmission line that is electrically one-quarter wavelength long at 7.2 MHz?

A. 10 meters.
B. 6.9 meters.
C. 24 meters.
D. 50 meters.

The coax cable is to be one-quarter wavelength with a presumed velocity factor of 0.66 (most common for the dielectric in use), so multiply 984 × 0.25 × 0.66. Now divide the result by the frequency in Mhz, 7.2, and you end up with 22.55 feet. Dividing by 3 to get an approximate length of 7.5 meters, leads you to select ***6.9 meters*** from the available answers. (See formula at E9F05) **ANSWER B.**

E9F10 What impedance does a 1/8 wavelength transmission line present to a generator when the line is shorted at the far end?

A. A capacitive reactance.
B. The same as the characteristic impedance of the line.
C. An inductive reactance.
D. The same as the input impedance to the final generator stage.

If a ***1/8 wavelength transmission line is shorted out at the far end***, it will look like an ***inductive reactance at the generator***. **ANSWER C.**

E9F11 What impedance does a 1/8 wavelength transmission line present to a generator when the line is open at the far end?

A. The same as the characteristic impedance of the line.
B. An inductive reactance.
C. A capacitive reactance.
D. The same as the input impedance of the final generator stage.

If a ***1/8 wavelength transmission line is open*** at the far end, it will look like a ***capacitive reactance*** at the generator. **ANSWER C.**

E9F12 What impedance does a 1/4 wavelength transmission line present to a generator when the line is open at the far end?

A. The same as the characteristic impedance of the line.
B. The same as the input impedance to the generator.
C. Very high impedance.
D. Very low impedance.

If the far end of a ***1/4 wavelength transmission line is open***, it will look like a ***very low impedance*** at the generator. **ANSWER D.**

E9F13 What impedance does a 1/4 wavelength transmission line present to a generator when the line is shorted at the far end?

A. Very high impedance.
B. Very low impedance.
C. The same as the characteristic impedance of the transmission line.
D. The same as the generator output impedance.

If the far end a ***1/4 wavelength transmission line is shorted,*** it at will look like a ***very high impedance*** at the generator. **ANSWER A.**

E9F14 What impedance does a 1/2 wavelength transmission line present to a generator when the line is shorted at the far end?

A. Very high impedance.
B. Very low impedance.
C. The same as the characteristic impedance of the line.
D. The same as the output impedance of the generator.

If a ***1/2 wavelength transmission line is shorted*** at the far end, it will have a ***very low impedance***, essentially a short. **ANSWER B.**

E9F15 What impedance does a 1/2 wavelength transmission line present to a generator when the line is open at the far end?

A. Very high impedance.
B. Very low impedance.
C. The same as the characteristic impedance of the line.
D. The same as the output impedance of the generator.

If that ***1/2 wavelength transmission line is open*** at the far end, it will have a ***very high impedance***, essentially an open circuit. **ANSWER A.**

Transmission Line Mnemonic

Here's a silly way to remember how the impedance along an open-ended transmission line changes as you move away from the open end in 1/8 wave increments:

I Like Zucchini Crumpets… I L Z C

1. Beginning at the open end: Infinity
2. One-eighth wavelength back: L (inductive)
3. One more eighth wavelength back: Zero
4. One more eighth wavelength back: C (capacitive)

One more eighth wavelength back: starts all over again!
This means you can start at any point and figure out where you're going to be at any distance back, such as: LZCI, ZCIL, CILZ.

Of course you remember ELI the ICE man! Just don't confuse the I used here to represent Infinity with the normal use of I for current.

E9F16 Which of the following is a significant difference between foam dielectric coaxial cable and solid dielectric cable, assuming all other parameters are the same?

A. Foam dielectric has lower safe operating voltage limits.
B. Foam dielectric has lower loss per unit of length.
C. Foam dielectric has higher velocity factor.
D. All of these choices are correct.

What you look for in coax cable is a little different now that you are on HF. On HF bands, coax cable losses are significantly less than those experienced when operating on VHF and UHF. Popular RG-8 foam coax, on the 15 meter band, represents only 0.7 dB loss per 100 feet. Although lighter in weight, foam dielectric has lower safe operating voltage limits and a slightly-higher velocity factor than solid dielectric. Some land mobile rated (LMR) coax cable that is about the same size as RG-8 exhibits wonderful qualities of being lightweight, having a non-contaminating jacket to resist sunlight cracks, a common 0.66 velocity factor, and slightly less attenuation on high frequency bands. It handles a little like a cold garden hose but, nonetheless, LMR type coax has plenty of benefits. ***All of the answer choices are correct for the difference between foam dielectric coax and solid-dielectric coax*. ANSWER D.**

Skin effect saves copper! Notice that there is no solid copper in the center conductor. This is high-power foam dielectric hard line for microwave use.

E4A07 Which of the following is an advantage of using an antenna analyzer compared to an SWR bridge to measure antenna SWR?

A. Antenna analyzers automatically tune your antenna for resonance.
B. Antenna analyzers do not need an external RF source.
C. Antenna analyzers display a time-varying representation of the modulation envelope.
D. All of these choices are correct.

It is likely that no other amateur radio accessory has offered ham radio experimenters a bigger breakthrough in measuring antenna and feed line SWR than the ***antenna analyzer***. This allows you to check the antenna feed point impedance and resonance without putting a big test signal out on the air. Since the ***micro transmitter is built into the analyzer, you don't need an actual radio to test your antenna system*. ANSWER B.**

E4B11 How should an antenna analyzer be connected when measuring antenna resonance and feed point impedance?

A. Loosely couple the analyzer near the antenna base.
B. Connect the analyzer via a high-impedance transformer to the antenna.
C. Loosely couple the antenna and a dummy load to the analyzer.
D. Connect the antenna feed line directly to the analyzer's connector.

When Gordo is working on a high-frequency antenna, he carries his trusty, portable antenna analyzer and a short coax jumper cable to connect his analyzer directly to the antenna input thus eliminating the radio from the testing system and protecting its valuable circuitry. The ***antenna analyzer*** contains a signal generator that takes the place of the transmitter for testing an antenna or feed line. It should be ***connected directly to the feed line of the antenna*. ANSWER D.**

E4A08 Which of the following instruments would be best for measuring the SWR of a beam antenna?

A. A spectrum analyzer.
B. A Q meter.
C. An ohmmeter.
D. An antenna analyzer.

Gordo always takes his ***portable antenna analyzer*** whenever he goes out for any ham radio station checkout. Whether he's at the top of a sailboat mast, up a tower, or down inside a boat's lazarette, his portable analyzer can handle everything from automatic antenna tuners to marine VHF antennas to a big 3-element beam up 90 feet in the air. **ANSWER D.**

E9G01 Which of the following can be calculated using a Smith chart?

A. Impedance along transmission lines.
B. Radiation resistance.
C. Antenna radiation pattern.
D. Radio propagation.

The ***Smith Chart*** is an invaluable graph for ***calculating the impedance along Amateur Radio transmission lines***. Many times you will see a completed, computer-generated Smith Chart illustrating the resonance of a commercial-quality VHF or UHF collinear base station or repeater antenna. **ANSWER A.**

E9G02 What type of coordinate system is used in a Smith chart?

A. Voltage circles and current arcs.
B. Resistance circles and reactance arcs.
C. Voltage lines and current chords.
D. Resistance lines and reactance chords.

Looking at a ***Smith Chart***, you see ***two circles – resistance circles*** centered on a resistance axis – all tangent to an open circuit point (infinity resistance and infinity reactance), ***and*** portions of ***reactance arcs*** fanning out from the resistance axis and passing through the same open circuit point. Circles for positive inductive reactance are to the right of the resistance axis; circles for negative capacitance reactance are to the left of the resistance axis. **ANSWER B.**

E9G03 Which of the following is often determined using a Smith chart?

A. Beam headings and radiation patterns.
B. Satellite azimuth and elevation bearings.
C. Impedance and SWR values in transmission lines
D. Trigonometric functions.

Commercial-quality coaxial cable, coiled on a 500-foot spool, may also contain a computer-generated ***Smith Chart*** that shows ***impedance and SWR values***.
ANSWER C.

E9G04 What are the two families of circles and arcs that make up a Smith chart?

A. Resistance and voltage.
B. Reactance and voltage.
C. Resistance and reactance.
D. Voltage and impedance.

Resistance and reactance values comprise the "two families of circles" on the Smith Chart, as shown the figure on the next page. Circles for positive inductive reactance are to the right of the resistance axis; circles for negative capacitance reactance are to the left of the resistance axis. **ANSWER C.**

Smith Chart—Finding Antenna Impedance

Finding Antenna Impedance, Z_A, when Fed with Known Length of Transmission Line (λ in Wavelengths) of Given Impedance, Z_o.
(Impedance and wavelength of transmission line are at the frequency of transmission)

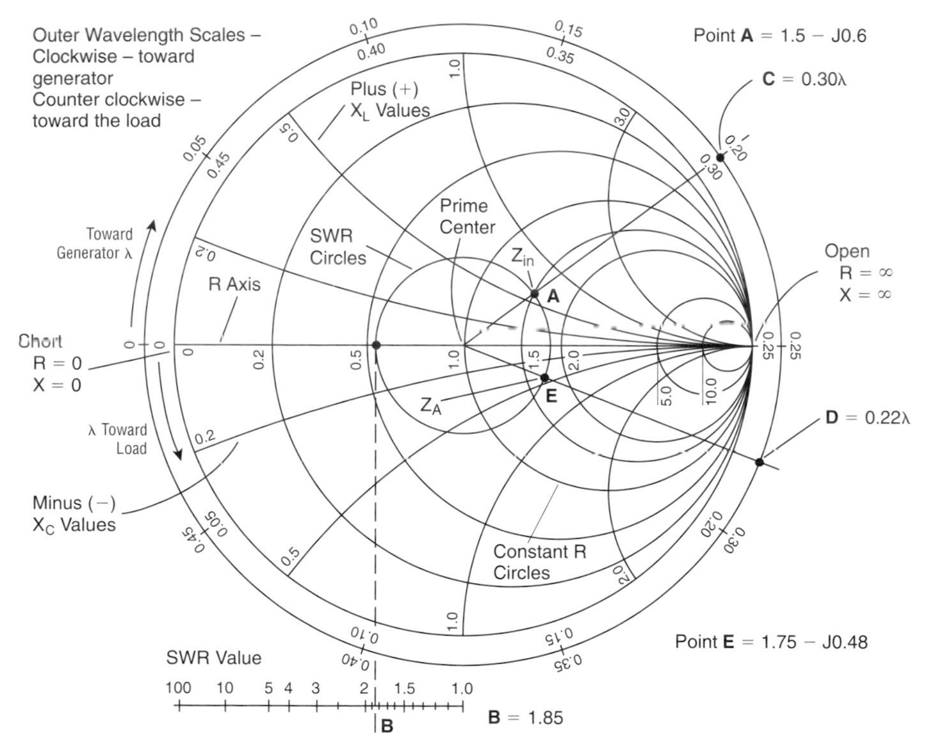

Known Values: $Z_o = 50$, $\lambda_Z = 0.42\lambda$

Steps

1. Connect transmission line to antenna.
2. Measure the input impedance Z_{in} to transmission line. (Example 75 + J30)
3. Normalize Z_{in} by dividing by Z_o.

$$Z_{in} = \frac{75}{50} + J\frac{30}{50} = 1.5 + J0.6$$

4. Plot 1.5 + J0.6 on chart (point A is at intersection of a resistance circle of 1.5 and a positive reactance curve of 0.6)
5. Draw SWR circle with prime center as center and radius through A. Project tangent to SWR circle at point where circle intersects the resistance axis down to linear scale and read value of SWR (Point B = 1.85).
6. Draw a line (radial line) from prime center through Z_{in} at A and project to wavelength scale. Use "toward load" scale because you are going from Z_{in} at generator to the load presented by Antenna. (Point C = 0.30λ)
7. To find the load impedance line on the wavelength (towards load) scale, add the transmission line length to Point C value. (Point D = 0.42 + .30 = 0.72λ)
8. Since the wavelength scales are only plotted for 0.5λ, the values repeat every 0.5λ. Therefore, subtract 0.5 for 0.72 to arrive at 0.22. (Point D = 0.22 toward load)
9. Draw another radial line from prime center to Point D. The intersection with the SWR circle is the load impedance of the antenna. (Point E = 1.75 – J0.48). You may have to interprolate between lines.
10. This is still a normalized value so multiply by $Z_o = 50$ to obtain true value or antenna impedance. $Z_A = 50$ (1.75 – J0.48) = 87.5 – J24. Which is an impedance of 87.5 ohms of resistance and 24 ohms of capacitive reactance.

The Importance of Knowing the Smith Chart

The Smith Chart is specifically designed to tell you what is happening within a transmission line. It won't actually tell you anything about the current distribution on an antenna itself, but if we apply a few "smarts" we can often make an educated guess as to what's happening "out there" at the far end of the coax based on the Smith Chart. Let's summarize the issue with a few principles we can rely on:

1. *If there is any reactance on the antenna, we will not have a 1:1 SWR. It is a common "newbie" misconception to conclude that if you have 25 ohms of resistance and 25 ohms of reactance the total "impedance" is 50 ohms, so the SWR will be flat. Wrong... dead wrong. The only time you will have no reflected power is if there is no reactance and the resistive component is matched.*

2. *Almost always, as you move off the resonant frequency of an antenna, the REACTANCE changes much faster than the RADIATION RESISTANCE. This is why SWR is a sensitive indicator of resonance, and a much poorer indication of "antenna impedance" (a bit of a misnomer).*

3. *The Smith chart can be normalized to any transmission line impedance you like. This is often forgotten. A 600 ohm transmission line with a 600 ohm load resistance has an SWR of 1:1... just like a 50 ohm transmission line with a 50 ohm load. This also tells you that you can feed a 50 ohm antenna DIRECTLY with 1/2 wavelength (or any multiple thereof) of 600 ohm ladder line and it will present a perfect 50 ohm impedance on the input end! This is in spite of the fact that the ladder line itself has a 12:1 SWR.*

4. *Impedances near the left side of the Smith Chart will always be more stable than those near the right side. A center fed full wave antenna will have an extremely high impedance (between 5-10 thousand ohms) at the feed point, with a reactance that changes drastically (flopping between nearly infinite capacitive reactance to nearly infinite inductive reactance) with a minuscule change of frequency. This manic depressive impedance can be totally calmed down with a quarter wave transmission line or "Q-section" which will present a low, broadband impedance at the input side of the line. This is how the end fed Zepp antenna (and its red-headed stepchild, the J-pole) works.*

5. *Here's an interesting fact closely related to item #2: Reactance has zero effect... zip, zilch, nada... on the ability of an antenna to radiate! There is absolutely nothing sacred about a self-resonant antenna, nor does it give us any advantage whatsoever. The reason we need the system (not the antenna itself) to be resonant is so that we can effectively transfer power. See all the questions on power factor. The truth of the matter is that the power radiated by an antenna is determined by the current and the radiation resistance only! Reactance, no matter how much, does not even enter the equation! Need proof? Let's talk about AM broadcast antennas which must, by law, be nearly 100% efficient. There isn't one AM broadcast antenna out of 1,000 that's anywhere near self-resonant! AM towers are generally made of 60 foot sections of tower, so if you're within 60 feet of resonance, you're usually doing okay. They rely on highly efficient Antenna Tuning Units (ATUs) at the base of the towers, basically transmatches on steroids, to cancel out any reactance and match the feed point impedance (which can also be far different from 50 ohms). The amateur radio community has finally, if reluctantly, caught on to this truth with the rising popularity of the 43 foot non-resonant vertical, used across the entire HF band very effectively. It is important to have an efficient matching network for these to perform as advertised.*

E9G05 What type of chart is shown in Figure E9-3?

A. Smith chart.
B. Free space radiation directivity chart.
C. Elevation angle radiation pattern chart.
D. Azimuth angle radiation pattern chart.

The ***Smith chart*** is an excellent way of visualizing the contours of resistance or reactance. The chart is used for circuit analysis and provides a graphic presentation for this purpose. **ANSWER A.**

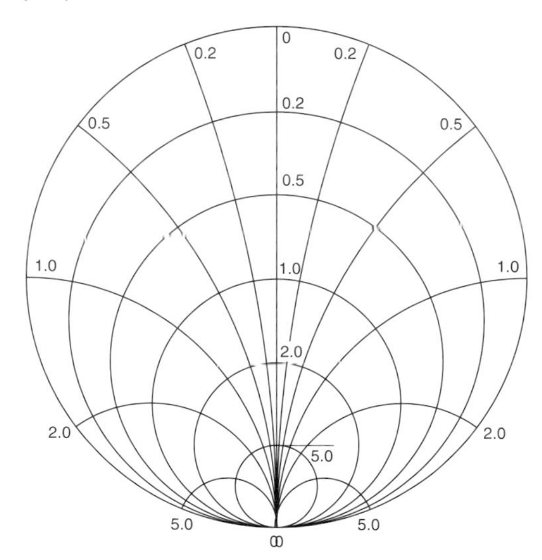

Figure E9-3

E9G06 On the Smith chart shown in Figure E9-3, what is the name for the large outer circle on which the reactance arcs terminate?

A. Prime axis.
B. Reactance axis.
C. Impedance axis.
D. Polar axis.

The large outer circle is the reactance axis. Within the reactance axis are curved reactance circles that are tangent to the resistance axis. Where the reactance circle intersects with the reactance axis, you will see an assigned value of reactance. Inductive reactance is to the right; capacitive reactance is to the left. **ANSWER B.**

E9G07 On the Smith chart shown in Figure E9-3, what is the only straight line shown?

A. The reactance axis.
B. The current axis.
C. The voltage axis.
D. The resistance axis.

On a ***Smith chart***, the ***straight line is*** always the ***resistance axis***. **ANSWER D.**

E9G08 What is the process of normalization with regard to a Smith chart?

A. Reassigning resistance values with regard to the reactance axis.
B. Reassigning reactance values with regard to the resistance axis.
C. Reassigning impedance values with regard to the prime center.
D. Reassigning prime center with regard to the reactance axis.

For the Smith chart coordinate system to apply to all types of transmission lines, the coordinates are "customized" to a particular transmission line by dividing test measurements by the characteristic impedance of the transmission line. This *process of reassigning resistance and reactance values with regard to the prime center is called "normalizing."* **ANSWER C.**

E9G09 What third family of circles is often added to a Smith chart during the process of solving problems?

A. Standing wave ratio circles.
B. Antenna-length circles.
C. Coaxial-length circles.
D. Radiation-pattern circles.

When a computer is used to analyze an antenna design, the software will draw a *third family of circles* on the Smith chart *showing standing-wave ratios*. The SWR circle is one of several *radially scaled parameters* available on most commercially printed Smith Chart forms. **ANSWER A.**

E9G10 What do the arcs on a Smith chart represent?

A. Frequency.
B. SWR.
C. Points with constant resistance.
D. Points with constant reactance.

The arcs (curved lines) on a Smith chart *are portions of reactance circles*. Remember that the large outer circle bounding the coordinate portion of the chart is the reactance axis. **ANSWER D.**

E9G11 How are the wavelength scales on a Smith chart calibrated?

A. In fractions of transmission line electrical frequency.
B. In fractions of transmission line electrical wavelength.
C. In fractions of antenna electrical wavelength.
D. In fractions of antenna electrical frequency.

The *wavelength scales* on the Smith chart are calibrated in *portions of transmission line electrical wavelength*. **ANSWER B.**

E9A18 What term describes station output, taking into account all gains and losses?

A. Power factor.
B. Half-power bandwidth.
C. Effective radiated power.
D. Apparent power.

It's important to always calculate *effective radiated power* when you set up your ham station. The table below shows you the dBs you could lose with cheap coax cable, and the valuable dBs you could gain by using a quality collinear VHF or UHF antenna. The decibel makes working out such problems a simple matter of addition and subtraction. It really helps to make friends with the dB! **ANSWER C.**

Coax Cable Type, Size and Loss per 100 Feet

Coax Type	Size	Loss at HF 100 MHz	Loss at UHF 400 MHz
RG-6	Large	2.3 dB	4.7 dB
RG-59	Medium	2.9 dB	5.9 dB
RG-58U	Small	4.3 dB	9.4 dB
RG-8X	Medium	3.7 dB	8.0 dB
RG-8U	Large	1.9 dB	4.1 dB
RG-213	Large	1.9 dB	4.5 dB
Hardline	Large, Rigid	0.5 dB	1.5 dB

E9A15 What is the effective radiated power relative to a dipole of a repeater station with 150 watts transmitter power output, 2 dB feed line loss, 2.2 dB duplexer loss, and 7 dBd antenna gain?

A. 1977 watts.
B. 78.7 watts.
C. 420 watts.
D. 286 watts.

For quick calculations combine gains as positive and losses as negative. Our total loss is about 4.2 dB (just add the losses together), and our total gain is 7 dB. This is an ***approximate 3 dB overall gain which*** almost ***doubles your 150 watts to an approximately 286 watts.*** The rule of thumb for antennas is that for every 3 dB of gain, radiated power almost doubles. But don't get too excited just yet, because you need to *double* the size of an antenna for every additional 3 dB of gain! The number one rule of thumb in ham radio is NFL... No Free Lunch! **ANSWER D.**

E9A16 What is the effective radiated power relative to a dipole of a repeater station with 200 watts transmitter power output, 4 dB feed line loss, 3.2 dB duplexer loss, 0.8 dB circulator loss, and 10 dBd antenna gain?

A. 317 watts.
B. 2000 watts.
C. 126 watts.
D. 300 watts.

Our total loss is 8 dB, and our total gain is 10 dB gain. This gives us a net gain of 2 dB; therefore, from Decibels chart in the appendix, we have a gain of 1.585 times. With 200 watts input, ***the ERP is 317 watts (200 × 1.585)***.
Now let's check our answer by working backwards through the formula for calculating gain or loss when we know input power and ERP.

$$dB = 10 \log_{10} \frac{P_1}{P_2}$$

where: P_1 = Power Output
P_2 = Power Input

Therefore,

$$dB = 10 \log_{10} \frac{317}{200}$$

$$dB = 10 \log_{10} 1.585$$

The $\log_{10}$ of 1.585 is 0.2

$$dB = 10 \times 0.2 = 2$$

Since the output power is 317 watts and the input power is 200 watts, there is a net gain of 2 dB, which was the total gain that we calculated by combining the gains and losses of the system. **ANSWER A.**

E9A17 What is the effective radiated power of a repeater station with 200 watts transmitter power output, 2 dB feed line loss, 2.8 dB duplexer loss, 1.2 dB circulator loss, and 7 dBi antenna gain?

A. 159 watts.
B. 252 watts.
C. 632 watts.
D. 63.2 watts.

Our total loss adds up to a 6 dB loss, and our total gain is 7 dB, for an actual gain of 1 dB. This would be a slight increase to our 200 watts output in effective radiated power. So from the answers given, ***252 watts ERP*** is the most reasonable selection. **ANSWER B.**

E0A04 When evaluating a site with multiple transmitters operating at the same time, the operators and licensees of which transmitters are responsible for mitigating over-exposure situations?

A. Only the most powerful transmitter.
B. Only commercial transmitters.
C. Each transmitter that produces 5 percent or more of its MPE limit at accessible locations.
D. Each transmitter operating with a duty-cycle greater than 50 percent.

If you are setting up a new repeater station on a mountaintop with many other transmitters, and ***if your station produces 5% or more of the maximum permissible*** exposure at that location, it will be ***your responsibility to mitigate*** this over the MPE situation. **ANSWER C.**

E0A11 Which of the following injuries can result from using high-power UHF or microwave transmitters?

A. Hearing loss caused by high voltage corona discharge.
B. Blood clotting from the intense magnetic field.
C. Localized heating of the body from RF exposure in excess of the MPE limits.
D. Ingestion of ozone gas from the cooling system.

Always stay away from the "business end" of a dish antenna, and watch out for any misaligned wave guides that could leak microwave radiation. ***Microwave energy heats the body*** with RF exposure in excess of maximum permissible exposure limits. This is what happens inside the microwave oven in your kitchen. Steer clear of any beam of microwave energy coming from the antenna system. **ANSWER C.**

Don't stand in front of any microwave antenna! Operate in a location where no one can stand in front of it.

E0A10 What toxic material may be present in some electronic components such as high voltage capacitors and transformers?

A. Polychlorinated Biphenyls.
B. Polyethylene.
C. Polytetrafluoroethylene.
D. Polymorphic silicon.

You know the dangers of power pole transformers that contain ***PCBs (PolyChlorinated Biphenyls)***. **ANSWER A.**

E0A09 Which insulating material commonly used as a thermal conductor for some types of electronic devices is extremely toxic if broken or crushed and the particles are accidentally inhaled?

A. Mica.
B. Zinc oxide.
C. Beryllium Oxide.
D. Uranium Hexafluoride.

Be careful around large antennas that may contain beryllium oxide on the metal internal contactors. These antennas are terrific performers and are safe, but if you decide to take one apart, remember that ***beryllium oxide is extremely toxic if inhaled***. Beryllium oxide is also used in some high powered ceramic transmitting tubes as well as most ceramic package RF power transistors. Never disassemble any of these! **ANSWER C.**

E0A05 What is one of the potential hazards of using microwaves in the amateur radio bands?

A. Microwaves are ionizing radiation.
B. The high gain antennas commonly used can result in high exposure levels.
C. Microwaves often travel long distances by ionospheric reflection.
D. The extremely high frequency energy can damage the joints of antenna structures.

Gordo runs 10 GHz microwave with a relatively large 20 dB dish antenna. He's always cautious to never let anyone stand in the antenna main lobe when transmitting. He doesn't want to ***"cook" his neighbors with his microwave transmissions***! **ANSWER B.**

This sign says all!

E0A03 Which of the following would be a practical way to estimate whether the RF fields produced by an amateur radio station are within permissible MPE limits?

A. Use a calibrated antenna analyzer.
B. Use a hand calculator plus Smith-chart equations to calculate the fields.
C. Use an antenna modeling program to calculate field strength at accessible locations.
D. All of the choices are correct.

When laying out your ham radio installation, remember it is the antenna that emits the most RF energy. ***Antenna modeling on your computer can help you calculate field strengths*** at any location where someone might get into the antenna pattern. Antenna modeling programs are available on the internet. **ANSWER C.**

San Bernardino Microwave Society members test their microwave stations before participating in the ARRL microwave contest.

E0A08 What does SAR measure?

A. Synthetic Aperture Ratio of the human body.
B. Signal Amplification Rating.
C. The rate at which RF energy is absorbed by the body.
D. The rate of RF energy reflected from stationary terrain.

SAR stands for *Specific Absorption Rate, the rate at which RF energy is absorbed by the human body*. **ANSWER C.**

E0A02 When evaluating RF exposure levels from your station at a neighbor's home, what must you do?

A. Make sure signals from your station are less than the controlled MPE limits.
B. Make sure signals from your station are less than the uncontrolled MPE limits.
C. You need only evaluate exposure levels on your own property.
D. Advise your neighbors of the results of your tests.

Your *neighbor's house is considered "uncontrolled"* when calculating maximum permissible exposure limits. Always make sure you do the calculations before putting up your new antenna system, and make absolutely sure you are always *less than the uncontrolled MPE limit*. **ANSWER B.**

E0A06 Why are there separate electric (E) and magnetic (H) field MPE limits?

A. The body reacts to electromagnetic radiation from both the E and H fields.
B. Ground reflections and scattering make the field impedance vary with location.
C. E field and H field radiation intensity peaks can occur at different locations.
D. All of these choices are correct.

Radio transmissions contain both an electric field (E) and a magnetic field (H). These field radiations may peak at different distances from the antenna. In addition, ground reflections and scattering will vary the field intensity. The human body reacts differently to E and H fields. So take *all of these factors* into account when do your calculations for *both magnetic and electric fields*. **ANSWER D.**

E0A07 How may dangerous levels of carbon monoxide from an emergency generator be detected?

A. By the odor.
B. Only with a carbon monoxide detector.
C. Any ordinary smoke detector can be used.
D. By the yellowish appearance of the gas.

While you might be able to smell generator exhaust, what you can't smell or see is a generator byproduct, carbon monoxide. Maybe you are working Field Day inside your motor home, and you have the generator exhaust well clear of the vehicle chassis. However, any small leak in the structure could allow the CO emissions to seep into your position. *Only a carbon monoxide detector* will alert you that you are running out of good air! **ANSWER B.**

Keep generators well away from any operating point during a field operation, and never run a generator where odorless carbon monoxide can creep into an operating position.

4

Taking the Extra Class Examination

Get set to pass the Element 4 examination and gain additional band privileges as an Extra Class operator! If you are upgrading from General to Extra Class, you will gain a whopping 500 kHz of *additional* frequencies on the 75/80, 40, 20, and 15 meter worldwide bands. If you are upgrading from an Advanced Class license, you will gain an *additional* 250 kHz of band width on these same worldwide bands. Your Extra Class license gives you full privileges and full frequency coverage on every amateur band assigned by the Federal Communications Commission.

In this chapter, we'll tell you how to prepare yourself for the exam, what to expect when you take the exam, and how you'll get your Extra Class license.

EXAMINATION ADMINISTRATION

Three Extra Class licensees who are also accredited Volunteer Examiners are required to administer your Element 4 exam. Many of you preparing for your Extra Class examination are already part of a Volunteer Examination team. If you are, it's perfectly acceptable to take your exam at an official examination session held by your team.

Volunteer Examiner teams offer examinations on a regular basis at local sites to serve their communities. Generally, the VEC's closely coordinate their activities with one another, so you should be able to find a nearby test site and exam date that is convenient for you. You can obtain information about VECs and exam sessions in your area by checking with your local radio club, ham radio store, or local packet bulletin boards. A list of VECs that was current at the time of publication is given in the Appendix (see page 283).

Once you have found your local VE team, contact them to select a test date and location and pre-register for your examination. They will hold a seat for you at the next available session. Don't be a no-show, and don't be a surprise-show. Call them ahead of time and pre-register!

The Volunteer Examiners are not compensated for their time and skills, but they are permitted to charge you a fee for certain reimbursable expenses incurred in preparing, administering, and processing the examination. The fee is adjusted frequently and currently is about $14.00. When you call to make your exam reservation, ask the VE the current amount of the exam fee.

EXAM CONTENT

The questions, answers, and distractors for each question of the Extra Class written examination are public information. The question pool included in this book contains all 712 possible questions that can be used to make up your 50-question Element 4 written examination. The VEC is not permitted to change any wording, punctuation or numerical values included in the questions, answers, or distractors. The VEC can change the A-B-C-D order of the answer choices if it wishes.

Also, the VEC is required to select one question from each syllabus topic under each subelement so you should expect an exam that contains one question from each syllabus topic within each subelement. Look again at the question pool syllabus on page 286 to see how the test will be constructed. Remember, if there is a particular topic within a subelement that you just can't seem to grasp, there will be only one question from that topic. You will still pass the exam if you miss that one question.

Want to find a test site fast?
Visit the W5YI-VEC website at www.w5yi.org, or call them at 800-669-9594.

WHAT TO BRING TO THE EXAM SESSION

Here's what you'll need to bring with you for your Extra Class examination:

1. The fee of approximately $14.00.
2. The original plus a copy of your current General or Advanced Class license. If your new license has not yet been granted, make sure you bring the original plus two copies of your Certificate of Successful Completion of Examination (CSCE) indicating your most current license status.
3. A photo identification card – your driver's license is ideal for this purpose.
4. Some sharp pencils and fine-tip pens. It's good to have a backup.
5. Calculators may be used. However, the examiners may erase the memory before your exam begins.
6. Any other items that the VEC asks you to bring.

If you have a physical disability that requires additional assistance by your examination team, let them know ahead of time so they can prepare accommodations for your special requirements. This could be reading equipment, a volunteer examiner reader, access for a wheelchair, or any other special requirements that they may be able to accommodate for you. It is best for everyone if your requirements are known in advance. Be sure to contact the volunteer examiner team leader before your test session.

Arrive at the testing site with plenty of time to check in and complete the necessary forms that are required before you can begin your exam.

COMPLETING NCVEC FORM 605

When you arrive at the examination site, one of the first things you will do is complete the NCVEC Form 605. This form is retained by the Volunteer Exam Coordinator which transfers your printed information to an electronic file and sends it to the FCC for your new or upgrade license. Your application may be delayed or kicked-back to you if the VEC can't read your writing. Make absolutely sure you print as legibly as you can, and carefully follow the instructions on the form.

NCVEC QUICK-FORM 605 APPLICATION FOR
AMATEUR OPERATOR/PRIMARY STATION LICENSE

SECTION 1 - TO BE COMPLETED BY APPLICANT

PRINT LAST NAME	SUFFIX (Jr., Sr.)	FIRST NAME	INITIAL	STATION CALL SIGN (IF ANY)
MAILING ADDRESS (Number and Street or P.O. Box)				SOCIAL SECURITY NUMBER (SSN) or (FRN) FCC FEDERAL REGISTRATION NUMBER
CITY	STATE CODE	ZIP CODE (5 or 9 Numbers)		FAX NUMBER (Include Area Code) OPTIONAL
DAYTIME TELEPHONE NUMBER (Include Area Code) OPTIONAL	E-MAIL ADDRESS (OPTIONAL)			

I HEREBY APPLY FOR (Make an X in the appropriate box(es))

Type of Applicant:

☐ INDIVIDUAL

☐ EXAMINATION for a **new** license grant

☐ CHANGE my **name** on my license to my new name

Former Name: ______ (Last name) (Suffix) (First name) (MI)

☐ CHANGE my mailing address to **above** address

☐ CHANGE my station **call sign** systematically

Applicant's Initials: To Confirm

NCVEC Form 605

Name

If you are upgrading from a current license, it is very important to compare the information on your present license with what you are writing on NCVEC Form 605. Make sure that everything on NCVEC Form 605 is identical to how your present license reads. Fill in your last name, first name, middle initial, and suffix such as junior or senior. You must stay absolutely consistent with your name on any future Form 605s for upgrades or changes of address. If you start out as "Jack" and end up "John," the computer will throw out your next application. If you decide to use a nickname, this is okay – as long as you have legal identification that supports this same nickname. It really is best to stick with the name that is on most of your personal pictured IDs, such as your Driver's License.

Social Security Number & FRN

You are required to write in either your Social Security Number or your FCC Registration Number (FRN) in the designated box. Your current license should show your FRN. If you do not have an FRN and you prefer not to disclose your Social Security Number, you should obtain an FRN before the exam session so that you can complete your Form 605. To obtain an FCC Registration Number, visit the following website and follow the instructions there:

☞ **https://apps.fcc.gov/coresWeb/publicHome.do**

Address

Have you moved? Check your current license. Did you write in the same exact address? If so, you are good to go. If your address is different, check the "change" box and list your new address.

e-mail Address

Your e-mail address is important! This is how the FCC will notify you of your new license and call sign under the Universal Licensing System. Make sure to write your e-mail address legibly. When you supply an e-mail address, the VEC that processes your application will include it as part of your application to the FCC. In turn, the FCC will and you an e-mail with a link to your new license grant as confirmation of your new or upgraded license. Once you have your upgrade or new call sign, you will be able to file interactive applications with the FCC directly via computer, including change of address, change of name, and license renewals without having to do any paperwork.

Phone Numbers

There are two boxes for phone numbers, one for a daytime contact, and the second for your FAX number, if you have one. If not, put down your home and cell phone numbers just in case the VEC or VE team needs to contact you because they can't read your writing.

Signature

Sign your name as legibly as possible and include all of the letters that you printed as your name at the top of the form. Don't just put down a squiggle or an initial. You need to sign your name all the way out, including all of the letters that were in your printed name.

Final Check

Finally, double-check that your handwriting is legible. If a single letter in your name can't be read clearly and is misinterpreted, subsequent electronic filings may get returned as no action. Make sure your Form 605 is as clear as a bell to your Volunteer Examination team, who will then forward it to their VEC.

Your Examiners' Portion

The VE team will carefully review your NCVEC Form 605 to ensure that they can read your handwriting and that everything looks okay. They will then enter this information into their computer database, and will most likely file your test passing results electronically to their Volunteer Examiner Coordinator. The VEC will then verify the information and electronically file your results with the FCC.

RUMOR MILL

While standing in line ready to enter the exam room, rumors usually begin to fly. Someone who is ill-informed may say they are still requiring the 20-wpm code test for Extra. Nonsense!

Someone may remark that all of the test questions have been changed, and this "new" 2016 test is different than the book. Nonsense! Again, the questions in this Extra Class book are valid from July 1, 2016, through June 30, 2020.

Someone comes out of the exam room indicating the Extra Class test had numerous questions not found in his book. Again, ridiculous! This book has every single question exactly as they will appear on the exam. Only the A, B, C, and D order of the answer choices may be changed.

TAKING THE EXAMINATION

Get a good night's sleep before the exam day. Continue to study the Extra Class formulas right up to the moment you go into the examination room. Don't believe that old saying that too much study will cause you to forget the subject material. If you purchased the audio CD course, continue to play them up to the last minute.

Listen carefully for the instructions of your Volunteer Examiners. Most examination sessions will conduct all levels of testing, so listen carefully to the announcements so you get the correct testing materials.

Make no marks on the examination sheet. If you don't know an answer, lightly shade in one of the boxes on the answer sheet for that particular question and make a big note on your scratch paper to go back and work out the correct answer. Using this technique will keep you from accidentally misaligning your answer boxes on the answer sheet. If you finish the exam at question #49, chances are you skipped a question and all of your work is out of sync with your answer sheet. Double check that all 50 answer boxes indeed have a mark.

When you are taking the written examination, *take your time!* Some answers start out looking correct, but end up wrong at the end. Don't speed read the exam – getting careless after all this preparation could undo all the hard work you have put in on preparing for the Extra Class exam.

If you don't know an answer, eliminate the obviously wrong answers, and then take your best guess. NEVER LEAVE AN ANSWER BLANK! Leaving an answer blank is counted as a wrong answer, so you might as well make your best guess. We also find that going back and second-guessing an answer will sometimes cause you to change to the wrong answer. Usually your first impression is the correct one. Another trick is to read your answer first and then see if it agrees with what the question asks. Start this technique at Answer 50, and go backwards.

AND THE RESULTS ARE...

When you are satisfied that you have completed the Extra Class exam, turn in all of your examination papers – including scratch paper and the original test paper. Thank your examiners for their courtesy and follow their instructions while they prepare your score. *Don't stand over the examiners* as they grade your exam.

When you pass, the examiners may or may not indicate how many questions you missed. DON'T ASK – the examiners may tell you it is against their rules to divulge which questions you got wrong. Although we're not aware of any rule that states they cannot tell you what you got wrong, it's usually just a case of not enough time to spend with any one applicant pouring over their exam performance. They usually will tell you your score, and passing the exam is exactly what you want to hear!

When you pass your Element 4 exam, the VEs will issue you an official CSCE – Certificate of Successful Completion of Examination. This is an important document that is official proof that you passed the exam. Check to see that it is signed by all three volunteer examiners. Be sure to sign your CSCE. Double-check that the date is correct. If the computer system should gobble up all of your test results, this is LEGAL PROOF that you indeed passed a particular examination element.

NOTE: The renewal period on your new Extra Class license is NOT 10 years. It remains the SAME renewal date that is now on your General or Advanced Class license. Make sure your address is up-to-date, too! An upgrade to your license class does NOT change the original grant date of your previous license class.

CHANGING CALL SIGNS

If you check the box "CHANGE my station call sign systematically" on the NCVEC Form 605, the FCC computer will assign you the next Group "A" call sign. A Group "A" call sign will be in a 2 by 2 format with "A" as the first letter. The complete 2 by 2 format has two letters followed by a number indicating the part of the country where you live, then two more letters, assigned alphabetically. Once all the Group "A" call signs are assigned, the FCC will begin to assign Group "B" 2 by 2 call signs, which begin with "K", "N", or "W".

You must initial the request if you decide to systematically change your call sign. Once changed, you cannot easily change back to your previous call sign. The only way you can go back to a previously-held call sign is under the Vanity call sign program. So be thinking about it right now – do you want to stay with your present call sign, or do you want to change over to a new Extra Class 2-by-2 call sign?

05 APPLICATION FOR
MARY STATION LICENSE

INITIAL | STATION CALL SIGN (IF ANY)
SOCIAL SECURITY NUMBER (SSN) or (FRN) FCC FEDERAL REGISTRATION NUMBER
(5 or 9 Numbers) | FAX NUMBER (Include Area Code) OPTIONAL
NAL)

☐ **CHANGE** my **name** on my license to my new name
Former Name: (Last name) (Suffix) (First name) (MI)
☐ **CHANGE** my mailing address to **above** address
☐ **CHANGE** my station **call sigs** systematically
Applicant's Initials: To Confirm
☐ **RENEWAL** of my license. Exp. Date

APPLICATION | PENDING FILE NUMBER (FOR VEC USE ONLY)

Only check the "CHANGE" my station call sign box if you want the FCC computer to give you a new call sign. If you want to choose a call sign, apply for a vanity call.

VANITY CALL SIGNS

The ability to select a call sign of your choice began in 1996 and it has been immensely popular, especially with Extra Class amateurs who historically like to have short call signs. While only a little more than 10% of all radio amateurs ever make it to the top Extra Class level, they hold nearly half of all Vanity call signs issued!

Extra Class radio amateurs can choose an available call sign containing any Group, A, B, C or D Amateur format. Most try to get a 1-by-2 or a 2-by-1 format, which are only available to an Extra Class licensee. A 1-by-2 Group "A" call sign begins with either K, N or W (but not "A") followed by a numeral and two letters. For example: W5YI is a Group "A" call sign.

An example of a 2-by-1 call sign is AA1A. The two letter prefix must be from the AA to AL, KA to KZ, NA to NZ or the WA to WZ prefix block followed by a numeral and a single letter. To make matters more confusing, certain prefixes (AH, AL, KH, KL, KP, NH, NL, NP, WH, WL and WP) are reserved for amateurs who have a mailing address outside of the 48 contiguous U.S. states.

An easier way to get a Vanity call sign is to let the W5YI Group handle everything for you. They charge a small service fee and do all the paperwork and filing for you. You can obtain complete details on this service by going to:
☞ **http://www.w5yi.org/page.php?id=269**

You can find out all the rules (and there are many!) surrounding obtaining a Vanity call sign by visiting the W5YI Group website at ☞ **www.W5YI.org** and clicking on the "Vanity Call Signs" button near the top of the home page.

You can file the application yourself online. Additional information on Vanity call signs is available from the FCC's Amateur Radio website located at:

☞ **http://wireless.fcc.gov/services/index.htm?job=call_signs_3&id=amateur&page=1**

To make it easier for you and to insure you get the vanity call sign that you want, the W5YI Group offers a vanity filing service and will electronically file your application for you. For a nominal service fee, they will research when and if a specific call is available, file your application. You could end up with the exact call sign of your choice using their 99% success rate vanity filing service. You can enter your application and call sign choices right on their web site at ☞ **www.W5YI.org**, or call them at 800-669-9594.

GOING FURTHER – GROL + RADAR

The Extra Class license is the highest level license you can achieve in the Amateur Radio service. Did you know you can also take the commercial general radio operator (GROL) license exams? The commercial license allows you to work on marine and aeronautical two-way radio equipment. This license is often required by two-way radio employers as a term of employment. And guess what? The GROL technical element 3 exam uses many of the same questions as those you just studied for your amateur Extra Class license exam. In fact, except for some of the numbers, the commercial radiotelephone examination may use identical math problems that you just studied.

So, why not upgrade to a professional-level FCC license? You can order the *GROL + RADAR* book by calling the W5YI Group at 800-669-9594.

BECOMING A VOLUNTEER EXAMINER

When you successfully complete your Extra Class exam, ask your VE team for a volunteer examiner sign-up application. After only a short processing time, you'll join a cadre of other Extra Class hams who give something back to ham radio by helping with testing. With just a few hours a month, and some paper and computer processing, you will enjoy this rewarding "extra credit" as an Extra Class operator.

Accredited Volunteer Examiners also help develop new questions when the amateur radio question pools undergo periodic updating. Each pool gets updated once every four years with a public notice for radio experts to submit new or improved examination questions along with four suggested right and wrong answers. Your comments and suggestions are filed with the NCVEC Question Pool Committee. Join in – as an Extra Class operator. It's fun to see that you have contributed to the new question pool and the growth of the hobby.

Here's how to become a volunteer examiner:
Write the W5YI-VEC at: P.O. Box 200065, Arlington, TX 76006-0065,
or call them at 800-669-9594 during regular business hours to obtain a VE application.
You also may apply online at: ☞ **www.W5YI.org**.
Click on the "Become a Volunteer Examiner" button.

SUMMARY

You made it through to Extra Class! Congratulations! Allow me to send you a free certificate of achievement. Send a large envelope, self-addressed, with 12 first class stamps on the inside to me at Gordon West Radio School, 2414 College Drive, Costa Mesa, California 92626. It will take about 15 days to get the certificate back to you with a hearty congratulations on making it to the top.

Eric and I hope to hear you soon on the worldwide airwaves or see you at an upcoming hamfest.

Welcome to the top!

Gordon West
73

Gordon West,
WB6NOA

I know congratulations will be in order very soon when you pass your Extra Class exam with flying colors!

APPENDIX

U.S. VOLUNTEER EXAMINER COORDINATORS IN THE AMATEUR SERVICE

W5YI-VEC
P.O. Box 200065
Arlington, TX 76006-0065
800-669-9594
nb5x@w5yi.org
W5YI-VEC@w5yi.org

Anchorage ARC VEC
PO Box 670616
Anchorage, AK 99567-0616
907-688-0660
e-mail: jwiley@gci.net
Internet: www.kl7aa.net/vec/vecmain.html

American Radio Relay League (ARRL)
225 Main Street
Newington, CT 06111-1494
860-594-0300
860-594-0339
e-mail: vec@arrl.org
Internet: www.arrl.org

Central America CAVEC
1204 Governors Dr. SE
Huntsville, AL 35801-2737
256-653-5007
e-mail: cavec.org@gmail.com

Golden Empire Amateur Radio Society (GEARS)
P.O. Box 508
Chico, CA 95927-0508
530-893-9211
e-mail: myw6js@gmail.com

Greater L.A. Amateur Radio Group
P.O. Box 500133
Palmdale, CA 93591
661-264-1863
vec@glaarg.org
e-mail: wa6yeo@sbcglobal.net
Internet: www.glaarg.org

Jefferson Amateur Radio Club
P.O. Box 73665
Metairie, LA 70033
504-831-1613
e-mail: w5gad@w5gad.org
Internet: www.w5gad.org

Laurel Amateur Radio Club, Inc.
P.O. Box 146
Laurel, MD 20725-0146
301-937-0394
301-572-5124
e-mail: aa3of@arrl.net
Internet: www.larcmd.org

Milwaukee RAC VEC, Inc.
2505 S. Calhoun Rd., #203
New Berlin, WI 53151
e-mail: MRACVEC@gmail.com
Internet: www.w9rh.org

MO-KAN VEC Coordinator
228 Tennessee Road
Richmond, KS 66080-9174
785-615-1097
e-mail: wo0e@lcwb.coop

Sandarc-VEC
P.O. Box 2446
La Mesa, CA 91943-2446
619-465-3926
e-mail: n6nyx@arrl.net
Internet: http://www.sandarc.net

Sunnyvale VEC Amateur Radio Club, Inc.
P.O. Box 60307
Sunnyvale, CA 94088-0307
408-255-9000
e-mail: vec@amateur-radio.org
Internet: www.amateur-radio.org

W4VEC Volunteer Examiners Club of America
P. O. Box 482
China Grove, NC 28023-04822
336-249-8734
e-mail: raef@lexcominc.net
Internet: www.w4vec.com

Western Carolina Amateur Radio Society VEC, Inc.
7 Skylyn Ct.
Asheville, NC 28806-3922
828-253-1192
e-mail: wcarsvec@wcarsvec.net

2016-20 ELEMENT 4 Q&A CROSS REFERENCE

The following cross reference presents all 713 question numbers in numerical order included in the 2016-20 Element 4 Extra Class question pool, followed by the page number on which the question begins in this book. This will allow you to located specific questions by question number. Note: 1 question was deleted from the pool by the Question Pool Committee, resulting in an active pool of 712 questions. The deleted question does not appear in this book, but its number is listed here followed by the word deleted.

ELEMENT 4 (EXTRA CLASS) QUESTION POOL SYLLABUS

The syllabus used by the QPC for the development of the question pool is included here as an aid in studying the subelements and topic groups. Review the syllabus before you start your study to gain an understanding of how the question pool is used to develop the Element 4 written examination. Remember, one question will be taken from each topic group within each subelement to create your exam. The QPC topic groups are not the same as those used by Gordon West in this book.

E1 – COMMISSION'S RULES
[6 Exam Questions – 6 Groups]

E1A Operating Standards: frequency privileges; emission standards; automatic message forwarding; frequency sharing; stations aboard ships or aircraft

E1B Station restrictions and special operations: restrictions on station location; general operating restrictions, spurious emissions, control operator reimbursement; antenna structure restrictions; RACES operations; national quiet zone

E1C Definitions and restrictions pertaining to local, automatic and remote control operation; control operator responsibilities for remote and automatically controlled stations; IARP and CEPT licenses; third party communications over automatically controlled stations

E1D Amateur satellites: definitions and purpose; license requirements for space stations; available frequencies and bands; telecommand and telemetry operations; restrictions, and special provisions; notification requirements

E1E Volunteer examiner program: definitions; qualifications; preparation and administration of exams; accreditation; question pools; documentation requirements

E1F Miscellaneous rules: external RF power amplifiers; business communications; compensated communications; spread spectrum; auxiliary stations; reciprocal operating privileges; special temporary authority

E2 – OPERATING PROCEDURES
[5 Exam Questions – 5 Groups]

E2A Amateur radio in space: amateur satellites; orbital mechanics; frequencies and modes; satellite hardware; satellite operations; experimental telemetry applications

E2B Television practices: fast scan television standards and techniques; slow scan television standards and techniques

E2C Operating methods: contest and DX operating; remote operation techniques; Cabrillo format; QSLing; RF network connected systems

E2D Operating methods: VHF and UHF digital modes and procedures; APRS; EME procedures, meteor scatter procedures

E2E Operating methods: operating HF digital modes

E3 – RADIO WAVE PROPAGATION
[3 Exam Questions – 3 Groups]

E3A Electromagnetic waves; Propagation and technique: Earth-Moon-Earth communications; meteor scatter; microwave tropospheric and scatter propagation; aurora propagation

E3B Propagation and technique, trans-equatorial; long path; gray-line; multi-path propagation; ordinary and extraordinary waves; chordal hop propagation, sporadic E mechanisms

E3C Aurora propagation; radio-path horizon; less common propagation modes; propagation prediction techniques and modeling; space weather parameters and amateur radio

E4 – AMATEUR PRACTICES
[5 Exam Questions – 5 Groups]

E4A Test equipment: analog and digital instruments; spectrum and network analyzers, antenna analyzers; oscilloscopes; RF measurements; computer aided measurements

E4B Measurement technique and limitations: instrument accuracy and performance limitations; probes; techniques to minimize errors; measurement of "Q"; instrument calibration; S parameters; vector network analyzers

E4C Receiver performance characteristics, phase noise, noise floor, image rejection, MDS, signal-to-noise-ratio; selectivity; effects of SDR receiver non-linearity

E4D Receiver performance characteristics: blocking dynamic range; intermodulation and cross-modulation interference; 3rd order intercept; desensitization; preselector

E4E Noise suppression: system noise; electrical appliance noise; line noise; locating noise sources; DSP noise reduction; noise blankers; grounding for signals

E5 – ELECTRICAL PRINCIPLES
[4 Exam Questions – 4 Groups]

E5A Resonance and Q: characteristics of resonant circuits: series and parallel resonance; definitions and effects of Q; half-power bandwidth; phase relationships in reactive circuits

E5B Time constants and phase relationships: RLC time constants; definition; time constants in RL and RC circuits; phase angle between voltage and current; phase angles of series RLC; phase angle of inductance vs susceptance; admittance and susceptance

E5C Coordinate systems and phasors in electronics: Rectangular Coordinates; Polar Coordinates; Phasors

E5D AC and RF energy in real circuits: skin effect; electrostatic and electromagnetic fields; reactive power; power factor; electrical length of conductors at UHF and microwave frequencies

E6 – CIRCUIT COMPONENTS

[6 Exam Questions – 6 Groups]

E6A Semiconductor materials and devices: semiconductor materials; germanium, silicon, P-type, N-type; transistor types: NPN, PNP, junction, field-effect transistors: enhancement mode; depletion mode; MOS; CMOS; N-channel; P-channel

E6B Diodes

E6C Digital ICs: Families of digital ICs; gates; Programmable Logic Devices (PLDs)

E6D Toroidal and Solenoidal Inductors: permeability, core material, selecting, winding; Transformers; Piezoelectric devices

E6E Analog ICs: MMICs, CCDs, Multiplexers, Device packages

E6F Optical components: photoconductive principles and effects, photovoltaic systems, optical couplers, optical sensors, and optoisolators; LCDs

E7 – PRACTICAL CIRCUITS

[8 Exam Questions – 8 Groups]

E7A Digital circuits: digital circuit principles and logic circuits: classes of logic elements; positive and negative logic; frequency dividers; truth tables

E7B Amplifiers: Class of operation; vacuum tube and solid-state circuits; distortion and intermodulation; spurious and parasitic suppression; microwave amplifiers; switching-type amplifiers

E7C Filters and matching networks: types of networks; types of filters; filter applications; filter characteristics; impedance matching; DSP filtering

E7D Power supplies and voltage regulators; Solar array charge controllers

E7E Modulation and demodulation: reactance, phase and balanced modulators; detectors; mixer stages

E7F DSP filtering and other operations; Software Defined Radio Fundamentals; DSP modulation and demodulation

E7G Active filters and op-amp circuits: active audio filters; characteristics; basic circuit design; operational amplifiers

E7H Oscillators and signal sources: types of oscillators; synthesizers and phase-locked loops; direct digital synthesizers; stabilizing thermal drift; microphonics; high accuracy oscillators

E8 – SIGNALS & EMISSIONS

[4 Exam Questions —4 Groups]

E8A AC waveforms: sine, square, sawtooth and irregular waveforms; AC measurements; average and PEP of RF signals; Fourier analysis; Analog to digital conversion: Digital to Analog conversion

E8B Modulation and demodulation: modulation methods; modulation index and deviation ratio; frequency and time division multiplexing; Orthogonal Frequency Division Multiplexing

E8C Digital signals: digital communication modes; information rate vs bandwidth; error correction

E8D Keying defects and overmodulation of digital signals; digital codes; spread spectrum

E9 – ANTENNAS & TRANSMISSION LINES

[8 Exam Questions – 8 Groups]

E9A Basic Antenna parameters: radiation resistance, gain, beamwidth, efficiency, beamwidth; effective radiated power, polarization

E9B Antenna patterns: E and H plane patterns; gain as a function of pattern; antenna design

E9C Wire and phased array antennas: rhombic antennas; effects of ground reflections; e-off angles; Practical wire antennas: Zepps, OCFD, loops

E9D Directional antennas: gain; Yagi Antennas; losses; SWR bandwidth; antenna efficiency; shortened and mobile antennas; RF Grounding

E9E Matching: matching antennas to feed lines; phasing lines; power dividers

E9F Transmission lines: characteristics of open and shorted feed lines; 1/8 wavelength; 1/4 wavelength; 1/2 wavelength; feed lines: coax versus open-wire; velocity factor; electrical length; coaxial cable dielectrics, velocity factor

E9G The Smith chart

E9H Receiving Antennas: radio direction finding antennas; Beverage Antennas; specialized receiving antennas; longwire receiving antennas

E0 – SAFETY

[1 exam question – 1 group]

E0A Safety: amateur radio safety practices; RF radiation hazards; hazardous materials; grounding and bonding

THREE SIMPLE RULES FOR 60 METERS

When operating 60 meters, it is crucial that you transmit on the correct frequency within the authorized channel. Compliance with the rules when using Single Sideband phone or sound card digital modes is very straightforward. CW operation is slightly different. 60 Meter channels are assigned by the following table, where the frequency represents the VFO display (suppressed carrier) frequency of your transceiver, which is also the low frequency boundary of the channel:

Channel 1: 5330.5 kHz
Channel 2: 5346.5 kHz
Channel 3: 5357.0 kHz
Channel 4: 5371.5 kHz
Channel 5: 5403.5 kHz

If you heed the following three rules, you will be in compliance with the FCC regulations:

(1) Phone Operation: Set your transmitter to Upper Sideband (USB) and your VFO to display the frequency of the chosen channel from the table. Be sure your transmitter is set to limit the bandwidth to 2.8 kHz. Check your operator's manual if you are unsure about how to do this. Most modern rigs default to this setting, but READ THE MANUAL!

(2) Sound Card Digital Modes: Set your transmitter to USB, select the channel frequency on your VFO, and click on 1.5 kHz on your waterfall or spectrum display. This will place your transmitted signal properly in the CENTER of the channel.

(3) CW operation: Place your transmitted signal in the CENTER of the channel, 1.5 kHz above the assigned carrier frequency shown in the table. Depending on your particular transceiver, this MAY OR MAY NOT correspond to your VFO display frequency! READ YOUR OPERATOR'S MANUAL TO BE SURE.

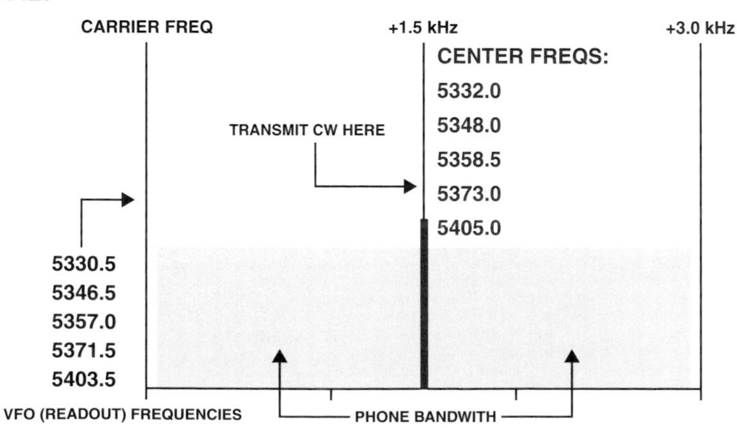

The diagram above shows the position of your transmitted signal within a 60 meter channel.

One final note: If you're using ***non-sound card digital modes***, such as "old school" RTTY with direct FSK modulation, you must assure that your transmitted signal is exactly centered in the channel, using any reliable means at your disposal.

SCHEMATIC SYMBOLS

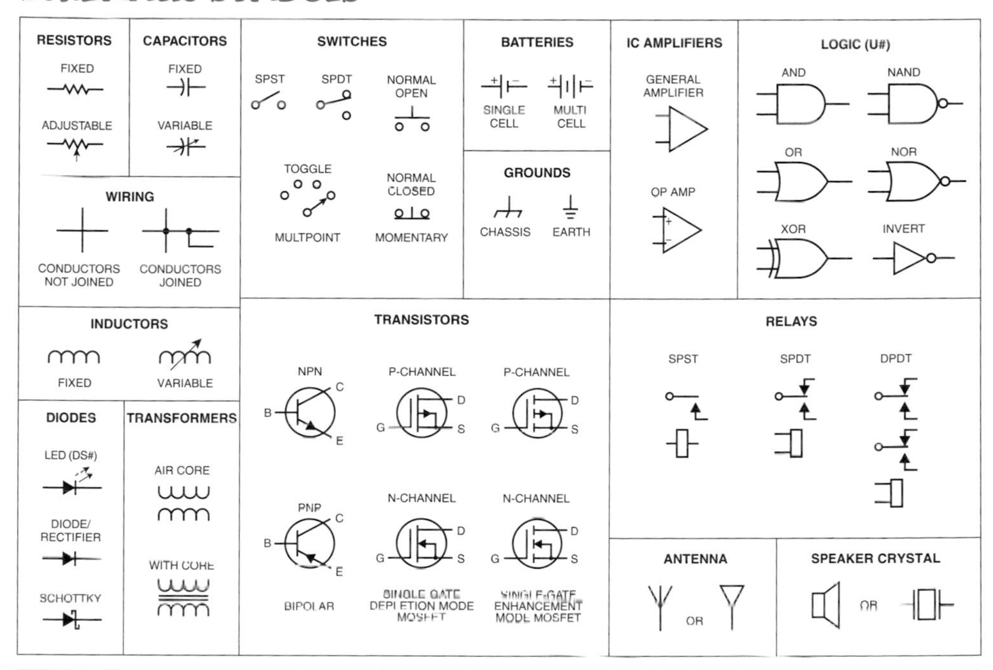

Decibels

It is important to know about decibels because they are used extensively in electronics. Look at the derivation for decibels and note it is a measure of the ratio of two powers, P_1 and P_2. Remember the power changes for different dB values. Also, since 6 dB is a four times change, 16 dB (6 dB + 10 dB) will be a 4 × 10 = 40 times change, and 26 dB (6 dB + 20 dB) will be a 4 × 100 = 400 times change. Thus, any dB value from 10 dB and above can be evaluted by using the power change of 1 dB to 9 dB and multiplying it by the 10 dB, 20 dB, 30 dB, 40 dB, 50 dB, 60 dB, etc. change. Thus, 57 dB (7 dB + 50 dB) will be a 5 × 100,000 = 500,000 times change.

dB	Power Change $\frac{P_1}{P_2}$	
1 dB	1 ¼X	Power change
2 dB	1 ½X	Power change
3 dB	2X	Power change
4 dB	2 ½X	Power change
5 dB	3X	Power change
6 dB	4X	Power change
7 dB	5X	Power change

dB	Power Change $\frac{P_1}{P_2}$	
8 dB	6 ¼X	Power change
9 dB	8X	Power change
10 dB	10X	Power change
20 dB	100X	Power change
30 dB	1000X	Power change
40 dB	10,000X	Power change
50 dB	100,000X	Power change
60 dB	1,000,000X	Power change

Derivation:

If $dB = 10 \log_{10} \frac{P_1}{P_2}$

then what power ratio is 60 dB?

$$60 = 10 \log_{10} \frac{P_1}{P_2}$$

$$\frac{60}{10} = \log_{10} \frac{P_1}{P_2}$$

$$6 = \log_{10} \frac{P_1}{P_2}$$

Remember: logarithm of a number is the exponent to which the base must be raised to get the number.

$$\therefore 10^6 = \frac{P_1}{P_2}$$

$$1{,}000{,}000 = \frac{P_1}{P_2}$$

$$\text{Or } P_1 = 1{,}000{,}000\, P_2$$

60 dB means P_1 is one million times P_2

SUMMARY OF QUESTION POOL FORMULAS

Page(s)	Formula	Where:
240,270	**Decibels** $dB = 10 \log_{10} \frac{P_1}{P_2}$	P_1 = Power (usually output) in **watts** P_2 = Power (usually input) in **watts**
215-216	**Counter Readout Error** Readout Error = f × a	f = Frequency in **MHz** being measured a = Counter accuracy in **parts per million**
131	**Intermodulation Product** $f_{IMD(2)} = 2f_1 - f_2$ $f_{IMD(4)} = 2f_2 - f_1$	f_{IMD} = Intermodulation product at particular f in **Hz** f_1 = Frequency of transmitted signal in **Hz** f_2 = Possible frequency of interfering signal in **Hz**
161	**Resonance** $X_L = X_C$ $2\pi f_r L = \frac{1}{2\pi f_r C}$ $f_r = \frac{1}{2\pi \sqrt{LC}}$	X_L = Inductive reactance in **ohms** X_C = Capacitive reactance in **ohms** f_r = Resonant frequency in **hertz** L = Inductance in **henrys** C = Capacitance in **farads** π = 3.14
162	**Phase Between V and I** ELI = Voltage (E) leads current (I) in an inductive (L) circuit ICE = Current (I) leads voltage (E) in a capacitive (C) circuit	E = Voltage in circuit I = Current in circuit L = Inductance in circuit C = Capacitance in circuit
176-177	**Half-Power Bandwidth** $BW_{-3dB} = \frac{f_r}{Q}$	BW_{-3dB} = Half-Power bandwidth in **kHz** f_r = Resonant frequency in **kHz** Q = Circuit **quality factor**
178-179	**Time Constants** $\tau = RC$ $\tau = \frac{L}{R}$	τ = Time Constant in **seconds** R = Resistance in **ohms** C = Capacitance in **farads** L = Inductance in **henries**

Time Constant Curve

		RC		RL	
Time Constant τ	% of Change in τ	% of Final Q or V on C When Charging	% of Initial Q or V on C When Discharging	% of Final I When Increasing	% of Initial I When Decreasing
1	63.2	63.2	36.8	63.2	36.8
2	23.3	86.5	13.5	86.5	13.5
3	8.5	95.0	5.0	95.0	5.0
4	3.2	98.2	1.8	98.2	1.8
5	1.1	99.3	0.7	99.3	0.7

SUMMARY OF QUESTION POOL FORMULAS (continued)

Page(s)	Formula	Where:
170	**Capacitors** Series: $C_T = \frac{C_1 \times C_2}{C_1 + C_2}$ Parallel: $C_T = C_1 + C_2$ **Resistors** $R_T = R_1 + R_2$; $R_T = \frac{R_1 \times R_2}{R_1 + R_2}$	C = Capacitance in **farads** R = Resistance in **ohms**
171-173	**Inductive Reactance** $X_L = 2\pi fL$	X_L = Inductive reactance in **ohms** L = Inductance in **henries** f = Frequency in **hertz** π = 3.14
171-173	**Capacitive Reactance** $X_C = \frac{1}{2\pi fC}$	X_C = Capacitive reactance in **ohms** C = Capacitance in **farads** f = Frequency in **hertz** π = 3.14
	$X_C = \frac{10^6}{2\pi fC}$ **Impedance in Series** $Z_T = Z_1 + Z_2$ **Impedance in Parallel** → $+Z_T = \frac{Z_1 \times Z_2}{Z_1 + Z_2}$	f = Frequency in **MHz** C = Capacitance in **picofarads** Z = Impedance in **ohms** π = 3.14 Special Case: When $Z_1 = R$ and $Z_2 = \pm jX$ $Z_T = \frac{RX \angle \Theta = \pm 90°}{\sqrt{R^2 + X^2} \angle \arctan \pm X/R}$
169	**Rectangular Coordinates** $Z = R \pm jX$ **Polar Coordinates** $Z = Z \angle \pm\Theta$	Z = Impedance in **ohms** R = Resistance in **ohms** +jX = Inductive reactance X_L in **ohms** −jX = Capacitive reactance X_C in **ohms** Θ = Phase angle in **degrees** arctan = angle whose tangent is
	Conversion: $Z = R \pm jX$ to $Z \angle \pm\Theta$: $Z = \sqrt{R^2 + X^2}\ Z \angle \arctan \frac{X}{R}$ $Z \angle \pm\Theta$ to $Z = R \pm jX$: $R = Z \cos \Theta$; $\pm jX = Z \sin \Theta$	
165	**Phase Angle Between V and I** $\tan \phi = \frac{X}{R}$	ϕ = Phase angle in **degrees** X = $(X_L - X_C)$ = Total reactance in **ohms** R = Resistance in **ohms**

SUMMARY OF QUESTION POOL FORMULAS (continued)

Page(s)	Formula	Where:
169	Do Addition and Subtraction in Rectangular Coordinates $Z_1 = R_1 + jX_1 \quad Z_2 = R_2 - jX_2$ **Addition** $Z_T = (R_1 + jX_1) + (R_2 - jX_2)$ $Z_T = (R_1 + R_2) + j(X_1 - X_2)$ **Subtraction** $Z_T = (R_1 + jX_1) + (R_2 - jX_2)$ $Z_T = (R_1 - R_2) + j(X_1 + X_2)$	Do Multiplication and Division in Polar Coordinates $Z_1 = Z_1 \angle\Theta_1 \quad Z_2 = Z_2 \angle\Theta_2$ **Multiply** $Z_1 Z_2 = Z_1 \times Z_2 \angle\Theta_1 + \Theta_2$ **Divide** $\frac{Z_1}{Z_2} = \frac{Z_1 \angle\Theta_1}{Z_2 \angle\Theta_1} = \frac{Z_1}{Z_2} = \angle\Theta_1 - \Theta_2$
175	**Circuit Quality Factor** $Q = \frac{R}{X_L}$ $Q = \frac{R}{2\pi f_r L}$	Q = Circuit **quality factor** $X_L = 2\pi f_r L$ X_L = Inductive reactance in **ohms** f_r = Resonant frequency in **Hz** L = Inductance in **henrys** R = Resistance in **ohms** $2\pi = 6.28$
184-185	**True Power** $P_T = P_A \times PF$ $P_A = E \times I$ $PF = \cos\phi$	P_T = True power in **watts** P_A = Apparent power in **watts** ϕ = Phase angle in **degrees** PF = Power factor E = Applied voltage in **volts** I = Circuit current in **amps**
216	**Inverting IC Op-Amp Gain** $G = -\frac{R_f}{R_1}$	G = Gain in **non-dimentional units** R_f = Feedback resistance in **ohms** R_1 = Input resistance in **ohms**
190	**Turns for L Using Ferrite Toroidal Cores** $N = 1000\sqrt{\frac{L}{A_L}}$	N = Number of turns needed A_L = Inductance index in **mH per 1000 turns** L = Inductance in **millihenrys**
189	**Turns for L Using Powdered-Iron Toroidal Cores** $N = 100\sqrt{\frac{L}{A_L}}$	N = Number of turns needed A_L = Inductance index in **µH per 100 turns** L = Inductance in **microhenrys**

SUMMARY OF QUESTION POOL FORMULAS (continued)

Page(s)	Formula	Where:
	Peak, Peak-to-Peak and RMS Voltages $^{**}V_P = \frac{V_{PP}}{2}$ $\quad ^{*}V_{RMS} = 0.707\ V_P$ $^{**}V_{PP} = 2V_P$ $\quad ^{*}V_P = 1.414\ V_{RMS}$	V_P = Peak voltage in **volts** V_{PP} = Peak-to-peak voltage in **volts** V_{RMS} = Root-mean-square voltage in **volts** * Equations for sinewaves ** Equations for symmetrical waves
	Peak Envelope Power $PEP = P_{DC} \times \text{Efficiency}$ Amplifier Class / Efficiency: C — 80% B — 60% AB — 50% A — <50%	PEP = Peak envelope power in **watts** P_{DC} = Input DC power in **watts**
102	**Modulation Index** $\text{Modulation Index} = \frac{\text{Deviation of FM Signal (in Hz)}}{\text{Modulating Frequency (in Hz)}}$	
103	**Deviation Ratio** $\text{Deviation Ratio} = \frac{\text{Maximum Carrier Frequency Deviation (in kHz)}}{\text{Maximum Modulation Frequency (in kHz)}}$	
97	**Bandwidth for CW** $BW_{CW} = \text{baud rate} \times \text{wpm} \times \text{fading factor}$	BW_{CW} = Necessary bandwidth in **hertz** wpm = Morse code signal rate in **words per minute** Fading factor = Constant of 5 for **CW**
91	**Bandwidth for Digital** $BW = \text{baud rate} + (1.2 \times f_s)$	BW = Necessary bandwidth in **hertz** f_s = Frequency shift in **hertz** Baud rate = Digital signal rate in **bauds**
246	**Antenna Beamwidth** $\text{Beamwidth} = \frac{203}{(\sqrt{10})^X}$	Beamwidth = Antenna beamwidth in **degrees** A_G = Antenna gain in **dB** $X = \frac{A_G}{10}$

SUMMARY OF QUESTION POOL FORMULAS (continued)

Page(s)	Formula	Where:
261	**Physical Length vs. Electrical Length** $L = \frac{984\lambda V}{f}$	L = Physical length in **feet** λ = Electrical length in **wavelengths** V = **Velocity factor** of feedline f = Frequency in **MHz**

Feedline	Velocity Factor
Coax	0.66
Parallel	0.95
Twin-Lead	0.80

SCIENTIFIC NOTATION

Prefix	Symbol	Multiplication Factor	
exa	E	10^{18}	= 1,000,000,000,000,000,000
peta	P	10^{15}	= 1,000,000,000,000,000
tera	T	10^{12}	= 1,000,000,000,000
giga	G	10^{9}	= 1,000,000,000
mega	M	10^{6}	= 1,000,000
kilo	k	10^{3}	= 1,000
hecto	h	10^{2}	= 100
deca	da	10^{1}	= 10
(unit)		10^{0}	= 1

Prefix	Symbol	Multiplication Factor	
deci	d	10^{-1}	= 0.1
centi	c	10^{-2}	= 0.01
milli	m	10^{-3}	= 0.001
micro		10^{-6}	= 0.000001
nano	n	10^{-9}	= 0.000000001
pico	p	10^{-12}	= 0.000000000001
femto	f	10^{-15}	= 0.000000000000001
atto	a	10^{-18}	= 0.000000000000000001

AUTHORIZED FREQUENCY BANDS – AMATEUR SERVICE (for U.S. Amateur Stations operating from ITU-Region 2–North and South America)

Current License Class[1]		Technician	Tech. w/Code	General		Extra Class
Grandfathered[2]	Novice	Technician	Technician Plus	General	Advanced	Extra Class
160				1800-2000 kHz/All	1800-2000 kHz/All	1800-2000 kHz/All
80 / 75	3525-3600 kHz/CW		3525-3600 kHz/CW	3525-3600 kHz/CW 3800-4000 kHz/Ph	3525-3600 kHz/CW 3700-4000 kHz/Ph	3500-4000 kHz/CW 3600-4000 kHz/Ph
40	7025-7125 kHz/CW		7025-7125 kHz/CW	7025-7125 kHz/CW 7175-7300 kHz/Ph	7025-7125 kHz/CW 7125-7300 kHz/Ph	7000-7300 kHz/CW 7125-7300 kHz/Ph
30				10.1-10.15 MHz/CW	10.1-10.15 MHz/CW	10.1-10.15 MHz/CW
20				14.025-14.15 MHz/CW 14.225-14.35 MHz/Ph	14.025-14.15 MHz/CW 14.175-14.35 MHz/Ph	14.0-14.35 MHz/CW 14.15-14.35 MHz/Ph
17				18.068-18.11 MHz/CW 18.11-18.168 MHz/Ph	18.068-18.11 MHz/CW 18.11-18.168 MHz/Ph	18.068-18.11 MHz/CW 18.11-18.168 MHz/Ph
15	21.025-21.2 MHz/CW		21.025-21.2 MHz/CW	21.025-21.2 MHz/CW 21.275-21.45 MHz/Ph	21.025-21.2 MHz/CW 21.225-21.45 MHz/Ph	21.0-21.45 MHz/CW 21.2-21.45 MHz/Ph
12				24.89-24.99 MHz/CW 24.93-24.99 MHz/Ph	24.89-24.99 MHz/CW 24.93-24.99 MHz/Ph	24.89-24.99 MHz/CW 24.93-24.99 MHz/Ph
10	28.0-28.5 MHz/CW 28.3-28.5 MHz/Ph		28.0-28.5 MHz/CW 28.3-28.5 MHz/Ph	28.0-28.3 MHz/CW 28.3-29.7 MHz/Ph	28.0-28.3 MHz/CW 28.3-29.7 MHz/Ph	28.0-29.7 MHz/CW 28.3-29.7 MHz/Ph
6		50-54 MHz/CW 50.1-54 MHz/Ph	50-54 MHz/CW 50.1-54 MHz/Ph	50-54 MHz/CW 50.1-54 MHz/Ph	50-54 MHz/CW 50.1-54 MHz/Ph	50-54 MHz/CW 50.1-54 MHz/Ph
2		144-148 MHz/CW 144.1-148 MHz/All	144-148 MHz/CW 144.1-148 MHz/All	144-148 MHz/CW 144.1-148 MHz/Ph	144-148 MHz/CW 144.1-148 MHz/All	144-148 MHz/CW 144.1-148 MHz/All
1.25	222-225 MHz/All	[3] 222-225 MHz/All	222-225 MHz/All	222-225 MHz/All	222-225 MHz/All	222-225 MHz/All
0.70		420-450 MHz/All	420-450 MHz/All	420-450 MHz/All	420-450 MHz/All	420-450 MHz/All
0.33		902-928 MHz/All	902-928 MHz/All	902-928 MHz/All	902-928 MHz/All	902-928 MHz/All
0.23	1270-1295 MHz/All	1240-1300 MHz/All	1240-1300 MHz/All	1240-1300 MHz/All	1240-1300 MHz/All	1240-1300 MHz/All

METERS

[1] Effective 4-15-00 [2] Prior to 4-15-00 [3] Effective 2/1/94 219-220 MHz is authorized for point-to-point fixed digital message forwarding systems.
[4] 60 meter operation restricted to 5 channels with center frequencies of 5332.0, 5348.0, 5358.5, 5373.0 and 5405.0 kHz, 100 watts PEP. See FCC 97.303.

Note: Morse code (CW, A1A) may be used on any frequency allocated to the amateur service. Telephony emission (abbreviated Ph above) authorized on certain bands as indicated. Higher class licensees may use slow-scan television and facsimile emissions on the Phone bands; radio teletype/digital on the CW bands. All amateur modes and emissions are authorized above 144.1 MHz. In actual practice, the modes/emissions used are somewhat more complicated than shown above due to the existence of various band plans and "gentlemen's agreements" concerning where certain operations should take place.

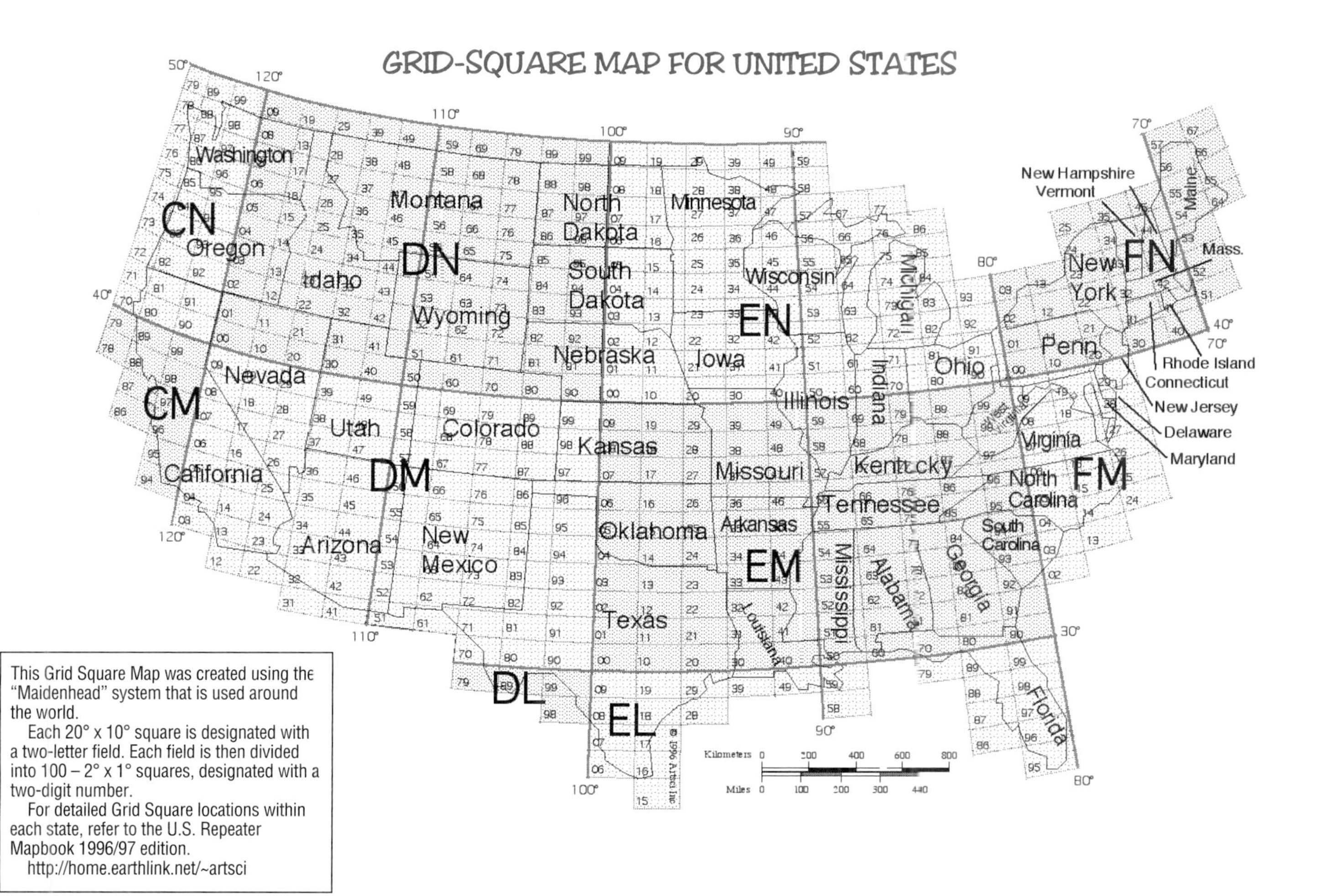
GRID-SQUARE MAP FOR UNITED STATES
CN
CM
DN
DM
DL
EN
EM
EL
FN
FM
Washington
Oregon
California
Nevada
Idaho
Montana
Wyoming
Utah
Arizona
Colorado
New Mexico
North Dakota
South Dakota
Nebraska
Kansas
Oklahoma
Texas
Minnesota
Iowa
Missouri
Arkansas
Louisiana
Wisconsin
Illinois
Michigan
Indiana
Kentucky
Tennessee
Mississippi
Alabama
Georgia
Florida
Ohio
Penn.
New York
Virginia
North Carolina
South Carolina
New Hampshire
Vermont
Maine
Mass.
Rhode Island
Connecticut
New Jersey
Delaware
Maryland
50°
40°
30°
120°
110°
100°
90°
80°
70°
Kilometers 0 200 400 600 800
Miles 0 100 200 300 440
© 1996 Artsci Inc
This Grid Square Map was created using the "Maidenhead" system that is used around the world.
Each 20° x 10° square is designated with a two-letter field. Each field is then divided into 100 – 2° x 1° squares, designated with a two-digit number.
For detailed Grid Square locations within each state, refer to the U.S. Repeater Mapbook 1996/97 edition.
http://home.earthlink.net/~artsci
Reprinted by Permission of Artsci Inc.,P.O. Box 1428, Burbank, CA 91507

ITU Regions

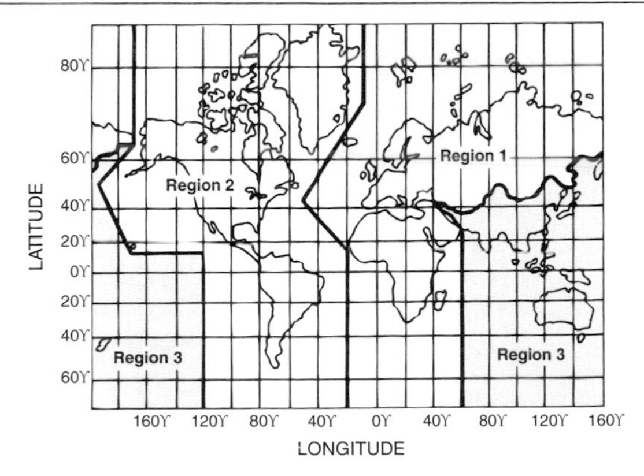

List of Countries Permitting Third-Party Traffic

Country	Call Sign Prefix	Country	Call Sign Prefix	Country	Call Sign Prefix
Antigua/Barbuda	V2	Gambia	C5	Peru	OA-OC
Argentina	LO-LW	Ghana	9G	Philippines	DU-DZ
Australia	VK	Grenada	J3	Pitcairn Island	VR6
Belize	V3	Guatemala	TG	St. Kitts/Nevis	V4
Bolivia	CP	Guyana	8R	St. Lucia	J6
Bosnia-Herzegovina	E7	Haiti	HH	St. Vincent & Grenadines	J8
Brazil	PP-PY	Honduras	HQ-HR	Sierra Leone	9L
Canada	VE, VO, VY	Israel	4X, 4Z	South Africa	ZR-ZU
Chile	CA-CE	Jamaica	6Y	Swaziland	3DA
Colombia	HJ-HK	Jordan	JY	Trinidad/Tobago	9Y-9Z
Comoros	D6	Liberia	EL	Turkey	TA-TC
Costa Rica	TI, TE	Marshall Islands	V7	United Kingdom	GB
Cuba	CM, CO	Mexico	XA-XI	Uruguay	CV-CX
Dominica	J7	Micronesia	V6	Venezuela	YV-YY
Dominican Republic	HI	Nicaragua	YN	ITU – Geneva	4U1ITU
Ecuador	HC-HD	Panama	HO-HP	VIC – Vienna	4U1VIC
El Salvador	YS	Paraguay	ZP		

Countries Holding U.S. Reciprocal Agreements

Antigua, Barbuda
Argentina
Australia
Austria
Bahamas
Barbados
Belgium
Belize
Bolivia
Bosnia-Herzegovina
Botswana
Brazil
Canada[1]
Chile
Colombia
Costa Rica
Croatia
Cyprus
Denmark
Dominica
Dominican Rep.
Ecuador
El Salvador
Fiji
Finland
France[2]
Germany
Greece
Greenland
Grenada
Guatemala
Guyana
Haiti
Honduras
Iceland
India
Indonesia
Ireland
Israel
Italy
Jamaica
Japan
Jordan
Kiribati
Kuwait
Liberia
Luxembourg
Macedonia
Marshall Is.
Mexico
Micronesia
Monaco
Netherlands
Netherlands Ant.
New Zealand
Nicaragua
Norway
Panama
Paraguay
Papua New Guinea
Peru
Philippines
Portugal
Seychelles
Sierra Leone
Solomon Islands
South Africa
Spain
St. Lucia
St. Vincent and Grenadines
Surinam
Sweden
Switzerland
Thailand
Trinidad, Tobago
Turkey
Tuvalu
United Kingdom[3]
Uruguay
Venezuela

1. Do not need reciprocal permit
2. Includes all French Territories
3. Includes all British Territories

POPULAR Q SIGNALS

Given below are a number of Q signals whose meanings most often need to be expressed with brevity and clarity in amateur work. (Q abbreviations take the form of questions only when each is sent followed by a question mark.)

QRG Will you tell me my exact frequency (or that of _____)? Your exact frequency (or that of _____) is _____ kHz.
QRH Does my frequency vary? Your frequency varies.
QRI How is the tone of my transmission? The tone of your transmission is _____ (1. Good; 2. Variable; 3. Bad).
QRJ Are you receiving me badly? I cannot receive you. Your signals are too weak.
QRK What is the intelligibility of my signals (or those of _____)? The intelligibility of your signals (or those of _____) is _____ (1. Bad; 2. Poor; 3. Fair; 4. Good; 5. Excellent).
QRL Are you busy? I am busy (or I am busy with _____). Please do not interfere.
QRM Is my transmission being interfered with? Your transmission is being interfered with _____ (1. Nil; 2. Slightly; 3. Moderately; 4. Severely; 5. Extremely).
QRN Are you troubled by static? I am troubled by static _____ (1-5 as under QRM).
QRO Shall I increase power? Increase power.
QRP Shall I decrease power? Decrease power.
QRQ Shall I send faster? Send faster (_____ WPM).
QRS Shall I send more slowly? Send more slowly (_____WPM).
QRT Shall I stop sending? Stop sending.
QRU Have you anything for me? I have nothing for you.
QRV Are you ready? I am ready.
QRW Shall I inform _____ that you are calling on _____ kHz? Please inform _____ that I am calling on _____ kHz.
QRX When will you call me again? I will call you again at _____ hours (on _____ kHz).
QRY What is my turn? Your turn is numbered _____.
QRZ Who is calling me? You are being called by _____ (on _____ kHz).
QSA What is the strength of my signals (or those of _____)? The strength of your signals (or those of _____) is _____ (1. Scarcely perceptible; 2. Weak; 3. Fairly good; 4. Good; 5. Very good).
QSB Are my signals fading? Your signals are fading.
QSD Is my keying defective? Your keying is defective.
QSG Shall I send _____ messages at a time? Send _____ messages at a time.
QSK Can you hear me between your signals and if so can I break in on your transmission? I can hear you between my signals; break in on my transmission.
QSL Can you acknowledge receipt? I am acknowledging receipt.
QSM Shall I repeat the last message which I sent you, or some previous message? Repeat the last message which you sent me [or message(s) number(s) _____].
QSN Did you hear me (or _____) on _____ kHz? I heard you (or _____) on _____ kHz.
QSO Can you communicate with _____ direct or by relay? I can communicate with _____ direct (or by relay through _____).
QSP Will you relay to _____? I will relay to _____.
QST General call preceding a message addressed to all amateurs and ARRL members. This is in effect "CQ ARRL."
QSU Shall I send or reply on this frequency (or on _____ kHz)?
QSW Will you send on this frequency (or on _____ kHz)? I am going to send on this frequency (or on _____ kHz).
QSX Will you listen to _____ on _____ kHz? I am listening to _____ on _____ kHz.
QSY Shall I change to transmission on another frequency? Change to transmission on another frequency (or on _____ kHz).
QSZ Shall I send each word or group more than once? Send each word or group twice (or _____ times).
QTA Shall I cancel message number _____? Cancel message number _____.
QTB Do you agree with my counting of words? I do not agree with your counting of words. I will repeat the first letter or digit of each word or group.
QTC How many messages have you to send? I have messages for you (or for _____).
QTH What is your location? My location is _____.
QTR What is the correct time? The time is _____.

Source: ARRL

ITU EMISSION DESIGNATORS

First Symbol – Modulation system

A – Double sideband AM
C – Vestigial sideband AM
D – Amplitude/angle modulated
F – Frequency modulation
G – Phase modulation
H – Single sideband/full carrier
J – Single sideband/suppressed carrier
K – AM pulse
L – Pulse modulated in width/duration
M – Pulse modulated in position/phase
N – Unmodulated carrier
P – Unmodulated pulses
Q – Angle modulated during pulse
R – Single sideband/reduced or variable level carrier
V – Combination of pulse emissions
W – Other types of pulses

Second Symbol – Nature of signal modulating carrier

0 – No modulation
1 – Digital data without modulated subcarrier
2 – Digital data on modulated subcarrier
3 – Analog modulated
7 – Two or more channels of digital data
8 – Two or more channels of analog data
9 – Combination of analog and digital information
X – Other

Third Symbol – Information to be conveyed

A – Manually received telegraphy
B – Automatically received telegraphy
C – Facsimile (FAX)
D – Digital information
E – Voice telephony
F – Video/television
N – No information
W – Combination of these
X – Other

The signal may be "assembled" from the different elements of the emission designators.
For example, an amplitude-modulated voice single sideband signal is J3E:
J 5 single sideband; 3 5 analog, single channel; E 5 telephony.

1. **CW –** International Morse code telegraphy emissions having designators with A, C, H, J or R as the first symbol, 1 as the second symbol, A or B as the third symbol.
2. **DATA –** Telemetry, telecommand and computer communications emissions having designators with A, C, D, F, G, H, J or R as the first symbol: 1 as the second symbol; D as the third symbol; and also emission J2D. Only a digital code of a type specifically authorized in the §Part 97.3 rules may be transmitted.
3. **IMAGE –** Facsimile and television emissions having designators with A, C, D, F, G, H, J or R as the first symbol; 1, 2 or 3 as the second symbol; C or F as the third symbol; and also emissions having B as the first symbol; 7, 8 or 9 as the second symbol; W as the third symbol.
4. **MCW (Modulated carrier wave) –** Tone-modulated international Morse code telegraphy emissions having designators with A, C, D, F, G, H or R as the first symbol; 2 as the second symbol; A or B as the third symbol.
5. **PHONE –** Speech and other sound emissions having designators with A, C, D, F, G, H, J or R as the first symbol; 1, 2 or 3 as the second symbol; E as the third symbol. Also speech emissions having B as the first symbol; 7, 8 or 9 as the second symbol; E as the third symbol. MCW for the purpose of performing the station identification procedure, or for providing telegraphy practice interspersed with speech, or incidental tones for the purpose of selective calling, or alerting or to control the level of a demodulated signal may also be considered phone.
6. **PULSE –** Emissions having designators with K, L, M, P, Q, V or W as the first symbol; , 1, 2, 3, 7, 8, 9 or X as the second symbol; A, B, C, D, E, F, N, W or X as the third symbol.
7. **RTTY (Radioteletype) –** Narrow-band, direct-printing telegraphy emissions having designators with A, C, D, F, G, H, J or R as the first symbol; 1 as the second symbol; B as the third symbol; and also emission J2B. Only a digital code of a type specifically authorized in the §Part 97.3 rules may be transmitted.
8. **SS (Spread Spectrum) –** Emissions using bandwidth-expansion modulation emissions having designators with A, C, D, F, G, H, J or R as the first symbol; X as the second symbol; X as the third symbol. Only a SS emission of a type specifically authorized in §Part 97.3 rules may be transmitted.
9. **TEST –** Emissions containing no information having the designators with N as the third symbol. Test does not include pulse emissions with no information or modulation unless pulse emissions are also authorized in the frequency band.

Glossary

Advanced: An amateur operator who has passed Element 2, 3A, 3B, and 4A written theory examinations and Element 1A and 1B code tests to demonstrate Morse code proficiency to 13 wpm.

Amateur communication: Non-commercial radio communication by or among amateur stations solely with a personal aim and without personal or business interest.

Amateur operator/primary station license: An instrument of authorization issued by the Federal Communications Commission comprised of a station license, and also incorporating an operator license indicating the class of privileges.

Amateur operator: A person holding a valid license to operate an amateur station issued by the FCC. Amateur operators are frequently referred to as ham operators.

Amateur Radio services: The amateur service, the amateur-satellite service and the radio amateur civil emergency service.

Amateur-satellite service: A radiocommunication service using stations on Earth satellites for the same purpose as those of the amateur service.

Amateur service: A radiocommunication service for the purpose of self-training, intercommunication and technical investigations carried out by amateurs; that is, duly authorized persons interested in radio technique solely with a personal aim and without pecuniary interest.

Amateur station: A station licensed in the amateur service embracing necessary apparatus at a particular location used for amateur communication.

AMSAT: Radio Amateur Satellite Corporation, a non-profit scientific organization.
(850 Sligo Avenue, Silver Spring, MD 20910-4703)

Antenna gain: The increase as a result of the physical construction of the antenna which confines the radiation to desired or useful directions. Usually specified in dB, referenced to the gain of a dipole.

ARES: The emergency division of the American Radio Relay League. See RACES

ARRL: American Radio Relay League, national organization of U.S. Amateur Radio operators. (225 Main Street, Newington, CT 06111)

ATV: Amateur fast-scan television.

Audio Frequency (AF): The range of frequencies that can be heard by the human ear, generally 20 hertz to 20 kilohertz.

Authorized bandwidth: The allowed frequency band, specified in kilohertz, and centered on the carrier frequency.

Automatic control: The use of devices and procedures for station control without the control operator being present at the control point when the station is transmitting.

Automatic Position Reporting System (APRS): Links a remote GPS unit to a ham transceiver which transmits the location to a base station.

Automatic Volume Control (AVC): A circuit that continually maintains a constant audio output volume in spite of deviations in input signal strength.

Beam or Yagi antenna: An antenna array that receives or transmits RF energy in a particular direction. Usually rotatable.

Block diagram: A simplified outline of an electronic system where circuits or components are shown as boxes.

Broadcasting: Information or programming transmitted by means of radio intended for the general public.

Business communications: Any transmission or communication the purpose of which is to facilitate the regular business or commercial affairs of any party. Business communications are prohibited in the amateur service.

Call Book: A published list of all licensed amateur operators available in North American and Foreign editions.

Call sign assignment: The FCC systematically assigns each amateur station their primary call sign.

Carrier frequency: The frequency of an unmodulated electromagnetic wave, usually specified in kilohertz or megahertz.

Certificate of Successful Completion of Examination (CSCE): A certificate verifying successful completion of a written and/or Morse code examination Element. Credit for passing is valid for 365 days.

Coaxial cable, Coax: A concentric, two-conductor cable in which one conductor surrounds the other, separated by an insulator.

Control point: Any place from which a transmitter's function may be controlled.

Control station: A station directly associated with the control point of a radio system, commonly used to control repeaters.

Control operator: An amateur operator designated by the licensee of an amateur station to be responsible for the station transmissions.

Coordinated repeater station: An amateur repeater station for which the transmitting and receiving frequencies have been recommended by the recognized repeater coordinator.

Coordinated Universal Time (UTC): Some-times referred to as Greenwich Mean Time, UCT or Zulu time. The time at the zero-degree (0∞) Meridian which passes through Greenwich, England. A universal time among all amateur operators.

Crystal: A quartz or similar material which has been ground to produce natural vibrations of a specific frequency. Quartz crystals produce a high-degree of frequency stability in radio transmitters.

CW: Continuous wave, another term for the International Morse code.

Dipole antenna: The most common wire antenna. Length is equal to one-half of the wavelength.

Direct Digital Frequency Synthesizer: Modern frequency synthesizer employing technology to reduce phase noise to an insignificant level.

Dummy antenna: A device or resistor which serves as a transmitter's antenna without radiating radio waves. Generally used to tune up a radio transmitter.

Duplex: Transmitting on one frequency, and receiving on another, commonly used in mobile telephone use, as well as mobile business radio on UHF frequencies.

Duplexer: A device that allows a single antenna to be simultaneously used for both reception and transmission.

Effective Radiated Power (ERP): The product of the transmitter (peak envelope) power, expressed in watts, delivered to the antenna, and the relative gain of an antenna over that of a half-wave dipole antenna.

Emergency communication: Any amateur communication directly relating to the immediate safety of life of individuals or the immediate protection of property.

Examination Element: The written theory exams and Morse code test required to obtain amateur radio licenses. Technicians are required to pass written Element 2 theory; Generals are required to pass written Element 3 theory plus Element 1 Morse code test at five (5) words-per-minute; and Extras are required to pass written Element 4 theory.

FCC Form 605: The universal application form used to renew or modify an existing license. (See NCVEC Form 605)

Federal Communications Commission (FCC): A board of five Commissioners, appointed by the President, having the power to regulate wire and radio telecommunications in the United States.

Feedline transmission line: A system of conductors that connects an antenna to a receiver or transmitter.

Field Day: Annual activity sponsored by the ARRL to demonstrate emergency preparedness of amateur operators.

Filter: A device used to block or reduce alternating currents or signals at certain frequencies while allowing others to pass unimpeded.

Frequency: The number of cycles of alternating current in one second.

Frequency coordinator: An individual or organization recognized by amateur operators eligible to engage in repeater operation that recommends frequencies and other operating and/or technical parameters for amateur repeater operation in order to avoid or minimize potential interferences.

Frequency Modulation (FM): A method of varying a radio carrier wave by causing its frequency to vary in accordance with the information to be conveyed.

Frequency privileges: The transmitting frequency bands available to the various classes of amateur operators. The privileges are listed in Part 97.7(a) of the FCC rules.

General: An amateur operator who has passed the Element 3 written theory examination and Element 1 code test to demonstrate Morse code proficiency to 5-wpm.

Ground: A connection, accidental or intentional, between a device or circuit and the earth or some common body and the earth or some common body serving as the earth.

Ground wave: A radio wave that is propagated near or at the earth's surface.

Half-Duplex: A method of operation on a duplex (separate transmit, separate receive) frequency pair where the operator only receives when the microphone button is released. Half-duplex does not permit simultaneous talk and listen.

Handi-Ham system: Amateur organization dedicated to assisting handicapped amateur operators. (3915 Golden Valley Road, Golden Valley, MN 55422)

Harmful interference: Interference which seriously degrades, obstructs or repeatedly interrupts the operation of a radio communication service.

Harmonic: A radio wave that is a multiple of the fundamental frequency. The second harmonic is twice the fundamental frequency, the third harmonic, three times, etc.

Hertz: One complete alternating cycle per second, abbreviated Hz. Named after Heinrich R. Hertz, a German physicist. The number of hertz is the frequency of the audio or radio wave.

High Frequency (HF): The band of frequencies that lie between 3 and 30 megahertz. It is from these frequencies that radio waves are returned to earth from the ionosphere.

High-Pass filter: A device that allows passage of high frequency signals but attenuates the lower frequencies.

Hysteresis: a lagging in the values of resulting magnetization in a magnetic material (as iron) due to a changing magnetizing force

Interference: The effect that occurs when two or more radio stations are transmitting at the same time. This includes undesired noise or other radio signals on the same frequency.

International Telecommunication Union (ITU): World organization composed of delegates from member countries, which allocates frequency spectrum for various purposes, including amateur radio.

Ionosphere: Outer limits of atmosphere from which HF amateur communications signals are returned to earth.

Jamming: The intentional, malicious interference with another radio signal.

Key clicks, Chirps: Defective keying of a telegraphy signal sounding like tapping or high varying pitches.

Lid: Amateur slang term for poor radio operator.

Linear amplifier: A device that accurately reproduces a radio wave in magnified form.

Long wire: A horizontal wire antenna that is one wavelength or longer in length.

Low-Pass filter: A device that allows passage of low frequency signals but attenuates the higher frequencies.

Machine: A ham slang word for an automatic repeater station.

Malicious interference: Willful, intentional jamming of radio transmissions.

MARS: The Military Affiliate Radio System. An organization that coordinates the activities of amateur communications with military radio communications.

Maximum authorized transmitting power: Amateur stations must use no more than the maximum transmitter power necessary to carry out the desired communications.

Maximum usable frequency (MUF): The highest frequency that will be returned to earth from the ionosphere.

Medium frequency (MF): The band of frequencies that lies between 300 and 3,000 kHz (3 MHz).

Mobile operation: Radio communications conducted while in motion or during halts at unspecified locations.

Mode: Type of transmission such as voice, teletype, code, television, facsimile.

Modulate: To vary the amplitude, frequency or phase of a radio frequency wave in accordance with the information to be conveyed.

Monolithic Microwave Integrated Circuit (MMIC): A four-legged device designed for RF amplification. Well-suited for VHF, UHF and microwave applications.

Morse code (see CW): The International Morse code, A1A emission. Interrupted continuous wave communications conducted using a dot-dash code for letters, numbers and operating procedure signs.

NCVEC Form 605: The form provided by VECs that is used to apply for an amateur radio license or license upgrade.

Noise level: The strengths of extraneous audible sounds in a given location, usually measured in decibels.

Novice operator: An FCC licensed, entry-level operator in the amateur service. Novices may operate a transmitter in the following meter wavelength bands: 80, 40, 15, 10, 1.25 and 0.23.

Ohm's law: The basic electrical law explaining the relationship between voltage, current and resistance. The current I in a circuit is equal to the voltage E divided by the resistance R, or $I = E/R$.

OSCAR: Acronym for "Orbiting Satellite Carrying Amateur Radio," the name given to a series of satellites designed and built by amateur operators of several nations.

Oscillator: A device for generating oscillations or vibrations of an audio or radio frequency signal.

Output power: The radio-frequency output power of a transmitter's final radio-frequency stage as measured at the output terminal while connected to a load of the impedance recommended by the manufacturer.

Packet radio: A digital method of communicating computer-to-computer. A terminal-node controller makes up the packet of data and directs it to another packet station.

Peak Envelope Power (PEP): 1. The power during one radio frequency cycle at the crest of the modulation envelope, measured under normal operating conditions. 2. The maximum power that can be obtained from a transmitter.

Phone patch: Interconnection of amateur service to the public switched telephone network, and operated by the control operator of the station.

Power supply: A device or circuit that provides the appropriate voltage and current to another device or circuit.

Propagation: The travel of electromagnetic waves or sound waves through a medium.

Q-signals: International three-letter abbreviations beginning with the letter Q used primarily to convey information using the Morse code.

QSL Bureau: An office that bulk processes QSL (radio confirmation) cards for (or from) foreign amateur operators as a postage saving mechanism.

RACES (radio amateur civil emergency service): A radio service using amateur stations for civil defense communications during periods of local, regional, or national civil emergencies.

Radiation: Electromagnetic energy, such as radio waves, traveling forth into space from a transmitter.

Radio Frequency (RF): The range of frequencies over 20 kilohertz that can be propagated through space.

Radio wave: A combination of electric and magnetic fields varying at a radio frequency and traveling through space at the speed of light.

Repeater operation: Automatic amateur stations that retransmit the signals of other amateur stations.

Repeater station: An intermediate station in a system which is arranged to receive a signal from a station, amplify and retransmit the signal to another station. Usually performs this function in both directions, simultaneously.

RST Report: A telegraphy signal report system of Readability, Strength and Tone.

S-meter: A voltmeter calibrated from 0 to 9 that indicates the relative signal strength of an incoming signal at a radio receiver.

Selectivity: The ability of a circuit (or radio receiver) to separate the desired signal from those not wanted.

Sensitivity: The ability of a circuit (or radio receiver) to detect a specified input signal.

Short circuit: An unintended, low-resistance connection across a voltage source resulting in high current and possible damage.

Shortwave: The high frequencies that lie between 3 and 30 megahertz that are propagated long distances.

Simplex: A method of operation of a communication circuit which can receive or transmit, but not both simultaneously. Thus, system stations are operating on the same transmit and receive frequency.

Single-Sideband (SSB): A method of radio transmission in which the RF carrier and one of the sidebands is suppressed and all of the information is carried in the one remaining sideband.

Skip wave, Skip zone: A radio wave reflected back to earth. The distance between the radio transmitter and the site of a radio wave's return to earth.

Sky wave: A radio wave that is refracted back to earth in much the same way that a stone thrown across water skips out. Sometimes called an ionospheric wave.

Spectrum: A series of radiated energies arranged in order of wavelength. The radio spectrum extends from 20 kilohertz upward.

Sporadic-E: Radio propagation caused by refraction of signals by dense ionization at the E-layer of the ionosphere.

Spread spectrum: Modulation method which employs lightning-fast frequency "hopping" to allow multiple users to share a band.

Spurious Emissions: Unwanted radio frequency signals emitted from a transmitter that sometimes causes interference.

Station license, location: No transmitting station shall be operated in the amateur service without being licensed by the FCC. Each amateur station shall have one land location, the address of which appears on the station license.

Sunspot Cycle: An 11-year cycle of solar disturbances which greatly affects radio wave propagation.

TAPR: Tucson Amateur Packet Radio Corporation, a non-profit research and development organization. (8987-309 E. Tangue Verde Rd. #337, Tucson, AZ 85749-9399)

Technician operator: An amateur radio operator who has successfully passed Element 2. This operator has not passed Element 1, the 5-wpm code test.

Technician with Code Credit: An amateur operator who has passed a 5-wpm code test in addition to Technician Class requirements.

Telecommunications: The electrical conversion, switching, transmission and control of audio signals by wire or radio. Also includes video and data communications.

Telegraphy: Telegraphy is communications transmission and reception using CW International Morse Code.

Telephony: Telephony is communications transmission and reception in the voice mode.

Temporary operating authority: Authority to operate your amateur station while awaiting arrival of an upgraded license.

Terrestrial station location: Any location within the major portion of the Earth's atmosphere, including air, sea and land locations.

Third-party traffic: Amateur communication by or under the supervision of the control operator at an amateur station to another amateur station on behalf of others.

Transceiver: A combination radio transmitter and receiver.

Transmatch: An antenna tuner used to match the impedance of the transmitter output to the transmission line of an antenna.

Transmitter: Equipment used to generate radio waves. Most commonly, this radio carrier signal is amplitude or frequency modulated with information and radiated into space.

Transmitter power: The average peak envelope power (output) present at the antenna terminals of the transmitter. The term "transmitted" includes any external radio frequency power amplifier which may be used.

Ultra High Frequency (UHF): Ultra high frequency radio waves that are in the range of 300 to 3,000 MHz.

Upper Sideband (USB): The operating mode for sideband transmissions where the carrier and lower sideband are suppressed.

Vanity call sign: Custom-chosen call sign instead of "next-in-sequence" assignment from the FCC.

Very High Frequency (VHF): Very high frequency radio waves that are in the range of 30 to 300 MHz.

Volunteer Examiner: An amateur operator of at least a General Class level who administers or prepares amateur operator license examinations. A VE must be at least 18 years old and not related to the applicant.

Volunteer Examiner Coordinator (VEC): A member of an organization which has entered into an agreement with the FCC to coordinate the efforts of volunteer examiners in preparing and administering examinations for amateur operator licenses.

Index

TAPE

FREE
CQ Amateur Radio Mini-Sub!

TAPE

Fill in your address information
on the reverse side of this page, fold in half,
tape closed and mail today!

TAPE

BUSINESS REPLY MAIL

FIRST-CLASS MAIL PERMIT NO. 2055 HICKSVILLE, NY

POSTAGE WILL BE PAID BY ADDRESSEE

CQ COMMUNICATIONS, INC.
17 WEST JOHN STREET, UNIT 1
HICKSVILLE NY 11801-9962

NO POSTAGE
NECESSARY
IF MAILED
IN THE
UNITED STATES